P9-DEH-737

TRIBAL BIGFOOT

TRIBAL BIGFOOT

David Paulides
Sketches by Harvey Pratt

ISBN 978-0-88839-687-7
Copyright © 2009 David Paulides

Cataloging in Publication Data

Paulides, David
    Tribal bigfoot / David Paulides ; sketches by Harvey Pratt.

Includes index.
Also available in PDF format.
ISBN 978-0-88839-687-7
    1. Sasquatch—North America. I. Title.

QL89.2.S2P386 2009   001.944   C2009-902493-4

All rights reserved. No part of this publication may be reproduced, stored in a retrieval system or transmitted, in any form or by any means, electronic, mechanical, photocopying, recording, or otherwise, without the prior written permission of Hancock House Publishers.

Editor: Theresa Laviolette
Production: Mia Hancock
Cover illustration: Harvey Pratt

Printed in Canada by
Island Blue/Printorium Bookworks

Published simultaneously in Canada and the United States by

HANCOCK HOUSE PUBLISHERS LTD.
19313 Zero Avenue, Surrey, B.C. Canada, V3Z 9R9
(604) 538-1114   Fax (604) 538-2262

HANCOCK HOUSE PUBLISHERS LTD.
#104 - 4550 Birch Bay-Lynden Road, Blaine, WA, U.S.A. 98230-9436
(800) 938-1114   Fax (800) 983-2262

website: www.hancockhouse.com   email: sales@hancockhouse.com

TABLE OF CONTENTS

DEDICATION: RAY & RUTH DELLA

I was raised in Cupertino, California, a suburb of San Jose. We had all the amenities that go with living in an urban environment while still being in a suburb. When I was quite young my folks wanted us to spend time away from the city. They spoke with my aunt and uncle about a weeklong fishing trip and the decision was made to take family vacations as a group, one large family group. Those vacations took us to the great outdoors. We went skiing and fishing in Lake Tahoe, spent summers staying in small cabins in Dunsmuir and fishing the Upper Sacramento and rafting the McCloud River.

As I got older, the men continued to vacation together and the women stayed closer to home. One of the best trips I ever took involved my Uncle Ray Della, my cousin, Randy, and my dad going steelhead fishing on the Smith River. I had the stars lined up on that trip and landed a 26-pound beauty; it's still mounted at my house. My dad caught a few, my uncle caught a few more, and my cousin caught zilch, but that didn't matter; Randy is a great fisherman and his positive attitude is contagious.

After long days on the river, a warm shower, and some clean clothes, the fun really started. Everyone exchanged their stories and fishing tactics, and tales of past trips kicked in. We always stayed in cottages either adjacent, or close, to each other; we always chided and kidded each other relentlessly, and continuous ongoing laughter was part of every trip. It's the type of activity and behavior that only longtime friends/family could ever appreciate and enjoy.

As I get older and reflect on what made me the person I am today, I think of family and the outdoors. It was Ray and Ruth agreeing to vacation with our family—it was their positive attitude about family, my uncle's laid-back demeanor, my aunt's nurturing character and great cooking—and it was the outdoors where I grew to appreciate the important side of living. The bond between the outdoors and family still resonates strongly in my life today, thanks to Ray and Ruth.

Happy 60th Anniversary, Ray and Ruth!

INTRODUCTION

"The deeper you look into the bigfoot issue, the more confusing the creature becomes."

—Ray Crowe, 2008.

Oregon researcher Ray Crowe made this statement to me when NABS was acquiring his research in the spring of 2008. When Ray told me this I immediately aligned it with statements from older adults who say, "As you get older you realize how much you really don't know." I do agree with Ray that the more we delve into the nuts and bolts of bigfoot research, the more we realize that the creature doesn't follow normal mammal behavior.

North America Bigfoot Search has expanded its reach to areas outside of California and into Middle America. There are stories in this book about Minnesota and Oklahoma that directly relate to our bigfoot friend in California. We have no doubt that bigfoot can differ in physical and behavioral activity throughout North America in the same manner that deer, bear, and even people differ in other areas. We have included Oklahoma and Minnesota in this book because of their relationships to Native American reservations and the physical descriptions supplied by the witnesses. Their descriptions of the hominid are too close to California depictions not to be included in the analysis.

We have several laboratory tests that are ongoing regarding potential DNA extraction. There have been hurdles in extracting DNA; with the valuable work and attention to detail that the PhDs have provided, we hope they will be able to overcome the obstacles and bring a viable resolution to the bigfoot DNA question.

This is the second book where we have utilized the services of one of the best forensic sketch artists in North America, Harvey

Pratt. NABS believes it is vitally important to put a face on bigfoot so that society can understand that this hominid is real, it does walk in our forests, and it does breathe our air. It does appear that it has many physical features that are very similar to a human's. Should we as society err on the side of caution, or ignore this hominid and continue to walk down a road of complete ignorance? Our first book, *The Hoopa Project,* gave several examples through the use of sketches that this creature has a consistent appearance that is not always as the Patterson–Gimlin creature is depicted. This creature has many more human qualities, and a possible human genetic link, that cannot be ignored. This book will explore the human/bigfoot genetic background from several angles, all of them intriguing.

This is also the second book where we have every witness that claims to have had a bigfoot sighting sign an affidavit, under penalty of perjury, to the facts surrounding their event. We guarantee that every witness in this book who has claimed a bigfoot sighting has signed an affidavit. We believe that we are the only bigfoot research organization in the world requiring an affidavit on every bigfoot sighting.

Any researcher who wants to understand the big picture needs to look deep into historical data. That data may include newspaper reports, books, and canvassing neighborhoods and interviewing elders in the community; all of this is chronicled in the following pages. I suggest that the reader take a deep breath when they read the historical newspaper articles about wild man; you may be shocked by what's been in print since the 1800s.

I have included a variety of reports from people who have directly observed bigfoot. The sightings are mixed with bigfoot incidents and physical evidence that we've obtained from on-site investigations.

The final chapter is probably the most controversial written about bigfoot in any book in many years. Our sketch artist has applied his years of experience to show a side of bigfoot that many have never seen. I would hope at the end of this book that you have a different perspective of this biped, a more human and compassionate perspective. There is absolutely no doubt in the minds of NABS researchers that the Northwest California, Minnesota, and Oklahoma biped, which people are claiming to be bigfoot, has a human side that has been ignored by many researchers. There is a chance that this creature is more aligned with human genetic coding than anyone has ever believed. Look at the sketches, interpret the historical data, and draw your own conclusions.

The bigfoot tribe may be the first indigenous people ever to have inhabited North America.

David Paulides
Executive Director
North America Bigfoot Search
www.nabigfootsearch.com
May 2009

1 HISTORICAL BIGFOOT PERSPECTIVE

"Upright posture was the hallmark of the first creatures on the human family tree."

—Meave Leakey (*National Geographic*, Volume 188, No. 3, September 1995)

BACKGROUND

When any researcher is delving into a topic to understand a historical perspective they will usually head for the periodicals, old newspapers, encyclopedias, and maybe religious readings and Native American stories. This chapter will deal with two of these sources, and attempt to offer the reader a valid historical source for bigfoot.

When I was a young boy my parents would take us fishing in the Sierra Nevada Mountains of California. The Sierras stretch from Los Angeles in the south to Yreka in the north. They are a huge range with several peaks cresting above 14,000 feet. The streams and rivers that flow out of these mountains harbor some of the best trout fishing in the western United States. The further we traveled back into the Sierras, the more mammals we would see, the fewer people we saw, and the more isolated was the trip. On one of these trips I remember traveling to the southern Sierra region southeast of Porterville. This region is between the Bay Area and Los Angeles and thus doesn't get the huge tourist numbers that areas like Lake Tahoe in the north or Mammoth Mountain in the south receive. I specifically remember fishing the upper stretches of the Tule River and having a fantastic trip with great fishing. What is memorable about the trip was the isolation, beauty, and abundant wildlife we observed. Later in life these memories were jolted when I learned about the Yokuts and Painted Rock.

THE YOKUTS & THE TULE RIVER RESERVATION

In 1864 the United States Congress authorized 55,000 acres or 85 square miles to be set aside in Southern San Joaquin Valley, to be called the Tule River Indian Reservation, as the permanent home for the Yokut Native American tribe. The exact location of the reservation is noted as being from the Kern Lake Region in the south to the mouth of the San Joaquin River in the north. In the late eighteenth century there were close to 18,000 Yokuts living in the region. In the early twenty-first century there were approximately 4,500 Yokuts living on the reservation.

In the mid 1800s the primary food source for the Yokuts was deer, quail, trout, ducks, wild oats, manzanita berries, pine nuts, rabbits, and squirrels. Food was plentiful, as they had several rivers in their area and they were at the base of the Sierra Nevada Mountains.

The tribe had several major occasions they celebrated: birth, death, girls' puberty, and marriage. A boy's puberty got no attention. A shaman (a type of medicine doctor) was a significant figure in the tribe, and some didn't live long. If the chief felt that the shaman wasn't acting in good faith or couldn't save people who were sick, he was killed.

The reservation is on the Tule River just southeast of Porterville. The Yokut tribe is isolated and makes a majority of its income from the Eagle Mountain Casino. Here you can play poker, blackjack, roulette, or slots. It's an odd site to see a Las Vegas-type casino in the desolation of the reservation.

On my first trip to the reservation I was drawn to the casino— the one location in the area where the public can travel without restrictions. The parking lot was full with tourists from the San Joaquin Valley who were all bussed in by huge, new, luxurious, air-conditioned motor coaches. They exited the air-conditioned bus and walked the ten feet into the air-conditioned casino. The day I was at the casino it was over 100 degrees outside—very dry and unbelievably inhospitable weather. I was able to catch a small view of the Tule River and see that it was running with a small flow. It was clean and had a beautiful backdrop of large boulders.

(Left) Painted Rock. (Right) Large bigfoot painted on rock. (Also in color section, page 305.)

The next day I had an appointment at the Yokut tribal chairman's office. I was directed by an assistant to explain the motives for my visit and to complete a special permit that would allow me access to the specific area I wanted to see. I explained that I had read about Painted Rock and the pictographs of a bigfoot family that were supposedly on it. I wanted to photograph the rocks and the associated drawings. The assistant stated that the chairman had granted me special access and that the assistant and security guard would escort me to the area of the rock.

It was a short drive from the tribal offices to Painted Rock. It didn't need to be identified or described; I immediately recognized it from photos. The location of the rock is directly adjacent to the Tule River in a gorgeous spot near a large pool. It's at the base of Mount Tilliman with the mountain facing on the opposite bank. You immediately get the feeling that if you had been part of the tribe hundreds of years earlier, this is the place you'd have wanted to live.

I was escorted to the far side of the rock and could readily see several colored drawings on the underside of the huge boulder. The drawings were protected from the weather by the situation of the rocks and the angle of the rock faces. The drawings were in remarkable condition, with no graffiti or other apparent damage. The area

14

(Left) Rock painting of bigfoot family. (Right) Beaver.
(The Painted Rock photos also shown in color section)

around the rocks showed that tribal members used the area for recreation, but it was obvious that this was a sacred area.

The assistant who escorted me was very helpful and explained the bigfoot relationship with the tribe. The Yokuts believed that bigfoot was their friend and kept grizzly bears, mountain lions, wolves, and other large mammals away from their people. They called bigfoot "hairy man," and that name has stuck through the generations. The elders would tell their children about the creature and advised them to be home before dark and stay close to home, as the hairy man might catch and eat small children. I was told that this was just a scare tactic used by elders to keep the kids nearby. My assistant told me that she did not know of any story where children were actually abducted. She confirmed that there are still stories of bigfoot visiting their valley and that it still makes the area its home.

The rock art on this site is extremely unique, for a variety of reasons. The colored drawings are beautiful and are in great condition, considering their age. What they depict is equally unique. One of the pictographs features an entire family of bigfoot, which includes the male, female, and child. The detail in the drawings is exceptional, as you can even detect five fingers in the bigfoot drawing.

The drawings also include a beaver, lizard, and a centipede that were drawn to perfection. There is white in the drawings, but the

15

primary color is red. The artist also placed horizons and sunsets in some drawings to bring out a sense of the setting. The largest bigfoot drawing goes from floor to ceiling and is massive. The other drawings of the smaller female and baby are much smaller. The large drawing of the male does look like a bigfoot, and you can absolutely get the sense of enormity of the hairy man.

I spent approximately three hours visiting the rock and the adjacent area. I can definitely understand the need to control access to this location, as it is of great historical value.

According to the Yokut tribe, an archeologist (Clewlow, 1978) estimated that the drawings were made around AD 500. Clewlow also stated that the drawings could possibly have been made as late as AD 1200 or as early as AD 1.

Native Americans were not known to waste their time drawing pictographs of fantasy creatures. The hairy man was placed amidst creatures that are all known and accepted animals of this region, so here we have physical evidence of this society identifying a creature as far back as 2,000 years that was part of their daily lives. Hairy man was important enough to the Yokuts to have occupied a majority of the space of Painted Rock and to be the biggest drawing in the collage.

RECENT SIGHTING

Bigfoot sightings in the Tule River Reservation area are still occurring. You can understand why tribal members would not seek out a website to report each incident of a sighting, as it is part of their heritage and not a huge issue. A search of Internet sites for bigfoot reports finds one sighting listed in the twenty-first century. In the spring of 2000, a counselor at the substance abuse center at the reservation happened to be on the building's outside deck and observed a bigfoot on the opposite side of the river. (I have visited the substance abuse center and it is directly adjacent to Painted Rock.) The witness described the creature as having little hair on its face, with its face being flat and almost ape-like, and described the hair on the creature as being a reddish-brown in color. This occurred on the hillside of Mount Tilliman (ref:: BFRO #9357).

This sighting is significant for many reasons; but most notably, both the description of the face and the witness statement that the creature had little hair on its face are similar to other sightings in Northern California. This is in direct contrast to the longest piece of bigfoot film—and considered the most important—the Patterson–Gimlin footage. In their footage they captured a creature that had full facial hair. Almost every artist rendering of bigfoot places considerable hair on the creature's face. I do not think this is actually the truthful rendering for most California bigfoot.

THE MORMON BIGFOOT

Historians who research events often will look to religious documents for information on old incidents, lifestyle issues, and even events of explorers and early settlers. In the "Journal of Mormon History" (Fall 2007, Mormon History Association) Matthew Bowman wrote an article about Cain, Mormons, and bigfoot. It's an interesting perspective from someone who has never conducted bigfoot research.

I am not an expert on religious history, but the reader will need a short lesson on Cain and Abel, so bear with me. A consistent theme in Genesis 41–16, Quran 526–32, or Moses 516–41, is that they all speak about Cain and Abel as the sons of Adam and Eve. Cain became a tiller of the land (a farmer) and Abel raised sheep (a shepherd).

At one point God asked Cain and Abel for a sacrifice. Cain offered fruit and grain from his fields, and Abel offered meat from his flock. God took Abel's meat and turned down Cain. Cain became upset because God had refused his offering, and got jealous over Abel's sacrifice. Cain's anger turned to rage and he killed his brother and hid the body. God could not find Abel's body and questioned Cain about where it was. At this point Cain made a statement that has become quite popular in today's world, "Am I my brother's keeper?"

God reportedly heard Abel screaming for help beneath the earth and found the body. God told Cain that he would be punished, and cursed hm to wander the Earth. Cain said he was afraid that other men would hear of his crime and kill him, so God remedied the situation by placing the "mark of Cain" on him so he could never be killed. He was then set free by God to wander the Earth for eternity.

Absent some finite details, this is the story of Cain and Abel according to three different religious documents. Keep these details handy as I explain the Mormon belief about bigfoot and Cain.

When I first heard that a very conservative publication like the "Journal of Mormon History" had written an article about bigfoot and Cain, I immediately ordered a copy. What follows is an abbreviated version of the article written by Matthew Bowman.

Mormon Apostle David W. Patten was living in Tennessee in 1835 and serving a mission for the Mormon Church. Patten was living with Abraham Smoot, who would later became quite important in Mormon history as the mayor of Salt Lake City and Stake President of the church. In 1838 Patton was killed in a battle in Missouri. In 1893 Smoot sent a letter to the then-head of the Mormon Church stating that a former roommate (Patten) had seen Cain while riding his mule to his residence.

> I suddenly notice a very strange personage walking beside me...for about two miles. His head was about even with my shoulders as I sat in my saddle. He wore no clothing but was covered with hair. His skin was very dark...He said that he had no home, that he was a wanderer of the earth. He said that he was a miserable creature that he had earnestly sought death."(pg. 62)

This sighting and statement are highly unusual for the late 1800s. The reader must remember that "bigfoot" and "sasquatch" were terms that were not coined until the late twentieth century. There were no writings about a creature such as Patten had seen. One could surmise that the creature must have had a face that was close to human or he would have noted the animal face. If the creature reached his shoulders while in the saddle, it would indicate it was close to eight or nine feet tall. Since Patten was a religious apostle, he knew the writings of the bible and knew the story of Cain and Abel and probably quickly associated the creature with the story. The most interesting part of the story is that this creature spoke English and was communicating with Patten, which would validate at a scientific level that this creature (bigfoot) was/is human.

South Weber Utah is in the high country east of Salt Lake and near Park City. This region is fairly desolate except for a small con-

18

tingent of local residents. Since the mid 1980s, residents and tourists have reported seeing a creature that matches the description given by Patten and matching the general accepted version of bigfoot. Since these sightings, the Mormons have associated the sightings with the appearance of Cain.

Author Bowman states, "The traditional Mormon image of Cain, a physical presence on Earth, an incarnation of supernatural evil sent by Satan, whose primary role was to undo the work of the church" (pg. 69). I have not ever heard of a bigfoot doing anything to destroy the work of any church, but I am willing to listen. Bowman doesn't cite anything directly related to this claim and does not make notations to events where follow-up could be accomplished.

"One 1998 story rejects the traditional curse entirely, instead explaining that Bigfoot was an Indian spirit that turns into a hairy Cain-like creature." (pg 79). There was no citation given for Bowman's statement, but I have heard this claim several times. This correlation between Native Americans and bigfoot is a constant theme when you delve deep into bigfoot research.

There are several aspects to the Mormon belief that Cain is bigfoot that are intriguing. Bigfoot has never been captured and a body has never been found. This is a very unusual circumstance considering that bigfoot has been seen in every state of the U.S. except Hawaii, and a bigfoot-type creature has been reported on almost every continent in the world. If humans have been able to capture every type of creature claimed to have been seen alive, then why not bigfoot? Does bigfoot have supernatural powers? If it does have these supernatural powers, how did it acquire the ability? If it doesn't have any unusual powers, is it smarter than a human and able to use its physical abilities to out maneuver man? If you scour bigfoot websites you can easily find thousands of reported sightings with relatively few pictures and videos. If something has been seen as many times as bigfoot, why aren't there more credible photos or extended film like the Patterson–Gimlin footage?

LEIF ERICKSON

Ray Crowe wrote about the exploits of Viking explorer Leif Erickson and his relationship to bigfoot in "The Track Record" #115 (pg. 16).

> It is a little known historical fact that the first Sasquatch encounter was perhaps observed by the Vikings who settled on the island of New Foundland in Eastern Canada. Leif Erickson, or Leif the Lucky (son of Erik the Red), and his crew of Norsemen landed on the rugged shores on "The Rock" some one thousand years ago, becoming the first Europeans to set foot in the western world. Leif kept a record of his journey across the Atlantic, from Iceland to Greenland, and of his experiences whilst in Newfoundland, the last point of land on his voyage. Among his accounts, Leif told of seeing huge hairy men who towered over him and his Berzerker crew (and the Vikings are known to have been large men). The "huge hairy men," according to Leif, lived in the woods and had a rank odor and a deafening shriek. Apparently, Leif had several sightings of the "huge hairy men" before departing the island.

DAVID THOMPSON

Thompson was known as an explorer, surveyor, and trader for the Northwest Trading Company and spent a majority of his time in the Western United States and Canada. When he did cross mountains or large barriers, he would employ locals to assist him. Many times he employed the services of local Native Americans or Canadians who knew the land and the obstacles. In total, Thompson mapped 3.9 million square kilometers and wrote 77 complete journals of his trips. T.C. Elliott wrote *The Journal of David Thompson,* putting Thompson's journal notes into a flowing story. The Oregon Historical Society printed the Elliott works in March 1914.

Thompson had a few notes in his journal that are of interest to any bigfoot researcher. In January 1811 he wrote,

I saw the track of a large animal, has four large toes, 3 or 4 inches long and a small nail at the end of each. The ball of his foot sank about 3 inches deeper than his toes. The hinder part of the foot did not mark well. The whole is about 14 inches long and 8 inches wide and very much resembles a large bear track. It was in the rivulet in about 6 inches of snow.

Forty years after making the notation in his journal, Thompson explained a little more about the finding when he wrote,

David Thompson stamp.

continuing on in our journey in the afternoon of January 7 1811, we came upon the track of a large animal, the snow about 6 inches deep on the ice; I measured it; four large toes each four inches in length to the end of a short claw; the ball of the foot sunk three inches lower than the toes. The hinder part of the foot did not mark well, the length 14 inches, by 8 inches in breadth, walking from north to south, and having passed about 6 hours. We were in no humour to follow him, the men and Indians would have him be a young mammoth and I held it to be a track of a large old grizzly bear; yet the shortness of the nails, the ball of the foot and its great size was not that of a bear, otherwise that of a large old bear, his claws worn away, the Indians would not allow.

Thompson's remarks about the dimension of the footprint parallel bigfoot tracks seen throughout North America. The 14- by eight-inch size is a fairly normal adult size print. The claw notation at the end of each toe could be an overgrown toenail. The length of the toes is not entirely unusual, as bigfoot tracks are noted for having a readily identifiable toe. Bigfoot in North America are normally accepted as five-toe prints, but I have personally talked to witnesses and have photos of four-toe prints that look very similar to the five-toe. Later in Thompson's 1851 narrative he states, "the sight of the track of that large a beast staggered me, and I often thought of it but could never bring myself to believe that such an animal existed."

21

Skeptics in today's society would state that someone planted the prints and some gullible person out walking in the woods followed them. In Thompson's era people were concerned about mere survival. If Thompson stated that the tracks continued for five miles in the middle of the Rocky Mountains near Jasper, Alberta, Canada, you can imagine that these had to be real, no doubt. Just another piece of evidence that bigfoot was around a lot earlier than 1967 when Patterson and Gimlin caught it on film.

It is interesting that Thompson thought enough about the sighting of the track to follow it a short distance and then take the time to measure it. This incident must have had a significant impact on the trip and his psyche. He even reviewed his experience 40 years later to ensure that his thoughts about the incident were accurate and understandable. Amazing.

REVEREND ELKANAH WALKER

The reverend, a missionary to the Spokane Indians in Washington, wrote a letter in April 1840 about a story that he had heard. The correspondence, part of the permanent archive at the Holland Library at Washington State University, is reprinted from *Washington Magazine* (October 1998). The narrative follows.

I suppose you will bear with me if I trouble you with a little of their superstition, which has recently come to my knowledge. They believe in the existence of a race of giants which inhabit a certain mountain off to the west of us...The account that they give of these Giants will in some measure correspond with the bible account of this race of beings.

They say their track is about a foot and a half long. They will carry two or three beams upon their back at once. They frequently come in the night and steal salmon from their nets and eat them raw. If the people are awake they always know when they are coming very near, by their strong smell, which is most intolerable. It is not uncommon for them to come in the night and give three

BEAR LAKE DEMOCRAT, PARIS, ID
OCT 20, 1883, BENOIT CREVIER

A Queer Family.

Consisting of a Mother and Three Children Who are Part Man and Part Animal.

Messrs. F. Rosengay, Allen M. Vandal and Edward P. Strong, for some time past have been traveling through portions of Florida, Alabama and Mississippi Inspecting pine lands. These gentlemen are the emissaries of a grand syndicate, or, at least, so they claim, which has been formed in Minnesota for the purpose of buying immense tracts of land in the south and forming a complete system of saw mills from which to furnish lumber for the world, but more especially for shipment into Mexico, where they claim they will always find a ready market at good prices.

These gentlemen are all men of means and very kindly in their deportment, making friends in every portion of the country they have visited, and bring with them undoubted credentials as to their honor and integrity; therefore some credit must be given to the following relation, which each of them declares to be nothing but facts.

It might be as well to state here that nearly on the line between St. Tammany and Washington parishes resides a man about 50 years of age, who answers to the name of "Crazy Aleck." This dilapidated specimen of crazy humanity resides in the most lonesome portion of the parish that it would be possible to find, seldom making his appearance among the settlers, and then only when forced to do so from hunger.

He is harmlessly insane and sometimes when he visits a place to secure a snack for something to eat will become quite communicative and tells some wonderful stories concerning his forest home; but where he resides no one knows or cares, and his queer stories were never heeded, being regarded as the fancies of a madman's brain.

But upon several occasions "Crazy Aleck" has told a story concerning "a woman and some bears" which sounded ridiculous and improbable in the extreme, but which, nevertheless, has proved to be true.

This is a narrative related by the three gentlemen first named:

Hearing, while in the lower portion of Mississippi, that some very desirable land lay in Washington and St. Tammany parishes, La., and that it could be procured very reasonably, our party started for that section. Upon arriving at the desired point we obtained accommodations at a farmhouse and the next day started on horseback on a tour of inspection; well, we had gone over considerable ground during the day and started on the return about 4 o'clock.

Mr. Vandal, who had become weary of riding, had asked one of us to lead his animal and that he would walk on while we stopped now and then to make observations and take notes, as was our custom upon such journeys. Vandal got some distance ahead of the two of us who had the horses, and suddenly we saw him come around a bend in the road a short distance ahead of us on a dead run; we started up our horses to meet him, and when he came up, almost out of breath, he said: "Tie your horses to a tree and come up the road; there's the greatest sight you ever saw."

We did as he requested and followed him up the road some 800 yards, when putting his fingers to his lips, to indicate that we must be silent and cautious, he started to creep through an undergrowth on the side of the road and we followed.

At this juncture we were much startled to hear a deep coarse laugh issue from a space apparently a short distance in advance of us. It was a human laugh, evidently, but still it was so hollow and unearthlike that it sent a chill of terror through us all; this laugh was followed by another, and another then of the same kind, and these were succeeded by a long hearty laugh, apparently that of a female.

Partly recovered from our astonishment, at a motion from Vandal we creeped on about thirty yards to the edge of the thicket, and there in an open space, a few yards beyond, we beheld the most astonishing sight that ever greeted mortal eye.

Seated on the trunk of a fallen tree was a dilapidated looking specimen of feminine humanity, whose small bit of clothing was naught but rags and whose hair hung about her shoulders in a tangled mass; she was apparently about sixty years of age, was of medium stature, and dark, as though from long exposure to the sun and weather. Near by her were three of the most hideous objects one could imagine, evidently playing, and she seemed to be watching them attentively.

To describe these three objects (you could not call them human) is very difficult. Their heads were shaped like human beings, the tops, ears, eyes and neck being perfect, but the mouth and nose coming together broad and thick, more on the order of a bear; the arms resembled perfectly those of men, but from the waist down they bore the shape and semblance of a bear, being covered entirely with long coarse, black hair. The feet were also similar to those of a bear, being flat, and seemingly armed with immense claws.

For some time we watched them with mute astonishment, as they wrestled and played, now dancing erect as a biped; and then on all four as a quadruped, and now, and then giving vent to their horrible laugh.

How long this would have continued we do not know, but just here Strong let his astonishment get the better of his discretion and changed his position, partly exposing himself to view, and at the same time making a slight noise. The noise attracted the woman's attention; and she evidently saw Strong, for uttering a cry of alarm she started through the undergrowth on the other side of the opening with the speed of a deer; closely followed by the three unearthly creatures, running on all fours.

As soon as we had recovered from our astonishment we attempted to follow them; but they were too swift of foot, and fearing we would become lost in this, to us strange section, we were forced to return.

The next day, in company with several other gentlemen, we scoured the section thoroughly, but could find no traces of the strange family.

"I have traveled far," said Mr. Vandal, "and seen many wonderful things, but this human-bear family, as I call them, surpasses my experience heretofore by far."—N. O.

Newspaper article "A Queer Family" reported October 20, 1883 in the *Bear Lake Democrat,* Paris, Idaho.

23

whistles and then the stones will begin to hit their houses. The
people believe that they are still troubled with their nocturnal visits. We need the prayers of the church at home…

A QUEER FAMILY

An article from the *Bear Lake Democrat,* Paris, Idaho, October 20,
1883 (shown on the following page), outlines what appears to be a
female with three bigfoot-like offspring in the swampy rural area of
Louisiana. They were heard to laugh. "It was a human laugh, evidently, but still it was so hollow and unearthlike that it sent a chill or
terror through us all." Their physical description followed, "Their
heads were shaped like human beings…but the mouth and nose coming together broad and thick…they bore the shape and semblance of
a bear, being covered entirely with long coarse, black hair." It's a
very odd story, but worth including in the historical perspective.

THEODORE ROOSEVELT

Most people interested in bigfoot are also intrigued with the outdoors.
If you have any fascination for what wildlife may offer to our souls,
then you probably know something about Theodore Roosevelt. If you
don't know much about this past U.S. president, you are about to learn
about an interesting person and a bigfoot relationship.

Theodore was born October 27, 1858 in New York City. To keep
dates in perspective, the Civil War began April 12, 1861. He loved
the outdoors and maybe his life in New York energized his need to
get away from the big city. From 1877 to August of 1886 Roosevelt
took trips and had adventures that were almost unheard of for a
nineteenth century gentleman. In addition to graduating from
Harvard University, he took a 30-mile trek through snow and ice in
Maine, climbed the Matterhorn, killed two deer with one shot, killed
a grizzly, hosted a hunt on Long Island and had his personal boat

stolen, tracked the thieves down and personally turned them over to law enforcement.

Roosevelt was a rare man and someone who could have been the Indiana Jones of his time. He loved the outdoors and would ask friends and acquaintances to join him on his trips. While his killing of animals is not politically correct for the twenty-first century, it does describe a man that wasn't afraid of much. He loved the exploration of new areas and the challenges that they offered. Just for the record, Roosevelt was sworn in as the 26th president on September 14, 1901. Even after winning the presidency, he still traveled to Africa and hunted rhino and Cape buffalo, and fished for piranha in South America. He died at age 60 on January 6, 1919.

In 1893 Theodore wrote about many of his exploits in a book titled, *The Wilderness Hunter,* a non-fiction work published prior to his run for president. One story in particular caught the attention of bigfoot researchers many years ago. The story had to do with rugged trappers working together in the mountains that divide the Forks of the Salmon and the headwaters of the Wisdon River. The two trappers weren't having a great year of trapping and decided to work a new area that had stories associated with it of being very wild and dangerous.

Roosevelt was told the story of the trappers by one of the men, Bauman. He was young at the time of the events, but was up in his years when he told the story. The region where the trappers were headed had a bad reputation because the year before a lone hunter went to the area and was purportedly killed and mutilated by a wild beast. He was found by two prospectors weeks later, his body in pieces (torn apart) and scattered in a small area.

Theodore
Roosevelt

Bauman and his friend weren't too concerned about the dead hunter and went directly to the area. The area they were entering was gloomy with a very thick forest and little sunshine. They got to their camp with a few hours of daylight and they went to canvass the

25

area after putting down their packs. They returned to find that something had gone through their packs, thrown items around the area, and even destroyed some of their equipment. They felt at the time that it was probably a hungry bear and didn't think much about it. Bauman was making dinner while his companion started to look closely at the footprints of what entered their camp. After examining many of the tracks made in the dust, the companion told Bauman that the creature that entered their camp was walking on two feet, an astounding claim. Bauman then looked at the tracks and said that they even looked a little like a human footprint. It was getting too dark to examine tracks, so the trappers went to bed.

After a few hours Bauman was wakened out of a deep sleep by something, he didn't know what. As he opened his eyes his nose caught the odor of a foul-smelling creature in the area. His eyes started to clear from being asleep when he saw a huge figure standing near his blanket. He grabbed his rifle and fired at the figure. He immediately heard a creature running away into the darkness at a furious pace, breaking tree limbs as it departed. The remainder of the night was uneventful.

The next day both trappers stayed together and set their traps and worked the area. After a long day of working they returned to their camp to find it was again torn apart. This time the creature walked through a soft marshy area where the prints were extremely clear. It was obvious that the creature was walking on two feet and its footprint was similar to a human's, but slightly different and a lot bigger.

After the trappers had turned to their sleeping blankets for the night, they heard a large creature walking on the opposite side of the creek. It wasn't far away, but it was just far enough that they couldn't see it. The creature was breaking branches and making noise that let them know it was definitely nearby. They didn't get a lot of sleep that night. They awoke the next morning and decided to gather their traps and leave the valley. They had agreed that while the trapping and wildlife in the valley were plentiful, the disturbances at night were getting to them.

Bauman volunteered to go out and quickly gather the three beaver traps while his partner stayed at camp and got their packs together, along with their pelts. He found the traps full and took time with each beaver to prepare the pelts and clean the furs. It was a long day before Bauman returned to the area of the camp. As he made his approach through the forest, it was abnormally quiet, no sounds. There was no wind, no sounds made by his walking, nothing as he approached the perimeter of their camp. He started to yell for his partner, but there was no response.

Bauman entered their camp and initially didn't see anything. He still called out for his friend, but there was no sound. At this point he started to walk the area and it was then that he saw his partner, dead, with a broken neck and four fang marks in his neck. The tracks of the creature that had harassed them were surrounding the body. It was obvious that the body had been moved around as though the creature had examined it. The creature had not eaten the body, and the only signs of attack were the fang marks and the broken neck.

Bauman was now an emotional wreck. He and his partner had high-powered rifles and yet his partner did not have time to fire his gun. Bauman was so upset, he left everything but his rifle and fled the area. He ran all the way to where their ponies were hobbled. He had reached the point of safety and then headed into a local town. I will end my version of the story at this point.

This story is fascinating for several reasons. It is believed that it occurred in the 1860s, approximately 100 years before bigfoot obtained international fame at Bluff Creek. The stalking of the men by standing on the opposite side of the creek is reminiscent of recent stalkings by bigfoot. The odor Bauman described is also indicative of a bigfoot presence. The trait of going into the camp when people are gone for the day is also something bigfoot has been known to do. The trappers identifying the creature as walking on two feet and having a track different than a bear's and closer to a human's, is something that is really unusual and strikes to the heart of the bigfoot identification.

Anyone who has studied bigfoot knows that the behavior of the creature in this incident parallels bigfoot on many grounds. But everyone has to remember that there are no recorded, validated killings of humans by bigfoot. I have personally read many third-person accounts of purported attacks by bigfoot, but none—repeat, none—have ever been validated.

WILD MAN

In any research project, one of the first items on the agenda is a historical search of the records for information about the subject being studied. Many times the information from 100 years ago is more accurate than the information gathered today. Newspaper articles from years ago represent data from a very confined area, usually the region where the newspaper was distributed. Information wasn't dispersed as quickly or as thoroughly as it is today, and many times people in different parts of California never got to read what was happening in another part of the state, and if they did, the news was weeks old. The literacy rate amongst state residents was a fraction of what it is in the twenty-first century, so that very often many people couldn't read an article when it was printed. Unless you lived in a fairly large city, news sometimes came weeks after an event, and sometimes a newspaper never arrived.

To think that bigfoot is a fabrication of twentieth-century film footage from Del Norte County would be a mistake. Granted, the term "bigfoot" is fairly new, having only been used to describe the creature since 1958. In the late 1800s and early 1900s bigfoot was known as "wild man," "hairy man," and a variety of other terms that proliferated across the United States and Canada.

News, November 18, 1908

WILD MAN OF THE MOUNTAINS ROBS TRAPS OF PREY AND COMMISSARY OF SUPPLIES

Peculiar Foot prints Left by the Man-Animal
as He Flees From His Would-Be Captor.

Special Correspondence.

Santa Monica. Nov. 18. — Is there a **wild man** roaming the mountains of the Santa Monica range? There is, according to the story by Bertrand Basey, who has just come down from the vicinity of Point Dume, where he has been acting as commissary for the contractors who engaged in the construction of the Malibu-Rindge railway.

Basey says that there were frequent, losses from the improvised store house. At first, these were attributed the thieving proclivities of coyotes and mountains squirrels, but as it was found to be impossible to trap any of these animals further investigation was made, with the result of learning that the theft was due to some cunning hand whose movements were directed by a reasoning mind. He kept a strict lookout and was soon rewarded by the sight of a brown being fashioned after the form of man, approaching the tent. The thing was on all fours, was devoid of clothing or such covering as might have been provided by the skins of animals, and had a face covered with hair.

Basey was about to fire at the intruder when he was deterred by the close resemblance of the uninvited visitor to a man. Thinking it might be a railroad laborer in disguise, he made a noise as if to frighten the thief away. Suddenly the wild man gave forth a guttural yell, rose upright on its hind legs and disappeared in the underbrush. Nothing more was seen or heard of the mysterious half-human beast, although the railroaders who went to the beach for a bath that morning are still wondering what manner of animal had been there before them to leave peculiar tracks in the sand. The tracks which they photographed were not unlike those that

29

would have been left by the hands and feet of a man were he provided with long claw-like nails for each of the five toes and the four fingers and thumb.

This article brings forward many interesting points that lead to this being a bigfoot. The article mentions impressions of the hands and feet and how they resembled those of a human, same as a bigfoot. Bigfoot has been known to take items when the opportunity presents itself, as it apparently did in Santa Monica. The description of a half human and half beast is a very common explanation for what people describe when they see bigfoot in the twenty-first century.

POINT DUME

Point Dume is a peninsula on the California Coast 14 miles west of Santa Monica very near Malibu. The region is now completely covered in houses and people, but in the 1800s there was significant open range starting at the Dume and heading north.

Daily Democrat, April 9, 1891
Woodland, California

WHAT IS IT?

An Unheard of Monstrosity
Seen in the Woods above Rumsey.

Mr. Smith, a well-known citizen of Northern Capay Valley, called on us to-day and tells us the following strange story, which we would be loath to believe if it were not for the fact that he is an old acquaintance of this office, and has always borne a spotless reputation. Several days ago, Mr. Smith together with a party of hunters, were above Rumsey hunting. One morning Mr. Smith started out early in quest of game, he had not gone far when his attention was attracted by a peculiar noise that seemed to come from an oak tree that stood near by. Looking up, Mr. Smith was startled to see gazing at him what was apparently a man clothed in a suit of shaggy fur. Having heard of wild men, Mr. (Smith) naturally was placed on his guard, but thinking that he would see "what virtue there was in kindness," he called to the supposed man to come down, as he was filled with nothing but the kindest motives. This speech did not have the desired effect, rather the opposite, for the strange thing gave grunts of unmistakable anger. Believing that discretion was the better part of valor, our informant stood not upon the order of his going, but went at once in a bee-line for the camp. After placing some distance between himself and the strange creature, the hunter turned around just in time to see it descend the tree. Upon reaching the ground, instead of standing upright as man would, it commenced to trot along on the ground as a dog or any other animal would do.

Smith then realized that it was no hermit he had seen, but some kind of monstrosity, such as he had never heard of, much less seen before. The hunter stood amazed and spellbound for a moment, but soon gathered his scattered sense again and was soon making his best speed to camp, where in a few breathless words, were telling his companions of what he had seen. They were disposed to laugh at him at first, but his sincere ness of manner and his blanched cheeks soon proved to them that he had seen something out of the usual order of things.

A hasty council was held, and the party decided to go in search of the monster, so taking their guns and dogs they were piloted by Mr. Smith to whom they soon came in sight of the unnamed animal. In the meantime it had commenced to devour the contents of Mr. Smith's game bag that he had dropped in his hasty retreat. The creature would plunge its long arms or legs into the bag and pulling forth the small game that was in it, transferred it to its mouth in a most disgusting manner. An effort was made to set the dogs upon it, but they crouched at their masters' heels and gave vent to the most piteous whines. The whines attracted the attention of the nondescript, and it commenced to make the most unearthly yells and screams, at the same time fleeing to the undergrowth, some half a mile distant, upon which the whole party

31

immediately gave chase. They soon gained upon the strange beast, and it, seeing that such was the case, suddenly turned, and sitting upon its haunches, commenced to beat its breast with its hairy fists. It would break off the great branches of trees that were around it, and snap them as easily as if they had been so many toothpicks. Once it pulled up a sapling five inches through at the base, and snapping it in twain, brandished the lower part over its head, much after the same manner a man would sling a club. The hunters seeing that they had a creature with the strength of a gorilla to contend with beat a hasty retreat to camp, which soon broke up, fearing a visit from their chance acquaintance.

Mr. Smith describes the animal as being about six feet high when standing, which it did not do perfectly but bent over, after the manner of a bear. Its head was very much like that of a human being. The trapeze muscles were very thick and aided much in giving the animal its brutal look. The brow was low and contracted, while the eyes were deep set, giving it a wicked look. It was covered with long, shaggy hair, except the head, where the hair was black and curly. Mr. Smith says that of late sheep and hogs to a considerable extent have disappeared in his vicinity and their disappearance can be traced to the hiding place of the "What Is It." Among those who have suffered are Henry Sharp, Jordan Sumner, Herman Laird, and J.C. Trendle.

Here is a chance for some energetic young man to start a dime museum and acquire a fortune with a very few years. Anyone wishing to learn more about this peculiar monstrosity can do so by calling on our informant who will no doubt take a delight in piloting them to the dangerous vicinity of the late scene of action.

• • •

Daily Democrat, June 25, 1891
Woodland, California

"THE WHAT IS IT"

Seen Once More—It Has Not Reformed Yet.

Once more the wild and wooly "What Is It" has been seen. It does not seem to have reformed as yet, as it is as frisky as ever. This time the person who saw it was a Mr. Herman Gilbert, who was up in the head of Capay Valley looking for a suitable piece of government land that he might homestead.

He says that he was near Rumsey, where he was stopping with some friends. On last Monday morning he started out with his brother-in-law, expecting to be gone a day or so, as he wished to combine business with pleasure. They came to a nice little valley about half a mile long on Tuesday afternoon, and as it was cool, well watered and full of nice green grass, they determined to pitch their tent there. This they did, and about half an hour later Mr. Gilbert went to the spring near by to water the horses, and was surprised to see around it tracks very much resembling that of a man, but thought nothing of it. Incidentally, when he returned, he mentioned it to his brother-in-law. He then, for the first time, heard of the terror, and suggested that the two return and track the mysterious animal to his lair. This they did, and as they followed the foot prints, they found that they led to the other end of the valley. Just as they came to the end of the defile and were about to turn down the mountainside, they heard a peculiar cry, half-human and half-brutish and quite near them. As may be supposed, they wended their way very carefully and slowly. Before they had gone half a mile, they came upon a path. The gentlemen were too sharp to walk in it, and followed the direction it took by walking in the underbrush near by. Just as they reached the bottom of the mountain, they came upon a deep ravine and there, walking up and down, they could be see "his nibs" himself. Mr. Gilbert says that the beast seemed to be mad a something and would beat its breast, which was covered with

33

gore, and the sound made thereby was like distant thunder. It had lost some hair since last seen, so the gentlemen should judge, for the cuticle was plainly discernable and was of a dark color, much like that of a horse.

Near by was a crude cave where the anomaly lived. About it could be seen bones from which flesh had been eaten. The stench arising from the decaying matter was horrible. The muscles of the creature were very powerful, and the animal made an exhibition of its strength once by lifting a huge rock that would weigh at least three hundred pounds and throwing it, without any apparent effort, a hundred feet. After watching, the "What Is It" for some time the men crept quietly back and as soon as possible left the locality, determined not to make any closer acquaintance with the Capay curiosity.

Both of the above articles, written about an area near Woodland, California, speak of a creature identical to bigfoot, and this was in the late 1800s. The throwing of a giant boulder is noted in sightings and incidents even today (refer to my first book, *The Hoopa Project*). The description is again "half-human" with great strength and a stench associated with its living area. Since the creature was in the area of the camp, the stench could have actually been the creature.

Woodland is northwest of Sacramento in the San Joaquin Valley. It is almost directly north of Davis, California. This area is now all agricultural with the nearest mountains being near Lake Berryessa, a very popular location for boating and fishing.

The first article's description of deep-set eyes, hair over the entire body, and a massive body is an almost perfect description of a bigfoot. When the hunters attempted to release their dogs on the creature the dogs cowered and wouldn't attack, a very normal response when people have accidentally walked up to a bigfoot with their dogs. There is something about the scent or some other factor that bigfoot releases that causes dogs not to want anything to do with the creature. It is a rare occurrence when a dog voluntarily attacks or even advances on a bigfoot.

Rumsey, California is approximately 25 miles northwest of Woodland and sits at the base of 3,000–4,000-foot mountains. It is

34

also only 14 miles from a very large body of water, Clear Lake. This is another area that is predominantly agricultural, but is much closer to mountains and forests. The area of Rumsey and Woodland get very hot in the summer with temperatures regularly in the high 90s to low 100s.

Morning Herald, November 10, 1870
Titusville, Pennsylvania

THE WILD MAN OF CALIFORNIA

A correspondent of the Antioch *Ledger,* writing from Grayson, California, under date of October 16, says: "I saw in your paper, a short time since, an item concerning the 'gorilla' which is said to have been seen in Crow Canyon and shortly after in the mountains of Orestimba Creek. You sneered at the idea of there being any such 'critters' in these hills, and were I not better informed I should sneer too, or else conclude that one of your recent prospecting parties had got lost in the wilderness, and didn't have sense enough to find his way back to Terry's. I positively assure you that this gorilla, or **wild man** as you choose to call it, is no myth. I know that it exists, and that there are at least two of them, having seen them both at once not a year ago. Their existence has been reported at times for the past twenty years, and I have heard it said, that in the early days, an ourang-outang escaped from a ship on the southern coast; but the creature I have seen is not that animal, and if it is, where did he get his mate? Import her as the Web-foot did their wives? Last Fall I was hunting in the mountains about twenty miles south of here, and camped five or six days in one place, as I have done every season for the past fifteen years. Several times I returned to my camp, after a hunt, and saw that the ashes and charred sticks from the fire-lace had been scattered about. An old hunter notices such things, and very soon gets curious to know the cause. Although my beddings and traps and little stores were not disturbed as I could see, I was anxious to learn who or what it was that so regularly visited my camp, for clearly the half-burnt sticks and cinders could not scatter themselves about. I saw no track near the camp, as the hard ground, covered with dry leaves, would show none. So I started on a circle around the place, and 300 yards off, in damp sand, I struck the track of a man's feet, as I supposed—bare, and of immense size. Now I was curious, sure, and resolved to lay for the bare-footed visitor. I accordingly took a position on a hillside, about sixty or seventy feet from the fire, and securely hid in the brush. I waited and watched. Two hours or more I sat there and wondered if the owner of the feet would come again, and whether he imagined what an interest he had created in my inquiring mind, and finally, what possessed him to be prowling about there with no shoes on. The fire-lace was on my right, and the spot where I saw the track was on my left, hid by brushes. It was in this direction that my attention was mostly directed, thinking the visitor would appear there, and, besides, it was easier to sit and face that way. Suddenly I was startled by a shrill whistle, such as boys produce with two fingers under their tongue, and turning quickly, I ejaculated, "Good God" as I saw the object of my solicitude, standing beside my fire, erect and looking suspiciously around. It was in the image of man, but it could not have been human. I was never so benumbed with astonishment before. The creature, whatever it was, stood full five feet high, and disproportionately broad and square at the shoulders, with arms of great length. The legs were very short, and the body long. The head was small compared with the rest of the creature, and appeared to be set upon his shoulders without a neck. The whole was covered with dark brown and cinnamon-colored hair, quite long on some parts, that on the head standing in a shock and growing close down to the eyes, like a Digger Indian's. As I looked, he threw his head back and whistled again, and then stooped and grasped a stick from the fire. This he swung round and round, until the fire on the end had gone out, when he repeated the maneuver. I was dumb, almost, and could only look. Fifteen minutes I sat and watched him, as he whistled and scattered my fire about. I could easily have put a bullet through his head, but why should I kill him? Having amused himself, apparently all he desired, with my fire, he started to go, and having gone a short distance, he returned, and was joined by another—a female, unmistakably—when they both turned and walked past me, within twenty yards of where I sat, and disappeared in the brush. I could not have had a better opportunity for observing them, as they (were) unconscious of my presence. Their only object in visiting my camp seemed to be to amuse themselves with swinging lighted sticks around. I have heard this story many times since then, and it has often raised an incredulous smile; but I have met one person who has seen the mysterious creatures and a dozen who have come across their tracks at various places between here and Pacheco Pass."

This article is fascinating, as it could have been written about a big-foot sighting in 2008. The witness was very wise to sit and wait for the creature's return. From the witness's own statement, the creature is not an orangutan, and the prints he found near the fire match those of a barefoot human, exactly how bigfoot tracks are described. The whistling that he heard is oftentimes heard subsequent to a sighting. When the male creature whistled several times, it would appear that it was an indicator that all was safe and the female or family could enter the camping area. This does indicate that the male leads the group into unknown areas. The description of short legs, large torso, no neck, and an appearance that the head was placed on the shoulders is also an exact match to many of the sketches we've done of the hominid.

The sight of the creatures playing with the fire is funny. It's obvious from our studies that bigfoot does like to play (Kirk Stewart sighting of bigfoot playing with another in a lake, chapter 9). Fire may be something they haven't seen often and is a new entity that they don't quite understand.

Grayson, California is located in the San Joaquin Valley 16 miles west of Modesto, just east of Interstate 5, very near the mountains on the western side of the valley. Pacheco Pass is the predominant means for vehicular travel between drivers traveling between Los Angeles and San Jose. The pass crosses a range of mountains that go to just over 4,000 feet and are mostly containing oak and madrone trees.

Independent, April 25, 1876
Helena, Montana

THE MISSING LINK.

Interview of a California Hunter
With a Gorilla-Like Wild Man.

A correspondent of the San Diego (Cal.) *Union* writes as follows concerning a "wild man" recently seen in the mountains of that country. "About ten days ago Turner Helm and myself were in the mountains about ten miles east of Warner's ranch of a prospecting tour, looking for the extension of a quartz lode which had been found by some parties some time before. When we were separated, about a half a mile apart—the wind blowing very hard at the time—Mr. Helm, who was walking along looking down at the ground suddenly Heard Somebody Whistle. Looking up he saw 'something' sitting on a large boulder about fifteen or twenty paces from him. He supposed it to be some kind of an animal and immediately came down on him with his needle gun. The object immediately rose to its feet and proved to be a man. This man appeared to be covered all over with coarse black hair, seemingly two or three inches long, like the hair of a bear; his beard and the hair of his head were long and thick; he was a man of medium size, and rather fine features—not at all like those of an Indian, but more like an American or Spaniard. They stood gazing at each other for a few moments when Mr. Helm spoke to The Singular Creature first in English, and then in Spanish, and then in Indian, but the man remained silent. He then advanced toward Mr. Helm, who not knowing what his intentions might be, again came down on him with the gun to keep him at a distance. The man at once stopped as though he knew there was a danger. Mr. Helm called to me, then The Wild Man went over the hill and was soon out of sight and made good his escape. We had frequently before seen this man's tracks in that part of the mountains, but had supposed them to be tracks of an Indian. Mr. Helm is a man of unquestioned veracity.

Warner Ranch is located approximately 30 miles from the Pacific Ocean between Los Angeles and San Diego. It is only five miles for the boundary of the Rincon Indian Reservation. The topography of this location is very similar to the Grayson sighting, same temperatures, same rolling hills, and same trees.

The description of the creature is an almost identical match for bigfoot: hair covering the body two to three inches long, an appearance of a man, all sounds very much like a bigfoot sighting that was documented in 1876. It appears the newspaper reporter felt that Mr. Helm was quite the honest person: "unquestioned veracity."

NORTH AMERICA SIGHTINGS

The earliest documented sighting I could find was in 1832 in the *Annual Register,* printed in Minnesota (see article next page). The article was found in the Track Record #110 (pg. 17). It is the first account I have ever seen where a creature described as bigfoot was called an Indian. This is a provocative identification because it is our belief that bigfoot associates closely with Native Americans, thus the many sightings around their reservations.

One of the earliest known newspaper articles regarding a creature that matched bigfoot's description came out of the *Memphis Enquirer* on May 9, 1851. (Quoted from "The Track Record" #85 [pg. 12]).

UNKNOWN CREATURE
SIGHTED IN ARKANSAS

Wild man of the woods. During March last, Mr. Hamilton of Greene County, Arkansas while out hunting with an acquaintance, observed a drove of cattle in a state of apparent alarm, evidently pursued by some dreaded enemy. Halting for the purpose,

MISCELLANEOUS ARTICLES.

American Expedition of Discovery.—The expedition sent out by the American government in 1820-21, to explore the rocky mountains, and north of the Numean line, has at length been heard of, after an absence of eleven years. The company landed at Green Bay, and wintered; went by Prairie du Chin to St. Anthony's Falls, Mississippi; went up St. Peter's 200 miles, in search of lead mines, where they discovered several very valuable ones; wintered there; went down the same river, and also the Mississippi to the mouth of the Missouri; thence up the Missouri to the foot of the Rocky Mountains; wintered there; and continued to the middle of August; then crossed the mountains, and were west eight years. While travelling by the Frozen Ocean, and having been over into Asia South, towards the head of Colombia river they were overtaken by a storm, and compelled to build houses, and stay there nine months, six of which the sun never rose, and the darkness was as great as during our nights. The snow, part of the time, was fourteen feet deep, and the company were compelled to eat forty-one of their pack horses to prevent starving, whilst the only food the horses had was birch bark, which the company cut and carried to them, by walking on the snow with snow-shoes. After crossing the mountains, they passed 386 different Indian tribes, some perfectly white, some entirely covered

with hair, who were among the most singular, and so wild that the company were compelled to run them down with horses. Of the company, five died by sickness, one by breaking a wild horse, one by the fall of a tree, and fifteen were killed by the Indians—total twenty two. Ten of the nineteen survivors are lame, some by accident.

Annual Register, (Minnesota) 1832, reproduced from "The Track Record" #110 (pg. 17).

they soon discovered as the animals fled by them, they were followed by an animal bearing the unmistakable likeness of humanity. He was of gigantic stature, the body being covered with hair and the head with long locks that fairly enveloped his neck and shoulders. The "Wildman," for we must so call him, after looking at them deliberately for a short time, turned and ran away with great speed, leaping from twelve to fourteen feet at a time. His footprints measured thirteen inches each.

This singular creature has long been known traditionally in St. Francis Green and Poinsett Counties. Arkansas sportsmen and hunters having described him so long as seventeen years since. A planter, indeed saw him very recently, but withheld his information lest he should net be credited, until the account of Mr. Hamilton and his friend placed the existence of the animal beyond cavil.

The description in the Memphis article certainly sounds like a bigfoot. If the writer of the article is accurate, then the creature was seen in this area as early as 1834, which is the earliest documented sighting of a bigfoot in a newspaper that I could find. The interesting angle of the article is the description of "unmistakable likeness of humanity"—it looks human! The descriptive term "Wildman" was one that was used throughout the 1800s to describe bigfoot.

The following article was quoted in "The Track Record" #119 (pg. 9).

Minnesota Weekly Record, January 23, 1869

A GORILLA IN OHIO

Gallipolis is excited over a wild man who is reported to haunt the woods near that city. He goes naked, is covered with hair, is gigantic in height and "his eyes start from their sockets." A carriage, containing a man and daughter, was attacked by him a few days ago. He is said to have bounded the father, catching him in a grip like that of a vice, hurling him to the Earth, falling on him and endeavoring to bite and to scratch like a wild animal. The struggle was long and fearful, rolling and wallowing in the deep mud, have suffocated, sometimes beneath his adversary, whose burning and maniac eyes glared into his own with murderous and savage intensity. Just as he was about to become exhausted from his exertions, his daughter, taking courage at the imminent danger of parent, snatched up a rock and hurling it at the head of her father's would be murderer, was fortunate enough to put an end to the struggle by striking him somewhere about the ear. The creature was not stunned, but feeling unequal to further exertion, slowly got up and retired into the neighboring copse that skirted the road.

An article dated April 29, 1871 from *The Petersburg Index*, in Petersburg, Virginia was titled "The Wild Man—What is He?" (see photos on pages 44 and 45). I'm not sure why there isn't more documentation on the incident reported in this piece, or why there is not a notation on what eventually happened to the captured creature. One of the more interesting portions of the article is noted on page one, column two, where it talks about the creature being seen simultaneously several hundred miles apart in multiple sightings. It would appear that the reporter felt there was only one creature and they found it astounding that it could be in two places at once. As we now know, there are thousands of bigfoot across the U.S. and multiple sightings during the same hour of any day would not be unusual.

A Canadian article from *The Hammond Times,* dated October 25, 1935 (opposite page) describes an incident 50 miles east of Vancouver near Harrison Mills. The applicable part of the article is in the last two paragraphs. NABS has taken many incident reports where bigfoot throws rocks and boulders when they want the recipients to leave an area, as is apparent in this article with the Indian. The description of "ferocious looking wild men, nine feet tall and covered from head to toe with thick black hair," sure sounds like a modern-day description of bigfoot.

REPORTS TELL OF CANADIAN MONSTER MEN

Settlers Fifty Miles From Vancouver Describe Hairy Giants

VANCOUVER, B. C.—(U.P.) — Sasquatch men, remnants of a lost race of "wild men" who inhabited the rocky regions of British Columbia centuries ago, are reported roaming the province again.

After an absence of several months from the district of Harrison Mills, 50 miles east of Vancouver, the long, wierd, wolf-like howls of the "wild men" are being heard again and two of the hairy monsters were reported seen in the Morris valley on the Harrison river.

Residents in the district tell of seeing the two giants leaping and bounding out of the forest and striding across the duck-feeding ground, wallowing now and again in the bog and mire and long, waving swamp grasses.

Reported Agile as Goats

The strange men, it was reported, after emerging from the woods, came leaping down the jagged rocky hillside with the agility and lightness of mountain goats. Snatches of their weird language floated on the breeze across the lake to the pioneer settlement at the foot of the hills.

The giants walked with an easy gait across the swamp flats and at the Morris Creek, in the shadow of Little Mystery Mountain, straddled a floating log, which they propelled with their long, hairy hands and huge feet across the sluggish glacial stream to the opposite side. There they abandoned the log and climbed hand over hand up the almost perpendicular cliff at a point known as Gibraltar and disappeared at the top of the ridge. They carried two large clubs and walked round a herd of cattle directly in their path.

Indian's Story Retold

The return of the giants to the legendary stronghold of the Sasquatch monsters recalls the narrow escape of an Indian at the same spot last March. A huge rock narrowly missed his canoe while he was fishing and looking up, he said he saw a huge and hairy monster stamping his feet and gesticulating wildly. The Indian escaped by cutting his fishing tackle and paddling away. The same Indian declares the Sasquatch twice have stolen salmon which he tied outside his house out of reach of dogs.

The latest appearance of the monsters was peaceful. They avoided the trails usually used by the people of the valley and molested neither cattle nor human beings.

People who have reported seeing the giants on their rare appearances described them as "ferocious looking wild men, nine feet tall and covered from head to toes with thick black hair."

News clipping from *The Hammond Times,* Oct 25, 1935, "Reports tell of Canadian Monster Men".

THE PETERSBURG INDEX.

VOLUME XI.   PETERSBURG, VIRGINIA, SATURDAY MORNING, APRIL 29, 1871.   NUMBER 52.

THE DAILY INDEX

THE DAILY INDEX.

SATURDAY, APRIL 29, 1871.

The Wild Man—What is He?

[From the N. Y. Times.]

As most of our readers are probably aware, there is at present roaming over the United States, and for aught we know, making occasional excursions into British America and Mexico, a singular creature known as the "Wild Man." We have before commented on his eccentricities, which are many and various. At regular intervals of time he appears in different parts of the country, creating always great excitement in the neighborhood, and a vast deal of discussion in the local Press. With one exception, which we shall presently mention, accounts uniformly agree as to his appearance. He is preternaturally hirsute and ferocious, swift, and strong. He has flaming eyes, and generally speaking, a horrent and uninviting aspect. The few rash mortals who have approached him near enough for conversation aver that he speaks an uncouth and unintelligible gibberish. So far the received descriptions of him are unanimous. The apparent discrepancy with regard to his stature, which varies in different narratives from about four feet to ten, is easily accounted for by the mental agitation under which these observations of his person must necessarily be made. Or he may have the power of extending or contracting himself at will. This would not be so strange as the faculty which gives rise to the exception above referred to, of changing his sex, so that after horrifying Alabama as a wild woman, he will immediately plunge Oregon into consternation as a wild man.

His seeming ubiquity is another perplexing attribute of this strange being. Superior to time and space, he thinks nothing of being seen simultaneously at points hundreds of miles apart. Quite as extraordinary is his peculiar power of eluding capture. Usually, all the able-bodied inhabitants of the district turn out en masse to hunt him with guns and dogs, but we have never heard of his being taken. After keeping up this game of hide and seek in one locality for a few days, he departs for another, with the same result. In all his wanderings, however, it is observable that he keeps aloof from the settlements, confining himself, for the most part, to that border of civilization where contact with nature keeps the imagination fresh and vivid, undulled by the prosaic realities of city life. After flitting about the continent in this manner for a while, he suddenly vanishes into impenetrable obscurity, where he rests quiet perhaps for months. Then he reappears, and the programme is renewed as before.

We have been led to return to this subject by the recent appearance at once of the real and a bogus wild man in Tennessee and Michigan respectively. The Michigan monster is, in some respects, creditably prepared, and shows ingenuity in its inventors. Its appearance is well calculated to deceive. To begin with, it is "fearfully deformed"—a sufficiently obvious trick, but still effective. Then its "hands are covered with long coarse hair; face grown full of rank whiskers," while its eyes "look like those of a wolf." These details, it must be admitted, show skillful handling, and only a practiced eye could detect the deception. But the one infallible test of the true wild man is here wanting—his defiance of pursuit. If his more than mortal speed fails to outstrip his would-be captors, he either subsides into the ground, or sim-

News clipping from *The Petersburg Index,* April 23, 1871, "The Wild Man— What is He?"

s quantities of furs, which are
n purchased for a mere trifle,
ined his cabins with them
hout, which rendered his rude
sry warm and comfortable.
ut the 14th of January, two
carpenters who had been out
suit of a gang of wolves, that
oved very troublesome, came
c camp and reported that they
en a huge monster in the for-
a branch of the Mississippi,
; the form of a man, but much
and stouter, covered with hair,
n frightful aspect. They sta-
at when seen, he was standing
g, looking directly at them,
e moment they raised their
ts, he darted into the thicket
sappeared. They saw him a-
about half an hour, apparent-
ching them, and when they
towards him he again disap-
d. Mr Lincoln was at first
ed to think lightly of this mat-
lieving the men might have
mistaken about the size and
of the object, or supposing it
have been a trick of the Indi-
frighten them. He was in-
d, however, by some natives,
uch a being had often been
in the St. Peters, and near the
of the Mississippi, and they
ed to guide a party of work-
o a bluff where it was thought
ght be found. The men were
dy for adventure, and arming
elves with rifles and hunting-
t, they started for the bluff un-
e direction of Mr Lincoln and
ndian guides. On the way
rure joined by several of the
es, and the whole party num-
twenty-three.
ey arrived at the bluff late in
lorneon of the 21st of January,

nd when within about a mile of the
cavern, the wild man crossed their
path, within twenty rods. They immediately gave chase again, and accidentally drove the creature from
the forest into an open prairie. At
length he suddenly stopped and turned upon his pursuers. Mr Lincoln
was then in advance. Fearing that
he might attack them, or return to
the woods, and escape, he fired upon
him, and lodged a charge of buck-
shot in the calf of his leg. He fell
immediately, and the Indians sprang
forward and threw their ropes over
his head, arms and legs, and with
much effort succeeded in binding
him fast. He struggled, however,
most desperately, gnashed his teeth,
and howled in a frightful manner.—
They then formed a sort of litter of
branches of limbs of trees, and placing him upon it, carried him to the
encampment. A watch was then
placed over him, and every effort
made that could be devised to keep
him quiet, but he continued to howl
most piteously all night. Towards
morning two cubs, about three feet
high, and very similar to the large
monster, came into the camp, and
soon as the monster saw them, he
became very furious—gnashed his
teeth, hollowed and thrashed about,
until he burst several of the cords,
and came very near effecting his e-
scape. But he was bound anew, and
after that was kept most carefully
watched and guarded. The next
day he was placed on the litter and
carried down to the mills on St. Peters.

er coat of hair than the rest of the
body—there is no appearance of eyebrows or nose; the mouth is very
large and wide, and similar to that
of a baboon. His eyes are dull and
heavy, and there is no indication of
cunning or activity about them.—
Mr Lincoln says he is beyond dis-
pute carnivorous, as he universally
rejects bread and vegetables, and
eats flesh with great avidity. He
thinks he is of the ourang outang species; but from what we have seen,
we are inclined to consider him a
wild animal, somewhat resembling a
man.

He is, to say the least, one of the
most extraordinary creatures that
has ever been brought before the
public, from any part of the earth,
or the waters under the earth, and
we believe will prove a difficult puzzle to the scientific. He lies down
like a brute, and does not appear to
possess more instinct than common
domestic animals. He is now quite
tame and quiet, and is only confined
by a stout chain attached to his legs.
It is Mr Lincoln's intention to submit these animals to the inspection
of the scientific for a few days, in or
der to ascertain what they are, and
after that to dispose of them to some
person for exhibition. Mr Lincoln
himself will return to the St. Peters
in the course of two or three weeks.

Page 2 of news clipping from *The Petersburg Index,* April 23, 1871, "The Wild Man—What is He?"

NORWALK, HURON COUNTY, OHIO—TUESDAY, JUNE 25, 1839.

From the Boston Times.

Wild Man of the Woods.

[The left-hand columns reproduce an 1839 account. Much of the fine print is illegible, relating the story of Mr Lincoln, Esq., agent of the New York Western Lumber Company, just returned from the St. river, near the head of steam navigation, on the upper Mississippi, bringing with him a living son of the Woods, with two cubs, supposed to be about months old. Lincoln went out to the northern agent of the New-York Company, in July last, with to establish extensive sawmills near the pine lands ...]

STEVENS POINT, WIS.,
SATURDAY, FEB. 15, 1908.

CAPTURED A WILD MAN.

Curious Find Recently Made at Velva, North Dakota.

The Journal is in receipt of a clipping from a Velva, N. D., paper from J. Thomas, who was formerly a resident of Keene, a son of Mrs. John Thomas, who still lives at Keene. It relates to the discovery of an alleged wild man near Velva not far from Mr. Thomas's home. It is stated for three years there have been rumors of this wild man being seen by persons of veracity but he had never been encountered at close range until a few days ago, when two cattle men who were out hunting, suddenly came upon him face to face as he emerged from a thicket of brush. One of them succeeded in throwing a lasso around him and before he could escape he was dragged to a tree and bound round and round with the lasso. Later he was bound hand and foot and carried to town on a dray, where he was imprisoned in a basement. His only clothing was a loin girdle of sheepskin tied with binder twine. He had not been shaved or had a hair cut in years and being a man of an extremely hairy variety he presented a very grotesque and wild appearance. His eye teeth are reported to be unnaturally elongated in the form of tusks. He refused to talk or eat anything, but drank water like a horse, half a pail at a time. The singular part of it is that this man has always been seen within two miles of the village of Velva.

News clipping from Stevens Point, Wis., Feb. 15, 1908 "Captured a Wild Man"

An 1839 news clipping from the *Huron Reflector of* Norwalk, Ohio reported a wild man capture.

46

An article dated Feburary 15, 1908 in *The Journal* in Stevens Point, Wisconsin, titled "Captured a Wild Man" discussed the discovery of a wild man in the town of Velva, North Dakota. I personally question one part of this story: "His eye teeth are reported to be unnaturally elongated in the form of a tusk." In the thousands of sighting reports I have read I have never seen this description of a wild man's teeth. Virtually every bigfoot sighting in existence does not have a description of the teeth. The description "He had not been shaved or had a haircut in years and being a man of an extremely hairy variety he presented a very grotesque appearance" fits bigfoot exactly and would be consistent with descriptions from the early 1900s and late 1800s. Again, the odd part of this story is that nothing can be found dated later that shows what happened with this capture.

There is much discussion among bigfoot researchers that the answers to bigfoot are probably in museums around the United States already. Where are the bones described in the above article? A DNA test completed on the bones today would reveal interesting details that may aid in identifying the victims and/or creatures. In 1908 nobody ever heard of DNA, thus the bones may be buried or in the back room of some museum. It would be great if museum curators throughout North America took a second look at their inventory to determine if they have giant bones in their inventory.

An article titled "Sounds Like A Bear Yarn" appeared in the *Twin City News* in Indiana on August 16, 1937. When a witness describes a creature as a "huge ape," reporters should pay attention. Just for starters, rarely would a bear tear apart a police dog; bears are normally afraid of dogs. The description of the ape and the fact that bigfoot has been known to kill dogs indicates this sighting was probably a bigfoot.

TWIN CITY NEWS

THE HAMMOND TIMES

Monday, August 16, 1937.

SOUNDS LIKE A BEAR YARN

BOONVILLE, Ind., Aug. 16. — (INS). — They're looking for a hairy ape near here today after a monster has terrorized this community.

Persons living near Cypress beach, a few miles south of here, have told tales of encountering a giant beast which specializes in blood-curdling screams in the middle of the night and leaves footprints larger than a human's.

Ralph Duff, a fisherman, first reported the animal about a year ago after his police dog was torn to shreds in an encounter with the beast.

This week-end Mrs. Duff said she heard a terrifying howl late in the night and saw a tower monster larger than a bear. When she screamed, the beast ran away.

Duff believes that the animal is a huge ape, which lives in one of the caves along the river, and has set a number of bear traps.

News clipping from *Twin City News,* Aug 16, 1937, "Sounds Like a Bear Yarn"

NEWARK ADVOCATE.

1883

A KENTUCKY WILD MAN.

A Man Covered with Thick Hair Who Refuses Bread, but Voraciously Eats Meat and Relishes Fruit.

Among the passengers the other night bound for New York from the west on the day express was a wild man, who occupied a seat in smoking car No. 158. He was accompanied by James Harvey and Raymond Boyd, his captors, both of whom belong in Paducah county, Kentucky. They had three second-class tickets to New York, which privileges them to three seats in the smoking car of any first-class train. They were on their way to Bridgeport, Conn., to make arrangements with P. T. Barnum to exhibit their prize in conjunction with his circus. When the day express arrived at the Broad Street station, James Harvey ran down the platform into the restaurant and purchased a box of sardines and some sandwiches for the wild man's supper. His companion remained in the smoker in charge of the wild man. He was dressed in citizen's dress, and wore big cloth shoes. His hair reaches nearly to his waist and falls over his shoulders, completely covering his back; his beard is long and thick, while his eyebrows are much heavier than those of an ordinary human being. There is nothing imbecile about the wild man's manners or actions. He cannot talk, and seldom makes any sound whatever except a low howl like a leopard. His actions are as much like those of the hyena in the Zoological garden as it is possible for anything in human form to be. Raymond Boyd, who seemed to have perfect control over the wild man, said his body was covered with coarse, brown hair as thick as the hair on a horse's hide. The palms of his hands looked like the paws of a bear, and his finger nails which were over an inch long, resembled the claws of an eagle.

He was first seen in Paducah county thirteen years ago, and was known as "Mum, the Hermit," because whenever anyone accosted him all he would say was, "Mum's the word." He lived in an old pine hut in the woods for about five years, and was seldom seen by anyone. Finally he abandoned the hut and took up his abode in a cave under a ledge of rocks known as "Lizard Rock." A little over six years ago two or three citizens of Paducah county, while out hunting, saw him run into his cave without a stitch of clothing on him. He was seen several times after that wearing no clothing. Three years ago it was discovered that a thick coat of hair had grown all over his body. Boyd and Harvey built a man-trap for him over three weeks ago, and placed a big piece of freshly killed beef in it. They watched the trap for three days before he entered it. He was not afraid of any bird or beast of prey, but ran terrified away from any human being who approached him. It took two days to accustom the man beast to their presence. The tinkle of a small dinner bell they used had a great influence over him. He watched the bell intently but would not touch it.

Some time ago a farmer missed a calf and two sheep, which had strayed off. They were tracked to "Mum's" cave; here all trace of them was lost, and it is supposed that he devoured them. In his cave, which he had occupied for the last seven or eight years, Boyd and Har-

News clipping from the *Daily Advocate,* Newark, 1883, "A Kentucky Wild Man"

vey found the skeletons of small animals and the skins of over fifty snakes. Some of the skins belonged to the most venomous species of reptiles. The floor of the cave was alive with red and green lizards and hundreds of toads hopped about. The wild man ate the box of sardines voraciously, and the two sandwiches which were handed him were greedily pulled apart. He ate the ham and threw the bread away. Whenever a train passed on the opposite track he crouched down in the corner of the seat terror stricken. After the train passed he would put his hand to his ear and listen with a look of animal cunning stealing out of his restless eyes, like a panther about to spring on its prey. Every time the engineer blew his whistle the wild man would grab the back of the seat with both hands and hold on until the whistle ceased blowing. Boyd had a little tin music box which he manipulated with a crank. The one tune of "Empty is the Cradle" was ground again and again to the great satisfaction of the ex-hermit, who sat and looked at it silently, but would not touch it.

When Conductor Harry Smith took out his glistening nickel-plated punch to cancel the tickets the wild man watched the punch intently until he heard it snap. Then he got down in the corner of the seat fairly shivering with fear, and set up a low howl supposing, evidently, that Conductor Smith was about to wing him. Boyd and Harvey said that there was a story to the effect that the wild man had originally come from North Carolina, and that during the war he had been a sharpshooter on Bald Mountain, and that shortly after the war he had murdered a whole family of settlers in the mountain and fled. Both Boyd and Harvey appear like shrewd fellows, and expect to make a fortune out of their prize. Their great anxiety and fear is that the authorities will interfere with them, and

claim that the man is simply a lunatic, and place him in some institution. They had the snake skins in a box in the baggage car, together with some other curiosities found in the cave. Boyd said that the wild man will not touch anything but fruit and meat, which he eats ravenously, and much the same as a wild beast. Cigar smoke bothered him a great deal and he kept driving it away from him with his clawy hands. When the train arrived in Jersey City the men took a carriage, and said they were going to take the New Haven night boat from the foot of Peck Slip, and avoid a daylight crowd in New York In case they cannot make satisfactory terms with Barnum, or some other prominent circus man, they intend exhibiting their prize themselves as soon as they can extensively advertise him beginning in New York City some time in May. In the meantime they are going to keep him in some secluded place on Long Island.— *Philadelphia Press.*

Page 2 of news clipping from the
Daily Advocate, Newark, 1883, "A
Kentucky Wild Man"

The following articles were taken from
"The Track Record" #52 (pg. 11).

THE MORNING NEWS, St. John, New Brunswick, Canada; June 18th, 1860. The inhabitants of Carrol (sic) county, Ohio, are very much excited about a male child, from seven to ten years old, that has been seen several times in the woods, but as yet has not been taken. It has approached children quietly, but flees from the approach of a man or woman. the place has been found where it had slept the preceding night and had eaten a frog. Several hundred persons, regularly organized, are out on the hunt.

TORONTO DAILY MAIL, Ontario, Canada, July 31st, 1883. A naked wild man is dashing about the country near Boerne, Texas. He runs with great speed, is tall and slender, with long locks flying in the wind. He has out-stripped chasing horsemen, but a party has been organized for his capture.

WEEKLY DISPATCH AND COUNTY OF ELGIN GENERAL ADVERTISER, St. Thomas, Ontario, Canada, July 7, 1870. A very-scantily-dressed negro has lately been terrifying the people of Magnolia, Louisianne (sic). When first seen in the neighborhood he was observed by a white man near Magnolia, seated upon a fallen tree, eating pine cones. On being approached he ceased to eat, threw himself on all fours, and began scratching up the earth like a terrier on the scent of a rat or other vermin, until he managed to get out of sight. When next seen it was eight miles below, near the railroad station at Chotaway. Every effort to get hem to talk to anyone. even of his own color failed, and on being approached he fled rapidly away until he was seen no more. He manifests no savage or brutal qualities, but seems to entertain an absolute dread of intercourse with human beings. He appeared to be about 25 years of age, well built and healthy. His finger nails have grown to an enormous length, resembling the claws of some wild feline animal. It is believed that he was originally a runaway, and that he has for years lived in the woods and swamps, and is not aware of the emancipation of his race. Some also believe that he is identical with the wild man described in Harper's Weekly as having been seen near Vicksburg a year or more ago.

HARPER'S WEEKLY, New York, New York, Aug. 1st, 1868, page 487. A very extraordinary "bear-story" is related by a Michigan paper. A little girl, whose home was near some woods, disappeared, and nothing could be learned of her whereabouts for thirty six hours. Bear-tracks were found in the vicinity, and it was supposed that she had been carried off and devoured. A gentleman, who was searching for her, thought he heard a child's voice among the bushes, and, calling her, asked her to come to him. She replied that the bear would not let her. The men then crept through the brush, and when near the spot where the child and bear were, they heard a splash in the water, which the child said was the bear. They found her standing upon a log extending about half-way across the river. The bear had undertaken to cross the river on the log, and being closely pursued, left the child and swam away. She had received some scratches, and her clothes were almost torn from her body; but apparently the bear had not designedly hurt her. The little one says the bear would put her down occasionally to rest, and would hang his head by her side and purr and rub against her like a cat. At night, she says, the old bear lay down beside her and put his "arms" around her, and hugged her to him and kept her warm, though she did not like his long hair.

HARPER'S WEEKLY, New York, NY, August 1st, 1868, page 487. The Nashville (Tenn.) "Banner" relates a wonderful story of a wonderful serpent, which has recently made a pretty public exhibition of himself, to the alarm of all beholders. Those who were favored with a sight of the monster allowed him, unmolested but carefully watched, to retreat to his hole - a process which occupied, with stoppages, some hours. The snake was estimated to be forty feet long, and about forty inches in diameter. After he retreated to his hole the entrance was blocked up with stones, and then a strong box fifty feet long was made, one open end of which was thrust into his snakeship's retreat. They hope to capture him. This is the substance of the story; and it is said that the creature has been seen several times before within the last twenty-five years.

HARPER'S WEEKLY, New York, NY, August 8th, 1868, page 507. The "big snake" of Tennessee has been almost caught! But, alas for the peace of the community, he has escaped, and is again at large. A trap had been set for him, consisting of an immense box, with its open end placed against the mouth of the hole into which his snakeship retired. He was seen - at least his head was - going into the trap; but the beholder was so terrified that he rushed away at full speed. The next person who ventured near the trap found it lying with one end still in the hole, but the other riven as if struck by lightning, some of the bars of iron being found fully fifty steps off. No one supposed for a moment he would be able to break the immense cage prepared for him; it would have held a lion. So says the Nashville "Banner". Latest intelligence. - The snake has at length been shot by a couple of young men! He measured 29 1/2 feet in length, and his greatest circumference was 30 1/2 inches. He was spotted like a rattlesnake, and a terrible-looking fellow.

Morning news.

Kentucky even had wild-man articles in 1883. The article "A Kentucky Wild Man" from the *Daily Advocate, Newark* (shown below on pages 50 and 51) describes a trip by train as a group transports an apparent bigfoot to a circus. The description is consistent with bigfoot: "his body was covered with coarse, brown hair as thick on a horse's hide. The palms of his hands looked like the paws of a bear and his fingernails which were over an inch long, resembled the claws of an eagle." I think this sounds pretty much like another bigfoot. The part of the story where they gave the creature a ham sandwich where he threw the bread away and just ate the ham, is indicative of many bigfoot incidents where the creature won't eat bread, yet may like salty meat.

The unusual angle on this story is I couldn't find anything in the history of the circus that displayed something close to this biped. Maybe the bigfoot was a little too wild for the "wild" animal trainers in the circus.

CONCLUSION

I have supplied numerous historical articles on a creature that obviously looks like, acts like, and is described as very similar to a bigfoot. Many of these articles are over 100 years old and have come from throughout North America. The idea that all these stories are fabricated would be ludicrous. Newspapers 100 years ago were grounded in a factual representation of the news and went to great lengths to ensure their accuracy. This should be your starting point in understanding the long history of bigfoot and how it has been an integral part of North American history.

In the twenty-first century, with increased communications and the Internet as a predominant communication tool, news can travel in seconds. In the 1800s and 1900s information moved slowly, and was sometimes difficult to obtain. It's amazing how the wild man descriptions parallel descriptions of today's bigfoot.

Sometimes studying history can give us a clear picture of an issue that blurs our perception today; maybe bigfoot is one of these history lessons.

2 THE BIGFOOT MAP PROJECT

In my years of researching bigfoot I had always hoped that I'd stumble onto a map of plotted sightings/incidents that someone had prepared on a small, confined area. I know there have been United States maps made showing locations of sightings, but little information is available on the incidents associated with those sightings. When NABS took on this phase of the project, we sat down and combined the information we had accumulated on the four counties of emphasis: Humboldt, Del Norte, Trinity, and Siskiyou, all in Northern California. We wanted to plot the sightings and incidents in the same method we used for *The Hoopa Project*. The scale of this effort is much, much larger and more cumbersome, but the results are that much more startling.

The initial phase of the project was to find a map that met our criteria and one that didn't require significant time filling in rivers, creeks, lakes that the map didn't possess. Here is a short list of what we needed on the map:

1. Roads, highway, freeways delineated by different colors and lines.
2. Major lakes and dams.
3. All rivers, creeks, and streams of minimal size.
4. Mountain peaks with elevation.
5. All wilderness areas with borders shaded in different colors.
6. National forest boundaries with colored boundaries.
7. County borders with the name of the counties listed at the boundary.
8. Airports must be listed.
9. Major US Forest Service roads must be listed with the corresponding number.
10. State and federal parks must be easily recognized and named.
11. Mileage distances between major points on roads need to be listed.
12. Map legend posted and easily read.

There was one map that met all criteria we identified—the California State Automobile Association (CSAA) Northwest California Map. We negotiated a licensing agreement with CSAA to utilize the map for our plotting requirements and they worked with us to bring the map to our exact specifications.

Once the map was produced, our team spent weeks meticulously plotting over 350 bigfoot sightings and incidents from the 1800s to 2008. The map is divided into counties, with each county having its own numbering system. Each county is listed on the opposite side of the map with all incidents/sightings listed by number, and each includes a sentence or two about the incident. This is the most comprehensive driving map of bigfoot sightings/incidents ever produced. The detail contained in the map and the size of the region covered made it impossible to include in the book. If we reduced the size of the map and reduced the area covered it wouldn't do the research justice. The map may be obtained at many of the locations that carry either of our books, or it may be purchased directly from North America Bigfoot Search at www.nabigfootsearch.com.

BY THE NUMBERS

The following is a list of accumulated data on the California counties covered that are important to our study of bigfoot:

Counties	Sightings/ Incidents	Population	Sightings per Population	Square Miles	Sighings per Square Mile
Humboldt	124	128,330	1 per 1035	3572	1/29
Del Norte	58	28,893	1 per 498	1007	1/17
Trinity	62	14,313	1 per 231	3180	1/51
Siskiyou	47	45,091	1 per 959	6287	1/134
Totals	291	216, 627	Avg = 1/744	14,046	Avg = 1/48

Interesting Facts

- Del Norte County has more sightings per square mile, by more than 50 percent, than any other county on the map.

- Trinity County has more sightings per citizen than any county listed. This is a slightly skewed number because of the large population base in Humboldt County.

- Humboldt County accounts for 43 percent of all bigfoot sightings/incidents even though it only accounts for 25 percent of the entire landmass for the four county regions.

WILDERNESS AREAS

Wilderness Areas	Elevation Range (ft.)	Acres	Square Miles	#Incidents	Sightings per Sq. Mile
Red Buttes	2800–6740	16,190	25	0	0
Marble Mountain	3100–8300	242,500	379	5	1/76
Trinity Alps	1360–9000	525,627	821	15	1/55
Mt. Shasta	4500–14,162	38,200	60	5	1/12
Castle Crags	2000–6500	10,500	16	2	1/8
Yolla Bolly	2600–8100	150,000	234	5	1/47
North Fork	1000–3600	8,000	12.5	0	0
Chanchelulla	3000–6400	8,200	13	0	0
Siskiyou	1900–7300	182,802	286	6	1/48
Russian	4800–8200	12,000	19	0	0
Totals		1,194,019	1,865.50	38	Avg = 1/49

Interesting Facts

- The smaller the wilderness area in acres (and square miles), the less likely there will be a bigfoot sighting or incident. Four of the smallest areas in size have no sightings or incidents, the exception being Castle Crags.

- The Trinity Alps Wilderness Area is more than double the square miles of its closest area in size, Marble Mountains. Even though it is only double the size of the Trinity Alps, it has three times the numbers of sightings/incidents.

RIVERS

The list below shows the number of bigfoot sightings and incidents within a 10-mile radius on all forks of the listed rivers.

Rivers	#Sightings/Incidents
Scott	4
Shasta	6
Van Duzen	7
Salmon	16
Mad	19
Eel	26
Sacramento/Shasta Lake	26
Smith	41
Klamath	106
Trinity	184
Totals	435*

Several of the bigfoot sightings and incidents are within a 10-mile radius of multiple rivers/lakes, thus the number exceeds the total number for the entire four county regions.

Interesting Fact

- The largest three rivers on the list (Trinity, Klamath, and Smith) are predominantly in the far northern section of the state. The Smith is the shortest of all three of the rivers in total distance, and this may account for the smallest number of sightings/incidents.

THE PACIFIC OCEAN

- The furthest point on the map from the Pacific Ocean is approximately 108 miles.

- Number of bigfoot sightings/incidents within a 40-mile radius of the Pacific Coast: 200.

- 37 percent of the map landmass holds 69 percent of the sightings/incidents (200/291).

There have been bigfoot sightings in the Pacific Ocean in an area covered on the map—Trinidad, California. Bigfoot was supposedly swimming between the large rocks that sit just offshore, possibly attempting to obtain bird eggs.

I have also personally taken reports of bigfoot seen with kelp draped over their shoulders walking back from the ocean. There are many reports on a variety of databases that show sightings quite near the coast, even on the roadway paralleling the ocean. There are a variety of reports stretching from British Columbia through California of bigfoot digging clams on the beach.

MAP PECULIARITIES

DEL NORTE COUNTY

One item that is striking regarding the central portion of the county is the lack of any bigfoot incidents along the South Fork Smith River between Hiouchi and the Gu Road. The South Fork Road parallels the river in this region, so the fact there are no incidents cannot be attributed to the absence of a roadway. There are residents along the South Fork who use the road daily. Granted, Highway 199 parallels the Smith River between Hiouchi and the Oregon border and this road gets ten times the traffic; however, I would think that there would still be some sightings in the South Fork area.

The lack of reported incidents on Highway 101 north of Fort Dick is understood once you visit the area. Open fields and mountains predominantly surround this stretch of roadway, and big trees are sometimes five miles east, which means there is no cover for the creature in this area.

SISKIYOU COUNTY

The county has a wide range of sightings without a predominant location where they congregate, other than eight incidents reported around the town of Happy Camp. One rationale for the Happy Camp incidents is that there are many berry bushes in the town and bordering the community. It sits almost perfectly between three different wilderness areas: Red Buttes, Siskiyou, and Marble Mountains, with the Klamath River flowing through the middle of town.

There is also an odd collection of incidents in the Mt. Shasta wilderness. There are five in the wilderness and another two relatively close by. This is a small wilderness area surrounded by cities and roadways on all sides. The Mount Shasta peak has had an interesting past, much associated with unidentified flying objects and other surreal incidents. There have always been claims that some type of civilization lives beneath the peak, protected from the outside world.

HUMBOLDT COUNTY

The biggest conglomeration of sightings on the map, maybe the biggest in the world, is located in the Hoopa Reservation stretching north through to Bluff Creek. There are almost 70 sightings/incidents in this alley as you travel north from the southern border of the Hoopa Reservation until you hit the Del Norte County border. It is unmistakable. There are large gaps noticeable with no sightings in the central part of the county, and then another small blip near Humboldt Redwoods State Park, also all along the Eel River.

NABS has done studies inside Redwood National Park and believes there is some significant bigfoot activity in that area. The ten sightings either in or directly near the park are an indicator of activity. The entire open area between the park and the Klamath River are areas of high interest.

TRINITY COUNTY

It appears that the towns in the county account for the small groupings of sightings: Hyampom has 13 in close proximity, Hayfork has six, and Wildwood has 10.

Anywhere near the Trinity River, along any fork or arm, is a good place to look for bigfoot. This fact jumps out at you while you study the map, and it can't be ignored.

Wildwood is a very small community, which shouldn't in itself justify the high volume of sightings in that region (12). There would appear to be something of interest about that area that attracts bigfoot. There are no major roadways anywhere near that could account for high levels of vehicular traffic, and thus higher bigfoot numbers. The only way a traveler even knows he or she is in the town is because the lone store in the region states "Wildwood Store" across the front door.

SIGNIFICANCE

There are two statistics that emerge from the map that need addressing. The one statistic that is the most significant is related to the Pacific Ocean; 69 percent of all the listed sightings/incidents logged on the map are within 40 miles of the coast. In my first book I wrote about the significance of sightings close to the ocean, and the importance precipitation has to the bigfoot equation in Northern California. It appears that 50 inches of annual rainfall is one item that bigfoot requires to maintain a habitat in the Pacific Northwest. Ironically, I was attending and presenting at a bigfoot conference this last summer (2008) and heard Dr. Jeff Meldrum talk about his research in China regarding the yeren. The region where they were studying the yeren has annual rainfall exceeding 50 inches. In northwest California the majority of rainfall occurs within 40 miles of the coast; as rain clouds approach the coastal range, they drop a majority of their water when they hit the mountains.

There are other factors that would cause bigfoot to live close to the Pacific Ocean. The area of Northern California that is covered on this map is also the only region of the state that has a substantial migration of salmon and steelhead into the north coast river system. That migration predominantly occurs in the 40-mile zone from the coast, and allows predators to catch the fish as they leave the main rivers and migrate up tributary streams and creeks, all close to the Pacific Ocean.

Although I haven't seen reports of California bigfoot searching for clams on beaches, it wouldn't mean that it doesn't happen. If bigfoot searches British Columbia beaches for clams, why wouldn't the same behavior be prevalent in California?

The second major point that comes from the map is that 184 bigfoot sightings/incidents are within a 10-mile radius of all forks/arms of the Trinity River. With 291 sightings on the map, 69 percent fall in the Trinity River category, an amazing statistic. The Trinity River doesn't occupy any portion of Siskiyou or Del Norte County, and these represent 52 percent of the total landmass on the map in California; yet 69 percent of the sightings are near a river in the lower 48. Amazing. We are still studying the significance of this specific finding and we'll keep you informed on the progress of our research.

3 ASSOCIATIONS

After spending thousands of hours committed to researching all angles of the bigfoot phenomena, there are a few associations that are readily identifiable. These items are specifically related to the Pacific Northwest and in no way can be attributed at this time to other regions of North America until an appropriate investigation is done. The same can be true for the sketches we have completed on bigfoot/wild man. I can say with reasonable certainty that the creature in this region of California does look like the sketches completed by Harvey Pratt.

ELEVATION

Bigfoot can be found at any elevation at anytime anywhere in California. That seems like a broad statement, but based on the databases of sightings from various bigfoot research organizations, bigfoot moves from place to place without cause or apparent rationale. This is not to imply that there isn't reasoning to their movement, it means that we aren't smart enough at this point to understand their thinking and their migratory routes, if they do migrate.

In *The Hoopa Project* I made an association between bigfoot and the elevation of 2,400 feet. The famous Patterson–Gimlin film footage was shot at 2,400 feet. I documented an abnormal number of bigfoot sightings and incidents that occurred at or near 2,400 feet. This is approximately the elevation that snow starts to stick on the ground in this area of California, but it also sticks at much lower elevations. We are not making a statement that most sightings occur at 2,400 feet, but we are stating that the range from 1,600–3,200 feet is a comfort area for bigfoot, and researchers should pay close attention to this when studying incidents and sightings.

One notable story caught my eye in "The Track Record" #62 (pg. 16). A woman named Maxine reported that she is a Klamath and Modoc Indian, and says she saw a bigfoot on Sage Hen Pass in 1981. She described it as reddish in color and saw it cross the highway in four easy strides. The creature was "as tall as the roadway snow poles, though there was no snow on the ground." This sighting is located on the eastern side of Shasta Lake in a region where there have been other notable bigfoot sightings. I was unable to obtain contact information for Maxine and thus didn't include this sighting on the map. I did include it in this section because Sage Hen Pass has a summit of 2,448 feet. I think this sighting is also important because a Native American made it.

More supporting evidence of the elevation association can be found in "The Track Record" #101 (pg. 11).

> Have one report from the early 1990s. Witness reported seeing a small group of sasquatch creatures carrying huge bones. Northern California, elevation 2,600 feet, dawn, summer. Witness declined to be identified. Approximately 4 Sasquatch creatures walking single file with the largest in the lead, huge bone balanced over its shoulder, approximately 4 feet in length. Two others carrying smaller bones, which although large were somewhat smaller.

The above article is interesting, but the pertinent fact is the elevation supplied by the witness. We don't think it was random that it falls inside our elevation association range.

I would encourage anyone who is interested in this association to purchase the Northern California bigfoot map (available on our website) that lists 350 sightings and their locations and elevations. The association will become very clear.

NATIVE AMERICAN RESERVATIONS

We have documented many sightings in and around the Hoopa reservation. In the middle of my first book I started to pay close attention to Native American reservations and bigfoot sightings. There are many instances across America where bigfoot is seen either in or near a reservation.

I have documented a reservation in central California that had bigfoot as part of their culture and they permanently put it on Painted Rock. Elders within the Tule River tribe have confirmed these beliefs. I've also been to the Leech Lake Indian Reservation in Minnesota and interviewed tribal members who have had firsthand sightings of bigfoot. This is a part of their culture and history. There are also stories I've written regarding the Lummi Indians in British Columbia who have had bigfoot sightings and encounters in and around the waters of the reservation. A close confidante and associate in our research directly related to bigfoot and Native Americans is Harvey Pratt, our sketch artist. Harvey has told me that bigfoot is a very large part of the Native American culture in Oklahoma, his home state. Members of his and other tribes around the state routinely tell Harvey about bigfoot sightings. Since he is intimately involved with his tribe and their culture, Harvey has a special view of the bigfoot and the Native American relationship that even I do not understand.

The following article comes from Ray Crowe's "The Track Record" #53 (pg. 7, para. 2).

Norma Truimble stopped by the shop to tell me that she had just heard from two Indian friends just down from Canada. They had stopped by the Lummi Indian Reservation near Bellingham, WA [Washington] and been told of a Bigfoot that had turned two mobile homes over about two weeks ago (end of December 1995) along the Nooksack River. The Nooksack has produced many Bigfoot reports in earlier days. Nobody was hurt and they were puzzled about the cause for the attack as the fishing had been poor and closed down, the runs had been nearly extinct and nobody had been out in canoes lately. Could the creature have been agitated over the lack of fish in the river and looking for someone, or something to take it out on?

The article is an interesting story as it occurred on a reservation and it involved an apparent attack on a Native American's residence. It would be interesting to know how long the trailers had been on that property; were they newly placed or had they just been placed since the last time the creature had been back? It reminds me of stories going back to the late 1950s and early 1960s along the road build out near Bluff Creek. There are several historical accounts of equipment being turned over at night when the crews left. This is just another example of bigfoot being associated with a reservation and Indians.

Native American reservations have been highlighted on the History Channel's *Monster Quest* in Oklahoma. The Cheyenne Arapahoe Nations have spoken of their belief in bigfoot and its visitations.

There are many Canadian Natives who have publicly stated that bigfoot has walked on their reservations, and John Green has documented their sightings on their reservations and in their rivers.

There is much talk about why bigfoot may be visiting reservations and possibly living near their boundaries. One answer may be that Natives have lived with bigfoot for hundreds of years and may hold it in reverence; they will never harm the creature. Many Americans have claimed they shot a bigfoot, but a Native American would never purposely do that, so bigfoot may feel comfortable and safe living near a reservation. The reservations are also protected zones where outsiders cannot travel without specific approval from the tribe. It may be the lack of traffic and absence of many people that is another rationale for bigfoot living close to reservations.

The following article exemplifies the relationship that Native Americans have with bigfoot. A sighting of a bigfoot by a Native American is not an uncommon occurrence. The following story was taken from "The Track Record" #52 (pg. 8). The sighting occurred in 1988.

> Earla Penn, a Quileute Indian from La Push, WA [Washington], says 8 years ago, in February, about 1:30 AM, she encountered a Bigfoot west of Kings Valley, OR [Oregon], in the Coast Range Mountains. In the car, she could hear something walking, and as it came closer to the car, she jumped out in time to see the silhouette of a really huge, 9 feet maybe, tall creature. She wasn't afraid, and waved at it. It stopped and looked at her, and then walked away without making any noises or leaving any smell. She told the police, and they looked for tracks, but she doesn't know if they found anything or not.

The following is another example of the association of bigfoot and

Native Americans. It comes out of Tacoma, Washington and is chronicled in the same issue of "The Track Record" #52 (pg. 4).

> Warm Springs Indian Garrett Suppah reports to tell of his Yakima Indian Grandfathers story of Bigfoot. Seems every fall a family of 6 passes near his place near Rattlesnake Creek, Klickitat County, Washington. They are heading west from a hilly-forested area east of him. He knows it's the same family because one of the females has tanned colored eyebrows that stand out.

If bigfoot is walking in areas where they can regularly be observed, they must have great faith that the residents will not do them harm. There must be some relationship between Native Americans and bigfoot that we are still struggling to understand.

In October 1996 a Navajo Indian dropped by Ray Crowe's bookshop in northern Oregon and was selling Native art from his tribe. He and Ray entered into a conversation about bigfoot and Ray asked if he had any stories. The Navajo Native stated that he had a friend who was cutting timber on the reservation. He later was at the base camp when he saw a large human-ape creature pick up a large tractor tire from one of the grading machines and throw it at him. This was in 1989 on the Navajo reservation in the Chuska Mountains on the Arizona side near the New Mexico state border. Ray documented this encounter in "The Track Record" #62 (pg. 5).

Another report came into Ray Crowe from Fran Krogstad from Filer, Idaho, dated October 5, 1996. The report was from a trip that she and her husband took to the southwest corner of Idaho.

> As is our lifestyle, we were out prospecting in the most remote county imaginable. We'd gone a few hours drive southwest of the Duck Valley (Indian Reservation), drove through a couple rivers to trails end, further broke through a couple miles of sage brush with the ATV and then hiked the rest. We brought plenty of noise with us, as we look back now.
>
> We came intending to stay a few days, if need be. Around 7:30pm we prepared camp. When we are alone like this I can't fall asleep

easily, but I was starting to fade about 9:30–10pm, when a powerful, lengthy scream roared forth from further up the canyon we
were in. At most this was 1/16 mile. I quickly realized this was not
a coyote. Could it be like a jet, definitely not! What about the wild
horses we'd seen that day? Well, even in most agonizing distress
it wouldn't be possible. With that process over, it still roared
unbroken. Interrupting as little as possible, I quickly asked Gene,
"You hear that?' As quickly he answered, "yes." The scream
tapered to an end. At this point we were resigned to the realization that we were being confronted by a Biqfoot.

I am going to jump ahead in Fran's story. She did talk about the
rationale for concluding it was a bigfoot. She had bought a tape of
bigfoot screams and the sound matched the scream. The noise
caused so much concern to Fran and Gene, they left the area the next
morning. I'll pick the story up when they left.

Next morning we drove back to civilization. There was road construction on the reservation. We passed time with the Indian lady
holding the stop sign. Thinking Indians are the most in touch with
such phenomena. I dared ask if she'd heard any talk of weird
sounds. Gene was very surprised (though I was not), that she
brought up Bigfoot immediately. She told us that, among other
things, that the past 15 years that they (Bigfoot) have been frequenting the reservation boundaries. We were eager to get home
and play our tape again.

There are many angles to the above letter that offer an interesting
insight into their story. To begin with, Fran was well versed in how
a bigfoot scream sounds; as she said, she had listened to a tape.
Also, they were in a very remote region, good for a bigfoot incident.
They were near an Indian reservation, even better for a bigfoot incident. The statement from the Indian woman confirmed that bigfoot
had been visiting the area of the Duck Valley Indian Reservation for
at least the past 15 years. Compelling…

In March 1996 Duane Richards was in Mandan, North Dakota
working on an oil rig. A Sioux Native American told him that some

65

Caucasians were trying to attract bigfoot by hanging used sanitary napkins in nearby trees. The lady didn't know if they were able to attract anything, but she definitely saw one of the creatures behind her house on the reservation. No description of the bigfoot was supplied with the article. This appeared in "The Track Record" #66 (pg. 7, para. 7).

I've heard many Native American stories about bigfoot over the years, but one consistently told is listed below. I know it sounds like a convenient explanation to many of the questions that surround bigfoot, but this story is so widespread that it deserves mention. The specific statement was forwarded from Native American PhD candidate Mel Brewster, on January 19, 1999, and is reproduced from "The Track Record" #83 (pg. 14). There are two parts of this statement that are notable, the first being Brewster's explanation of bigfoot as a spirit. The second is his explanation of his grandmother's story of sighting a bigfoot. The part of the story where the bigfoot is seen laughing is not unusual. Witnesses claim it appears to laugh, or makes sounds consistent with laughing, and this usually occurs when it's in the presence of other bigfoot.

I meant to tell you, there are some indications from the Numu (In Nevada, north of Reno) community that Pahazoo or some other creature is a spiritual creature. That meetings with it are spiritual and that the only way to get to him is spiritually. That is to contact him formally. Otherwise chance encounters with this spirit are impossible to follow up. A brief encounter with it spiritually is the most someone may have ever dreamed of or be able to handle in the past. Some people had and still have his medicine. What would somebody do? You would have to be completely submerged in tribal traditions and meditations that would allow ones self to communicate with such a powerful spirit.

This is heavy, you may be chasing a spirit that takes form on occasion, leaving hair, footprints, smells and so on (like its whinny high pitched but powerful roar)! Incredible-Uh na bu da ha!!

Otherwise, they must be part of the ancient (Num Wad). Part of the animal teaching stories. And also a spirit. My great great grandmother also states that they seen a huge man that was really tall and naked and hairy sitting on a branch over-looking the river, in Montana. Basically, it just looked at itself and laugh for hours. Story was passed down since late 1800's.

I'm sure you can meet that spirit but you will need to be by yourself away from the smell of lots of people. You have to find out how to get really spiritual but that is scary because it takes many years to prepare. And you will have to leave the Western tradition behind. Spooky!!

In an effort to offer a broad perspective to the Native American/bigfoot connection, I am including the following report from South Dakota, printed in "The Track Record" #91 (pg. 7). There are five different incidents listed (including a newspaper article) that all occurred at the Lakota Sioux Indian Reservation.

BFRO report, Dawn Harrack, August, 1999, Cheyenne River, Lakota Sioux Indian Reservation near Eagle Butte, South Dak. Four sightings were reported within 50 miles.

#1. Five people were fishing on Cheyenne River south of Faith saw a tall and hairy creature similar to a bear which was watching them from across the river, and it appeared to be waving at them. Witness T.C. said it was taller than some small cottonwoods by the river bank.

#2. A white man living in Faith saw a hairy creature cross his path and called the highway patrol to investigate. This report was heard on a scanner by several listeners.

#3. R.D., "Iron Lightning," was swimming north of Faith, and saw a hairy creature with a little one. Iron Lightning and others gave chase but were unable to get close.

#4. F.V.D., a white rancher and farmer from east of Faith, saw a hairy creature while making a dam at dusk. It made a noise and dove into a narrow creek bed.

The local Eagle Butte News carried a story from some years ago about sighting on the reservation at Green Grass, SD. There were actual foot print photos in the paper that were investigated by Officer S.P. of the Cheyenne River law enforcement agency.

Note, many UFO sightings have been noted in the 2100-foot elevation area around and south of Faith, S.D.

It's interesting that a UFO notation is listed at the bottom of this report, because many of the tribal members in Hoopa had told me that they routinely saw UFOs over Hoopa.

In "The Track Record" #90 (pg. 15), Ray Crowe documented a very unusual report involving a sound from a bigfoot that sounded like a language. I have talked to many Native Americans who were told by their elders that if they were ever to be confronted by bigfoot, to speak in their native language, apologize, and slowly back away while facing the creature. Damon Colegrove confronted a bigfoot after shooting a deer, and then spoke to it in Hoopa as he backed away *(The Hoopa Project)*. Hoopa Native Inker McCovey had told me several times that his people believed that the bigfoot understood their language *(The Hoopa Project)*. I also interviewed Pliny McCovey who resided in Hoopa and saw two creatures on a hillside in the bushes behind his house. As he crept up to them he could hear them speaking a language that he couldn't quite understand. He stated that it was definitely a language but he just couldn't understand it *(The Hoopa Project)*.

The following story aligns well with the stories coming out of Hoopa. Lamonica was able to obtain a tape recording from a bigfoot group, and he has attempted to identify the language for several years. The following explains his findings.

I am sorry if I am beginning to sound like a broken record, when it comes to this tape recording. It is just that it has been nearly five complete years since it happened. I am not one step closer to figuring it out today, than I was the day it occurred. I have talked to linguists at Kent State University. According to them, it sounds as if it may be a language. However, they could not tell what lan-

68

guage it was. The hottest lead I have gotten on this tape is from Hawk Spearman's Grandfather, Graycoat Spearman. He has said, without any inkling as to what I think it is, that he hears words on it that are in the Algonquian-Wakashan Native Language. That is about as much as I have learned regarding the origin of the spoken language on the tape.

If bigfoot were related in some way with Native Americans, then it would only follow that they understand their language and possibly started their own language. There are too many Native Americans claiming that bigfoot speaks a language close to their own for there not to be some truth to the claims.

Rob Alley wrote about more Native American associations that appeared in "The Track Record" #91 (pg. 11).

> While working as a physical therapist about the same time in the hospital in Powell River (British Columbia), my next door neighbor on the Indian Reserve, an obstetrician, told me that two nurses had to be treated by two different psychiatrists the year before following a car crash into the ditch. They had both recounted near identical versions of a sighting of a hairy manlike creature on the road which they had swerved to avoid.

No other information was available.

K. Foster found bigfoot tracks in southern Colorado in 1993. The discovery of the tracks prompted Foster to dig a little deeper and see if there was any historical information about bigfoot in the southern U.S. area. The following was written in the "The Track Record #95 (pg. 5).

> I found out that there is an old Pueblo (Jemez Tribe) Indian archeological site just across the border in New Mexico called 'Gee-tow-to-out-ay-new,' which is translated 'Where Giant Man Stepped.' I asked the Jemez Tribal Archaeologist about the name, and what he told me sent chills up my spine. He said, 'Oh, that is a name in reference to a tribe of hairy giant men that the Jemez have many tales about that lived in the ponderosa forests of that area.' He

also told me that the Jemez tales told of the fear the Jemez had of these "Devils" and that many of the tales told of the giants eating people. He said he could not share the stories in whole, because Jemez hold the stories sacred and are part of a ritual that only Jemez can take part in. I asked him, 'when did the Jemez build and name the ruin?' He said, 'we know for sure that it was in existence and had that name in 1540 A.D.

The Jemez story is a remarkable confirmation of a bigfoot related tale going back almost 500 years. This is also a credible report with information coming directly from the tribal archaeologist. It is also important to note that the archaeologist called the group a "tribe."

Another Native American story from Canada, reprinted from the "The Track Record" #108.

The Province—Wednesday, June 20, 2001, Huge Footprints Bring Myth to Ontario Peawanuck, Ont.

Footprints measuring 35 centimeters long and 12cm wide have been spotted in the Native community on the north shore of Hudson Bay, and the Chief of the Weensuk First Nation has only one explanation: Bigfoot. "Its definitely not a bear," Abraham Hunter said firmly. I looked at them. They were 6 feet apart, walking." News of the prints first spotted June 9 by a band member riding an ATV through the bush soon spread through the community of 250 and the hunt was on for the mythical ape-like beast. The tracks drew the curiosity of the Natural Resources Ministry who arrived June 14 to investigate and record the images of the footprints. "We were surprised," said ministry official Brett Kelly, who admitted no one could explain what caused the tracks.

"The Track Record" #111 (pg. 9, para. 7).

Roland Bourguignon, 8 Jan 2001, my second cousin's husband is a former Indian Chief of the community of Ogoki, about 400 miles or so north near Webequie (Ontario), which I told you earlier of a Bigfoot sighting. Well, apparently some of the fellows in the Ogoki area have mentioned sighting our big hairy friend. Will confirm these stories in a later letter, OK? Also, these places are only accessible by air, no roads, some have winter roads though. I often wonder how many more people have sighted our big hairy friend and have said nothing about it. Around here over the last number of years cougar sightings are popping up. Of course these skeptics are saying na, na. Will track down photos I witnessed, and can account of their screams and droppings being found. Lots of deer around here, more toward Fort Francis/ International Falls, MN also lots down near Sudburry to Manstoalen.

"The Track Record" #112 (pg. 2, para. 6).

Wynoochee Reservation, Washington, a man and wife walked up a logging road, the man on crutches. Bigfoot, apparently curious, came out to see the strange man on four legs. It stood and looked at them, eventually turning and walking away after the woman talked to it, "that's ok, we're not going to hurt you."

"The Track Record" #115 (pg. 2, para. 2). The following came from an individual living on the Lummi Indian Reservation.

3/10/02 Follow-up. Feel free to write it in the newsletter, it was Tuesday the 5th officially at 12:38am. There are reports here on the reservation monthly, but of course you're never sure how many are true. It appears that the Bigfoot is coming in specifically to get salmon out of the Sush-Nak River. Most sightings occur at the time salmon make the runs for spawning. Some of the locals once you earn their trust open up about their experiences. Jason Valenti and myself are going to begin geographically mapping out the area and start putting pins in the areas where Bigfoot have been spotted or heard. Hopefully we will see a pattern and set up cameras.

The next two pieces are from "The Track Record" #117. (pg. 11, para. 4). The first is an abduction story.

Dr. Edward Fusch tells of updates on his book concerning the Spokane and Colville Indian Reservation tales. Of the most interesting in his search for DNA from a lady who is ¼ Bigfoot. Seems a newly wed maiden was carried off, leaving a trail of white fragments from her slip, and escaped later when she was noted gathering wild potatoe roots and the creature slept, so, she ran and jumped on a horse behind the rider and left, but was pregnant. She said the Bigfoot made fire with flintstone and covered mouth of caves with skins it had stolen from Indians. Her son, Patrick had long arms (to his knees), long fingers, was 5'4" tall, was a pin head with a sloping forehead, large lower jaw, large wide mouth, straight protruding teeth, hump backed, ears peaked, a large head, was very ugly, but very intelligent. He is buried on the reservation having died at 30. He had 5 children, two daughters that lived.

• • •

(pg. 11, para. 15).

Tim Cassidy recently moved to Washington from Indiana and has been studying the 13,000 acre Lummi Indian Reservation. He says that Bigfoot reports number in the thousands. He has studied the history and the culture extensively and thinks 5 creatures were there. He tells of taking salmon out of a freezer, of an apple orchard where he has a couple of reports of drunken creatures from fermented fruit.

"The Track Record" #118 (pg. 3, para. 7).

Ophelia Jartay of Sweethome, OR [Oregon], Spanish/Native American was a guest of our VP (International Bigfoot Society) Patti Reinhold. Living north of Seattle at Marysville, she and her son went to the Tulalip Reservation where her brother lives, in 1989 on a Thursday night in late spring or early summer. There was a big bonfire party in the hills of the reservation on this clear night, but she had to leave early. It was a school night and she had to work the next day. At 10:30pm as she descended the windy road from the 5 acre site, half mile off the main road to her friends at the party she rounded a curve and saw a monster. It was just standing on the side of the road, arms hanging down, like at attention. As she passed by, coming within 10 feet, the monster hunkered over to look into the vehicle window to see her. It was matted and dirty looking, dark colored, eyes reflecting in the in the headlights. It turned its whole body to follow her movements as she passed. She was terrified, and couldn't say how tall or how heavy it was, and wasn't about to go back past it to warn her friends, calling later to tell of a humongous monster. They put out the fire and went in.

The notation by the witness of the hair on the creature being matted may be associated with the time of day. Many researchers believe that bigfoot is nocturnal. Noting the time, 10:30 p.m., it may have just woken up and been a bit disheveled.

The following two entries are from "The Track Record" #121. This first (pg. 6, para. 6) is one part of a series of stories by Don Monroe.

Today (9/1) again more Indians from the Fort Hall Reservation near Pocatello (Idaho) came in to tell me about a recent Sasquatch sighting in the Black Foot River area. Got it all on tape. Three more stories this past week as well.

No details of the event were included.

• • •

(pg. 12, para. 7). Small Bigfoot Jumps in Tree.

My friend told me that while doing an update on a report he was covering, he talked to an Indian lady that lived in that area (Idaho). She told my friend that when she was a girl she was walking in the woods one day and she heard a pack of dogs barking and running in her direction. So the girl hid herself behind some brush and watched as the dogs came closer. As she was waiting to see what was to come about, she also heard footsteps as if running fast. As she heard the dogs getting closer she could see what it was they were chasing. It was what she called a small child Bigfoot. The dogs were a pack of beagles. This small Bigfoot was running way ahead of the dogs but came to a tree that was pretty close to a drainage ditch with a pipe that went under the highway. The small Bigfoot got up to the tree and hid itself inside. Just then the dogs came and stopped at the tree, sniffed around the tree and the area. They lost the scent and got uninterested and turned back to where they came from.

The Indian lady said she waited for a while to see what the small Bigfoot was going to do. Then she said, it jumped down out of the tree looked around real fast and then ran into that pipe that went under the road. The lady knew very much about the Bigfoot. Her granddad had told her many stories of them.

The following entries are from "The Track Record" #122. The first (pg. 3, para. 8): bigfoot travels through the trees. These stories were first told to NABS by a Hoopa police officer who explained this behavior.

Tim had no reports from Crater Lake but did have a few other things of interest to relate. In 1998 on the Lummi Reservation, Washington, Chief of Police Charles Skeetum tracked a Bigfoot that was traveling through the trees. This was a younger creature, about 6 feet tall. Skeetum followed the creature for a quarter of a mile as it moved through the trees.

• • •

Nelchina Plateau, Alaska 1930's in the month of September Gilyuk, the cannibal giant, killed an Indian. His sign was a sapling twisted to shreds.

Alaska 1940, near the ghost town of Kaluka, Emily Supanich's mother was berry picking with others when "they came upon a large hairy creature that resembled a man covered with long black hair. They ran back to the village to tell people.

From the same issue (pg. 4, para. 2): Eric G. of Bend, Oregon and his friend Matt, an avid hunter, had joked about the possibility of bigfoot, but neither gave much credence to its existence, until they saw signs of bigfoot on the Warm Springs Indian Reservation.

Not just one or two, but roughly 60 yards of tracks from an animal that obviously walked on two feet, was very heavy based on the depth of the prints, and its feet were 18 inches long when this thing walked out of the river, then traveled upstream stepping over a downed tree that Matt and Eric had to leap over with two feet. This thing simply hopped over it landing on one foot and left an enormous print. It continued downstream, walking through the clearing where its prints were first spotted. Those tracks lead back into the thickness of the riverside pine forest. (At this point the tracks were in the dirt and forest carpet. They were not complete tracks, as only the heel of the foot was leaving evidence behind). The he noticed an almost perfect print, minus toes. It had stopped for a moment. Then Eric noticed the scat. "The Bigfoot you are looking for squats like a man when it poops. The scat was not directly between the prints but back from them 2-3 feet. I was able to stretch and almost put my feet in the prints, squat with my forearms on my knees and that is exactly where the poop was in relationship with the prints."

The reservation's Oliver Kirk was able to accurately guess where on the reservation this took place as he has close to 30 cases that he has documented in the area.

• • •

(pg. 10, para. 2).

Daris Swindler, for example, is not your typical Bigfoot believer. When he retired in 1991 after more than 30 years at the University of Washington, Swindler was an acclaimed expert in the arcane study of fossilized primate teeth.

[I'll jump ahead four sentences at this point.] Swindler continues to state, "Mythical giant apes lurk in the traditions of nearly every North American linguistic group and in legends handed down through the ages from Europe to Asia. Each year Bigfoot or similar creatures are reported by hundreds of hunters, hikers and motorists and others from central Asia to central Rockies.

• • •

(pg. 13, para. 3).

Bigfoot trivia, native Americans have more than 60 names for the animal legend known as Bigfoot. The most common are Sasquatch, Bigfoot and Yeti. The name Sasquatch is an Anglicization of the Coast Salish Indian word, "Sesque" of British Col. Seaquac means "wild man."

"The Track Record" #124 (pg. 15, para. 3).

Sasquatch also has a place in history of Nuu-chah-nulth people. Jessie Hamilton an elder with the Hupacasath First Nation, recalled a story of one Hupacasath ancestor who went up into the mountains where he encountered the creature, "One of our ancestors came across on in one of the mountains," she said. "He saw it and didn't know what it was." As the story has it, the man attempted to hide from the beast by wrapping himself in a blanket. While wrapped in the blanket the man was allegedly picked up by the great creature and twirled around before being set down. Hamilton said that the sightings of Sasquatch, or Cacuuqhsta as his name is in the Nuu-Chah-nulth language are not unheard of but are usually confined to the west side of the island (Victoria, BC). "He was seen by different tribes on the other side of the island," she said. But it surprised her that the creature has been spotted so low in elevation. She speculated that the beast might have been driven into lower elevations by a lack of precipitation in the mountains over the past couple of months. "It surprises me that he was seen so low. He may be looking for water.

"The Track Record" #125 (pg. 4, para. 3) printed a letter from a Northern California bigfoot organization.

When I came back from Korea I was stationed in Alaska with a special military unit. I was assigned as a military game warden and had the privilege of traveling all over Alaska. I was assigned to the admiral's staff. The admiral was an avid hunter and fisherman, especially fly-fishing. While assigned in Alaska, I was asked to do some research on the local commercial fishing. My assignments were out of Sitka and Ketchikan. Upon arrival in Ketchikan I was asked to go along with a commercial fisherman on a 32-foot commercial fishing boat. The captain fished for crab and halibut. The captain was a native Alaskan (Haida Nation). I spent four days with this captain. One night we were getting ready to set the anchor for the night. We used two anchors, one for the bow and one for the stern. This was due to extreme tidal changes. While we were setting the anchors, I noticed a large blackish animal walking the beach. At first I thought it was an extremely large black bear. I mentioned this to the captain and he casually looked up and stated it was a Sasquatch. I inquired what was a Sasquatch? I'd never heard of a Sasquatch. After we secured the boat for the night, I watched Sasquatch for about 15 minutes, and it was approximately 50-75 yards away. As it was walking along the beach it was digging and picking up clams, I later found out.

The captain of the commercial fishing boat told me that his nation has admired the Sasquatch for many years. Also, when we returned back to the harbor, he would introduce me to the Nation's shaman, if it were permissible. It was. The shaman told me numerous stories about the Sasquatch. One story that stands out in my mind or recollection was a story about a very old shaman, his grandfather, about a fight between a Sasquatch and a large bear. The shaman that I talked to was 102 years old at the time.

Unfortunately there were no more details about the sasquatch and bear encounter. The story does exemplify that even in Alaska the sasquatch is well known and accepted. From the story you can see that the sasquatch is not afraid of the First Nations people, while it would probably be afraid of Caucasians. It allowed itself to be viewed for an extensive period of time while it was on the beach.

"The Track Record" #127 (pg. 4, para. 11).

BIGFOOT ABDUCTS
NATIVE AMERICAN WOMAN

A second case is the Skanicam. It is one of the Indian ladies being kidnapped by a Bigfoot and made pregnant. The child, Patrick, married and had two children, one of which resides today in the Seattle/Olympia area (Dr. Fuch is trying to locate for DNA Sampling).

No further details on this incident, but we thought it was important because it is another article relating the abduction of a human female by a bigfoot.

The information received about sasquatch or bigfoot throughout the Pacific Northwest contains many reports from Native Americans throughout Canada and the United States. There isn't one country or a few tribes in either country reporting the sightings; it is consistent throughout each country.

QUARRIES & MINES

Areas that offer significant open space appear to be a draw for the hominid. Either a typical quarry or mining operations seem to attract the creature. There are many bigfoot sighting reports—too many to mention them all—in a variety of databases that identify mines as places where bigfoot has been seen. A mine would have many of the same features as a quarry, including dirt thrown out by the mine and an open area where the mine is located. It may also offer shelter under some circumstances.

Bigfoot may use the mines/quarries as something similar to a beacon or landmark visible from miles in any direction. While humans initially think about visual markers, for bigfoot these could also be odors/scents associated with locations, which we cannot smell. I've heard researchers hypothesize that the horrible smell coming from a bigfoot is related to sulphur and that sulphur is related to underground caverns found in caves or mines.

Rose French (chapter 10) saw her bigfoot within 200 yards of an old quarry that was very near her vacation cabin. She actually commented that she saw huge boulders at the quarry stacked on top of each other, and saw they had been moved from mound to mound, even though the quarry had been closed for years and there was no heavy equipment on site to move the boulders. The only way the boulders could be moved was by someone's, or something's, hands, and that something had to have tremendous strength.

Aaron Carroll saw bigfoot at the Tyson Mine in Del Norte County. Kirk Stewart later found footprints (stories in chapter 9) on a ridgeline directly above the mine. These were two distinctly different incidents years apart, yet both were related to Tyson Mine.

There were bigfoot sightings and footprints reported behind the airport in Happy Camp where there was an abandoned mine. Since the time of the sightings the mine has been covered, but during the sightings it was open, deep, and accessible.

It would make sense for a bigfoot to live in an abandoned mine. If they are nocturnal and have outstanding night vision, there would be no need for artificial light. It would be a place that would shelter them from the elements and insulate them from humans.

I decided to include a short story about a bigfoot seen in a mineshaft that was taken from the "The Track Record" #52 (pg. 7). This is just another example of the association between mines and bigfoot.

Larry Lund tells of Dr. Fahrenbach receiving a sighting report from the upper Clackamas River, OR that was very interesting. The man and wife were exploring an old mine-shaft January 14th. There are about 30 abandoned quicksilver mines in the vicinity, on the south side of the river, reached by cable or boat. This mine is near Lake Harriet, off Hy. 57, east of Ripplebrook Ranger Station, T6S R7E, Sec. 4. This particular horizontal shaft was boarded up, with a special door for entry that had to be crawled through (Larry wonders how a Bigfoot got in). Using big flashlights, the lady first, it was just a short distance from the entry of the 400 foot shaft where she saw a Bigfoot. She was so startled that she dashed back out with her husband (he didn't even see it), before she told him of the shaggy, silvery gray creature. It was sitting hunckered up, head between it's [sic] knees, and the arms swrapped around it, one had shading its eyes from the light. She described it in human terms, rather than those of a bear, and commented on the foul odor when they enetered. Later, her husband with some neighbors, re-entered the shaft, and explored it to it's [sic] end, without seeing anything significant, but the smell was gone. They did collect some bedding material that was due to be delivered to Dr. Fahrenbach on Feb. 7th (he examined it and it was thought to be a wood rat nest, and wood rat hairs were found). The bed was 4½ x 4 feet, and composed of sticks, moss, and ferns. There was snow on the ground, but no tracks were noted. At a nearby work center garden, stalks of Brussels sprouts had been eaten by something with big teeth (elk?).

K. Foster did some research and found a historical note about bigfoot and mines. It was covered in "The Track Record" #95 (pg. 5). There isn't a lot of detail but the association is there.

There were also some tracks found by miners in the 1920's and a short article in a newspaper about giant man tracks found at the entrance to a mine. The locals called the giant "Boji" for some reason, and several residents described a dark colored hairy giant "wild man" seen up in the Crestone, Colorado area in the 1920's, strange.

BEARS

In my first book I went into a great deal of explanation about the association between bigfoot and bears. There is no doubt that where there are a lot of bears, there is probably a very good chance it's good bigfoot country. They probably access the same food source and occupy the same place. It's also likely that they attempt to stay out of each other's way. There is nothing to gain and everything to lose by getting into an altercation. In British Columbia, bigfoot has a much tougher bear to deal with than the black or cinnamon bears in California; they must deal with the grizzly. A confrontation between a grizzly and bigfoot would no doubt cause serious injury to each, possibly fatal. It does not appear that bears are a normal food source for bigfoot, thus they may not feel that they are a physical threat. If bigfoot found a dead bear they might, out of pure opportunity, take portions to eat, but I have personally never heard a credible story of a bigfoot attacking a bear, killing it, and then carting it away.

In *The Hoopa Project* I wrote that there is no place in California with more bears per square mile than in Hoopa. It also has more bigfoot sightings than anywhere else in California, and is also where a food source for both bears and bigfoot grows in great quantities: berries.

BERRIES

There will never be an argument about whether berries are a substantial food source for all bears; bears seek berries out when they come into season, as they offer a significant source of vitamins and energy. There have also been many sightings of bigfoot either eating berries or being near berry bushes, and there have been many bigfoot tracks found near berry bushes. The oldest bigfoot sighting noted in this book—the Marble Mountain sighting in "The Hermit of Siskiyou" — talks about the creature at a berry bush eating berries.

My first book recounts a story where several kids and a mother watch a young bigfoot walk up to a berry bush and delicately pull the berries off the bush, one at a time (McCardie). There are many, many other sightings noted in various bigfoot databases about bigfoot eating berries.

It's always encouraging when another researcher finds supporting evidence for a theory you've developed, as Ray Crowe did with the berry connection. There were a series of sightings and incidents that the editor of "The Track Record" listed.

"The Track Record" #52 (pg. 8, para. 2).

Another story, this time told by a logger friend in Montana about 10-15 years ago, but he doesn't know where. His grandmother was tent camping and picking berries. Going to bed that night, after dark a noise was heard outside of the tent. Peeking out she saw a male, female, and an adolescent Bigfoot snooping around. The male reached into the dead campfire, dug his hand through the ashes of the dead fire pit, picked up a handful of cold ashes, smelled them, then threw them away. She was scared and didn't know what to do, so climbed back in her sleeping bag until dawn when the creatures were gone.

Note: There was a sighting from the 1800s that also indicated that bigfoot played with fire at a campsite, an interesting behavior observation.

"The Track Record" #100 (pg. 3, para. 7).

Jason commented that there was an Alaska report from 1975 printed on page 58 in the *Alaskan Gazeteer.* Maknek was out with his family and saw for a half hour, three hominids on the tundra picking tundra berries. They were on a hill 200 yards away and clearly seen. They didn't run off, but stayed on their task.

"The Track Record" #112 (pg. 14, para. 9).

A father and son were out picking huckleberries on a cutblock on Vancouver Island, the buckets were filling up so the father told the son to go take some of them down to the truck and empty them into a larger bucket. The son did as he was told, and on the way back through some heavy brush he noticed that a stump was beside the trail that was not there before. It turned out to be a Sasquatch kneeling with its hands over its knees, it rose up to look at the boy. The boy yelled out and the dad thought it was a bear and came running down towards the boy with a high powered rifle, yelling to the boy if he was ok. The boy yelled that he was ok, and the Sasquatch started to move at a very fast run through the young planted cutblock. The dad emptied his gun at the Sasquatch and it was screaming as it ran towards the other side of the block, 1964. I am assuming that the Sasquatch was screaming either because it was hit or that it wanted the shooter to know that it wasn't a bear, or else it was so scared it was yelling in fright.

This is an interesting story when you think about the psychology of the sasquatch. We've always heard that they are afraid of firearms, and that they know what they are. The boy was obviously not carrying a gun; he had berries. The sasquatch was probably in the area because of the berries. Was the sasquatch trying to mildy intimidate the boy out of his berry area or was he trying to interact, since records show they are not afraid of children? The sasquatch must not have known that the father was nearby with a firearm or the encounter with the boy probably would not have occurred. The chance that this sasquatch will casually walk up to a child again without thinking of possible consequences is remote.

This next story from "The Track Record" #117 (pg. 14, para. 10) starts on Lost Creek in Oregon (T3N R9W sec 26). A couple was camped in the area and found huge human-like footprints in the area of the creek bed. The couple would walk the area at night with flashlights and hear rummaging in the bushes, screams, etc. Another group of campers came by—satanic worshippers—and the bigfoot chased them off. The couple went to the bottom of Lost Creek Road and heard a strange dog bark, but they couldn't understand why a dog would be in that area. This was in September 1994. It was during this time that the couple was walking down a wash and saw a bigfoot stand up from behind logs. It was chocolate brown, thin lips, huge mouth, and large pupils. The bigfoot "stared down" the people so they decided to get to their truck quickly and get out of the area. Here is part of the original story.

I asked for more details and Brian said it had a scrunched up nose, a very wide chin and cheek and its head was more like a dome, rather than pointed like he had heard.

What other signs have you seen? I noticed that red huckleberries were harvested up to 9-10' and branches of trees had been broken off.

There are a few amazing parts to this story. We have long believed that the vast majority of bigfoot sightings are of creatures with domed heads, not pointed. Another point is the creature's scrunching of the nose. It was probably trying to reconcile the smell it was associating with the strange creatures and the trucks.

"The Track Record" #122 (pg. 7, para. 7).

> Near the ghost town of Kaluka, Alaska, Emily Sapunich's mother was berry picking with others when they came upon a large hairy creature that resembled a man covered black hair. They ran back to the village and told the people. The men went out and captured it and caged it. She said her mom fed it raw fish. After some time the hair began to fall out and it turned out to be a female with breasts. Not long after the hair began falling out the creature died. This is recorded in a letter to Roger Patterson by Emily Supanich.

Note: Roger Patterson and Bob Gimlin were, of course, the two individuals that shot footage of a bigfoot in Bluff Creek in October of 1967.

"The Track Record" #128 (pg. 10, para. 7) received an email concerning bigfoot and an individual's lifetime of observations of the creature in Washington and Oregon. He wrote about a trust he developed with a few of the bigfoot near the Rogue River where he regularly walked along and looked at the water. I'll start to quote the story mid paragraph.

> I purposely don't plaster cast but would not mind doing one just to have it. They don't like to know their tracks are being examined, it feels like you are stalking them. If I want to establish trust in an area I don't try to track them too much or mess with their tracks, it scared the female on the Rogue River when we were bent over her tracks. I heard her in 10' tall blackberry bushes rocking back and forth from foot to foot.

On July 30, 2008 out of Kenora, Ontario an article titled "Blueberry Pickers Report Seeing Sasquatch-like Creature" ran across Canadian news wires. The sighting occurred on the Grassy First Nation Reservation, where three Native Canadian women went out onto the reservation to pick blueberries. They all stated that they saw an eight-foot-tall, slender black creature just a few meters ahead of them. They all agreed that they had observed a sasquatch. Family members went back later and measured the footprint of the sasquatch and found a print 38 centimeters long by 15 centimeters wide (15 x 6 in).

WATER

We would all agree that animals need water from somewhere. How that water is obtained—off plants, through streams and rivers, in mines, or from fruits and vegetables—is the real question.

In my first book I reported a sighting from Julienne McCovey that told of her watching a female bigfoot taking leaves and then dragging them through her mouth in an apparent attempt to get moisture. This section really isn't about how bigfoot may obtain water; it is about the rationale of many sightings and incidents so close to bodies of water. In the bigfoot sightings map section I noted several facts about the map that showed a large percentage of the sightings occur close to water—an abnormal number of sightings.

In Robert Alley's book, *Raincoast Sasquatch* (2003, Hancock House) he describes several incidents where bigfoot is seen swimming in British Columbia and Alaska waters. On page 51, paragraph 2, he writes, "Its style of swimming is commonly noted as submerged, not on-the-surface style one might expect to hear for any ape or other primate, or the crawl style if one were to think of a human." This style of swimming falls into a category that caused NABS to reflect on the proximity bigfoot has to water and the benefits of that association.

In *The Hoopa Project* I had several paragraphs about a creature that the Hoopa people believe may occupy the Trinity River, Kamoss. They describe it as having a large head, and swimming in the river when the water is high. Few people have ever actually seen the creature and they believe that it lives a portion of its life in the ocean and then occupies the river during the winter months.

I asked several Hoopa tribal people if Kamoss could actually be bigfoot. Several of the tribal people gave me strange looks and stated, "Possibly." We know that bigfoot likes the water (Kirk Stewart sighting in chapter 9, and revelations in Alley's book). We know the creature can swim. We know that bigfoot doesn't occupy the same region for long periods of time. If bigfoot were to find the most efficient method to move across large areas within a region, floating the river downstream would be an ideal highway through counties. It would also account for continued sightings close to the Trinity and Klamath rivers. If bigfoot traveled predominantly at night, the likelihood of witnesses seeing the creature floating the river, or swimming the river submerged (as Alley describes the swimming in his book) would be unlikely. It would be a very efficient method to move great distances in a very stealthy manner. It would also be a very good method to sneak up and ambush prey going to the river at night to drink, similar to the way an alligator slowly moves up on animals drinking from a river bank.

In *The Hoopa Project* I described several incidents on the Trinity River where a bigfoot threw huge boulders into the river where residents were sitting. It was obviously trying to scare the individuals (Hank Masten story) away from the area. One has to ask why bigfoot wouldn't like having people near the river at night unless bigfoot wanted access to it, for whatever reason.

A website that monitors bigfoot sightings (BigfootEncounters.com) had a sighting in March 2007 off the coast of Trinidad, California in Humboldt County. Witnesses reported that the bigfoot was swimming off the coast in the ocean adjacent to large rock islands. It was unknown what the bigfoot were looking for, possibly bird eggs. The witnesses talked to locals and one person stated, "they (bigfoot) probably come ashore after dark." They were seen swimming around the rock outcroppings and a few people believed they might have been staying in caves on the backs of the rocks.

The idea that bigfoot swims in the Pacific Ocean is not an isolated idea. The Bigfoot Encounters website also lists a sighting in August 1995: "Vacationing truck driver and wife watch as Sasquatch holding seaweed in his hands disappears across highway into the woods from the beach with black sand." When they talk about seaweed in this location they are speaking about kelp. The properties found in kelp are very beneficial and healing. This is another example of bigfoot found near water. The likelihood that bigfoot was in the water is very high.

"The Track Record" #119 (pg. 12, para. 4).

> According to Bindernagel [Dr. John Bindernagel, a British Columbia wildlife biologist] and others, Bigfoot tends to live where there is water and wooded high grounds. Based on eyewitness accounts and footprint casts, Bigfoot is, well, big, standing up to ten feet tall and weighing more than 800 pounds.

RIDGEWALKER

Early in my research, several Yurok tribal members told me that bigfoot was a ridgewalker. I didn't quite understand what that meant and I asked for clarity. They stated that bigfoot likes to stay on ridge tops and can be seen walking the ridges more often than any other single location in the mountains.

I didn't think much about what I heard from the Yuroks about the ridgewalker until I grabbed a United States Geological Survey map for the region of Bluff Creek. Looking back to 1958 when the roads were initially being built into the area behind Orleans, California you can see that the road makers attempted to stay on the ridge tops as they built the Go Road towards the Siskiyou Wilderness Area. The roads they were building were almost all on the top of ridges. Crews reported hundreds of large, human-type footprints on the freshly cut roads when they returned each morning. The same association can be made for the parallel road on the opposite side of Bluff Creek at the Blue Creek Mountains. Again, this road was built on the top of a ridge from the area of Onion Lake north towards the wilderness area. Road builders again reported hundreds of tracks on their freshly cut road, and again this was on the top of the ridge.

Several years ago a family that was camping in the Marble Mountain Wilderness Area saw a small den that resembled other shelters thought to have been built by bigfoot. As they were examining the shelter, one of the kids started screaming that a bigfoot was walking down the ridge towards them. Well, the creature was probably a mile away and 1,000 feet above them on a ridge that surrounded the valley. The father had the videotape playing as he filmed the bigfoot slowly walking towards them on the ridge top. The video clearly shows something that doesn't look human, and a reenactment by the BFRO shows the creature was very large. There are no real conclusions that can be drawn from the tape, but we think that it's a clear example of ridgewalker behavior.

If you can think from purely a predator's position, the idea of staying on a ridge is beneficial from many angles. The bigfoot would be able to see threats coming from several directions almost simultaneously. The chance of sneaking up on a bigfoot while it's on a ridge is almost impossible. One of the main reasons that battlefield tactics in war always aim towards taking the high ground as fast as possible is so you can see your enemy advance. You would also be able to see food sources from many points, and would be ambushing from a superior position; chasing a deer downhill is much easier than chasing it uphill.

If a bigfoot wanted to communicate with another bigfoot, it would not be effective to yell, scream, or whistle from the bottom of a canyon; it would be easier from the top of a ridge where the voice would carry for miles and echo off canyon walls.

Most ridges are either very warm during the summer or very cold during the winter. They are almost always windy and rarely would there be dust or loose dirt present on trails; the chance of finding footprints in these areas would be almost nonexistent. They are also not areas where backpackers would spend the night, as campers mostly stay in the bottom of canyons, near water and away from the elements. Bigfoot would clearly have an advantage watching, smelling, and hearing activity below from a position on a ridge. If you were camping in a valley and even had a small fire, the smoke and the flame against a dark backdrop would be an easy identifier to bigfoot that he had company in his forest.

The Kirk Stewart incidents (documented in chapter 9) at his organic farm all occurred on the top of a ridge, not at the lower elevations where other crops were being raised. The portion of the Kirk Stewart story where they showed me large, human-type tracks walking directly down a ridgeline above the Tyson Mine fits this theory to perfection. There were thousands of open acres around this location, yet the hominid that made these tracks walked precisely down the highest point in this ridge (see photos for details).

Researchers and sportsmen with night vision and FLIR units should keep their eyes on the ridgelines at night and they may be pleasantly surprised at what they find.

"The Track Record" #124 (pg. 9, para. 1).

> Dear Ray Crowe and fellow Track Record readers. The prints pictured are from a recent Southern Oregon Bigfoot Society outing from last Sunday (Feb 9) on the backside of our research area, near the Oregon caves. We found the tracks on our way up a ridge.

The photos were very unclear in the above article, but this does go to show the ridge association.

Paul Freeman was an employee of the U.S. Forest Service in Walla, Walla Washington. He had seen indicators of bigfoot during his career with the Forest Service, and later found tracks and even made a sighting. He later resigned from the USFS and made a career of tracking and researching bigfoot. He was a regular guest with Ray Crowe at the Western Bigfoot Society meetings. Ray made the following notation regarding some of Paul's findings in "The Track Record" #72 (pg. 3).

> Paul often found beds of the creatures on the end of ridge tops, where they can see for a long distance, but not be seen. The beds are made of long grass, fir boughs and smelled with hair mixed with grass. They were about two feet thick and 8-9 feet long and 4-5 feet wide. Often there was scat in the nest (gorilla's [sic] poop in their own nest). Also the nests would have bones; coyote teeth, fish bones, mice and deer and elk bones.

This is a quality report from a known researcher regarding bigfoot frequenting ridges. The idea of living at the end of a ridge accomplishes many feats associated with living, tracking game, tracking predators (humans), and keeping track of any fires in the area.

Track Record #104 (pg. 12).

> Two weeks ago I took a load of carpet and pad up to a farm some thirty miles north of here (Iowa). There was so much snow in the driveway that my truck wouldn't take it, and I had to carry the stuff all the way to the house. A carpenter was there and helped me unload. As we walked up and back we made some small talk. He said that he had just moved to Missouri from Florida. I asked about his roots and he said he wasn't a native of Florida, he had grown up in Oregon near Crater Lake. I mentioned my interest in Sasquatch and immediately his eyes lit up. He said "Oh Really?" "Well I saw one of them." I asked him for his story. He said that one time when he was a teenager he was in the woods (don't know where) down in a valley, he looked up and at the top of the ridge/mountain and there was a hair covered "guy" following him. He said it stayed with him for over a mile before he lost sight of it. It spooked him. He said a lot of the guys he grew up with had stories of having seen things in the woods like that. He said you always see "them" watching you from above. Then you go up to look for them they are not to be seen.

Another validation for bigfoot being a ridgewalker, and more confirmation that people who live in areas where bigfoot lives have encounters and don't talk about them.

CHILDREN & WOMEN

There are many sightings of bigfoot in close proximity to kids. I have never heard of a report where bigfoot has taken, abducted, or otherwise molested a child in anyway. I documented an incident in *The Hoopa Project* where a bigfoot repeatedly made visits to the area where children were playing and camping. It appears to have felt completely comfortable and returned multiple times. In my

90

reading of thousands of sightings, there are many instances where kids see the creature while nobody else does. Al Hodgson (see his story in chapter 10) told me a story where he met a father with his young boys at the museum. The family lived in Blue Lake and the boys had seen a bigfoot in their yard as it had walked up to them while they were playing. Al didn't get the contact information for the family so I had no chance to interview them, but if they are reading this, I'd like to talk.

Bigfoot probably understands that children and women offer them no threat. Imagine the size of an adult bigfoot and then imagine a small child or small-framed woman—definitely no threat. There are also many accounts where adults shoot at bigfoot. We do believe that bigfoot has excellent communication skills, and there is no doubt that they explain threats to each other in a way we don't understand. Part of their explanation probably includes telling each other what damage rifles can do and that rifles are only in the hands of adults. There is really no other threat a human can leverage on a bigfoot other than shooting it, unless you're driving a logging truck.

It is our belief that bigfoot also has a very good sense of smell. Children and women smell much different than adult men, and bigfoot can probably distinguish that difference. In *The Hoopa Project* one of the sighting reports was of a bigfoot that continued to come into Raven Ullibarri's yard and take garbage. During our investigation it was determined that the creature came back when Raven was menstruating. She thought this through very carefully and was positive that each incident occurred during that specific time of the month. (Subsequent to this incident, NABS investigators have been very careful how females are questioned regarding bigfoot sightings in specific lone scenarios. Researchers need to use great caution when crossing this bridge with a female witness. Ensure you have a comfortable, professional environment and you have the trust of the witness; don't make this one of the first questions you ask.) Reference also the story earlier in this chapter from "The Track Record" where people in North Dakota tried to attract bigfoot by hanging used sanitary napkins on trees.

We encourage researchers to think this association over carefully. We believe that the possibility exists of surveillance of the creature if the proper situation is staged, for example, possibly stuffing a tent with clothes kids have worn (smell association). Place a long-running tape inside the tent of kids playing (sound association) and then with a long-range spotting scope and video, set up in a blind several hundred yards away while soaking yourself in some ingredient to allow you to blend into the environment. The best choice is to set up in a large tree 40–50 feet off the ground where the human scent can't be detected as easily (get off the ground). We realize this is a cumbersome process, and takes time and extreme effort, but if someone wants to capture bigfoot behavior on tape over a prolonged timetable, steps similar to this need to be implemented and followed. (A FLIR video camera and a fourth-generation night vision scope attached to a video tape recorder is a must.) This off-ground surveillance technique was attempted by Bob Titmus and Syl McCoy (see chapter 10, Mary McCoy) in a region just outside Hyampom. It was in the early 1960s when they developed the technique in an attempt to observe a bigfoot. They never saw a bigfoot, but they did see a lot of wildlife. I haven't heard of any researchers attempting this type of surveillance in recent times.

A woman from Modesto California emailed the "The Track Record" #118 (pg. 10, para. 8) describing the encounter she had with a bigfoot near Ebbetts Pass in Alpine County, California in 1973. The witness was with a male friend on a granite outcropping. They heard something circling their campfire just out of view. They could clearly tell it was a biped. They heard guttural breathing, large braches breaking, and other sounds consistent with bigfoot. After bedding down, the couple heard an ear-piercing scream—a cross between an elephant's and a lion's. Another nearby camper threw his items into his car and left. The woman reporting this tried to get her partner to leave, but he was paralyzed in fear. She was crying and was very frightened. Rocks started to roll down the hillside and the roars continued. Another car then came down the roadway and they could hear the creature run up the hillside; the rocks were vibrating as it ran.

George Haas interviewed the witness in 1978. She says, "He asked me if I happened to be having my period at the time of the trip. I replied yes, and he told me that a Bigfoot has been attracted to us because of that."

George Haas was one of the original bigfoot investigators from the 1960s out of Oakland, California. He did a lot of groundbreaking work.

"The Track Record" #123 (pg. 7, para. 3) received a letter from a reader; the following is the narrative.

> One of my customers called to go over a project today. He knew I was into the subject (Bigfoot). He let me know of an experience last year during deer season near the Saddle Mountain area. He was a bit uncomfortable since he didn't want it spread around. He and his hunting partner brought their wives up with them on the normal weekend hunting trip to the area. The guys left right before light leaving the ladies behind. As the ladies sat drinking coffee, they heard the camper door (which was locked) being jerked on and pushed on as if one of the guys was trying to get in. It happened very quickly and lasted no more than a few seconds. The ladies crossed the camper to look out the window to see who it was and said they saw a very large dark upright shape walking off into the woods. They could not believe what they had seen and were really excited about it. Since then they have not told barely anybody because of the negativity they normally would receive.

The description fits a bigfoot. Its unclear if the door was touched soon after the men left. Maybe the creature was waiting for the men to leave. When they left, the creature could probably smell women inside. It makes you wonder what would have happened if the men hadn't left and they had answered the door. No other details are available about this incident.

"The Track Record" #116 (pg. 11, para. 8) printed an article titled, "Is Bigfoot in Wyoming" by Minnie Woodring, Staff Writer. The article describes a 13-year-old boy calling a writer for the newspaper asking a series of questions about bigfoot. The boy explained that he and his 13-year-old friend were on horseback at his grandmother's house. He explained that they saw a creature covered with hair, 12 feet tall and then it came out from near some hay and started to walk towards them. The horses got spooked and started back towards the grandmother's house.

The parents contacted the local police. They responded and found footprints. The police reported that they had multiple reports in the past few weeks of a bigfoot-type creature, but this was the first time they had a multiple witness scenario.

The behavior of bigfoot in the above story is not surprising. They do not have an innate fear of children. They either know that they intimidate kids, or they merely want to make friends. It would be interesting to think what might happen if bigfoot understood kids and took a passive posture when kids confronted them, maybe sitting on the ground and looking friendly. If bigfoot ever got to the point where they made themselves less offensive to children, we may have some very interesting interaction. This sighting also relates to the "Horses" category listed below.

HORSES

Bigfoot has a very noticeable affection for horses and definitely does not associate their odor or sound with any threat. They are seen near horses at ranches on a regular basis. Roger Patterson and Bob Gimlin rode into Bluff Creek on horses and were able to get relatively close to bigfoot before they shot film footage. There are many other stories of researchers and backpackers in wilderness areas that have had bigfoot sightings while they are on horseback.

Many biologists feel that people who want to get close to wildlife need to get up off the ground and get their scent up in the air. Deer hunters routinely sit in chairs in trees in an attempt to get the deer closer to their location. Many times the deer will walk directly up to the tree and never smell the hunter above. The six feet that a horse puts you up off the ground may be one of the secret ingredients to getting up close to bigfoot. The sound of the horses' hooves on rocks may be a sound that bigfoot hears regularly from deer, elk, and moose hooves; another large, four-legged mammal coming down a trail may be overlooked as normal forest sounds.

Nabs found there was a strong association with horses in Oklahoma incidents involving Mary Lonebear and her family (chapter 12). There were several incidents involving bigfoot, their barn and horses.

SUMMARY

All of these associations are continuously being researched, and much work still needs to be done to understand the rationale between the associations. There must be links between several of these factors, and an understanding of those links is critical to understanding the behavior of the hominid.

It must also be understood that NABS is only drawing these associations with the bigfoot being studied in the Pacific Northwest, specifically the far northwestern corner of California. If there are human genes in these creatures, then their behavior and thought processes can be compared at some level with human. That one factor may be a basis for understanding a creature that has eluded scientific acknowledgement.

4 EXTREME SIGHTING LOCATIONS

During our research of bigfoot/sasquatch/wild man, NABS found some very unusual locations that reported sightings of this biped. I feel it's important to include some of these reports in this book so that the reader can understand how adaptable bigfoot can be. It's important to understand that these sightings have been made in regions where the general climate conditions are extreme, so any biped that can successfully relocate to such areas is extremely adaptable.

COLD WEATHER LOCATIONS

While winter months in the Yukon can regularly see temperatures 30–40 degrees below zero, with an additional factor for wind chill, I have spent many summer weeks in the Yukon and can guarantee that even in the middle of summer the weather has its extremes. The summer trips I have taken revolve around a beautiful lodge approximately 185 air miles northeast from Whitehorse. Any bigfoot researcher that wants to establish base camp deep in the Yukon and close to Kluane National Park and the Northwest Territories wants to utilize Inconnu Lodge (www.inconnulodge.com). The lodge is a five-star facility that has multiple types of aircraft to allow the client unlimited access to every conceivable location in the two territories. I have taken their Hughes 500 helicopter into extremely remote areas of the Northwest Territories where no plane could possibly land. We've flown over vast spaces of the Yukon while observing every conceivable type of wildlife you can imagine. On one trip, returning from a day of fishing with my kids, we saw caribou, moose, two grizzly bears, and a moose calf, all in just 45 minutes.

The Yukon's weather is known to change by the hour, and when

McEvoy Lake.

you're in that environment you need to dress for almost anything. I've been hiking a creek in the morning when it's hailing and 40 degrees, and later in the afternoon it's sunny and 80 degrees. The area is so vast, undeveloped, and wild that it's completely believable that a sasquatch/bigfoot clan could live and move throughout the two territories undetected for decades.

YUKON TERRITORY

VICTOR MOUNTAIN, 1901

Source: Western Canadian Sasquatch Research Organization (WCSRO)

A "Bushman" is seen crossing a clearing. The witness stated that the creature was very tall and had fur as black as the night. (Note: "Bushman" was an old northern Canadian word for sasquatch or bigfoot.)

NANITENEE CANYON, 1930S & 40S

Source: WSCRO

A Mr. Taylor (85 years old) told a radio show that while he was working he had seen "large beings full of hair" many times when he was trapping in the 1930s and 1940s in a river canyon in the Yukon called the Nanitenee. This place has been called "Dead Man's Canyon." Taylor states that they were large beings and very friendly. However, when he tried to get close to them they just casually walked away.

FARO, WINTER 1970

Source: WSCRO

Witness Terry Gaines observed a brown-haired sasquatch lurking near garbage bins. The creature moved off when the witness's dogs started to bark. Mr. Gaines claims that the creature had a bad limp and was heavily favoring its right leg.

SIX KILOMETERS OUTSIDE WHITEHORSE (ON EAST SIDE OF YUKON RIVER), AUGUST 1978

Source: BFRO

It was 8:00 a.m. and the witness had gone to his favorite fishing spot near a small utility shed. He had just started to set up on his spot when he looked 200 yards downriver and observed a bigfoot on two feet standing near a grove of trees. The creature was somewhat hunchbacked, black in color, thin, and was observed grabbing a branch on the tree. As the creature tried to hold onto the branch, it broke and the creature fell onto its butt on the ground. The witness got very scared and left the area immediately. He stated that he was only 15 minutes from civilization. The witness is now 42 and has the memory etched into his brain.

YUKON TERRITORY, SEPTEMBER 29, 2004

Source: Gulf Coast Bigfoot Research Organization

The witness has lived in Alaska and has seen grizzly bear and other large mammals many times. He stopped his vehicle around noon to relieve himself. He walked a slight distance into the woods and smelled an odor similar to something dead. He turned and saw a creature standing on two feet looking at him from 50 feet away. The creature was bigger than the witness and slowly started to walk towards him, swinging its arms as it made its way towards the witness. The witness ran to his car and left the area immediately. He felt he had just seen a bigfoot. (Note: The GCBRO blacked out the exact location of the sighting.)

SOUTH MACMILLAN RIVER, AUGUST 1984 (DAYLIGHT HOURS)

Source: Monster Quest, *Kiefer Irwin*

The witness is a professional tracker and is considered an expert in wildlife track identification. She found several human-like footprints in the mud near the river. The tracks measured 16 inches long, five inches wide and they were pressed two inches into the soil. The stride was 4–5 feet and five footprints in total were found.

ALASKAN HIGHWAY,
JUNE 2004

Source: Whitehorse Star *newspaper article by Chuck Tobin*

2 YUKONERS THINK THEY SAW
SOMETHING LIKE A SASQUATCH

Whatever two Teslin residents saw in the early morning mist, conservation officer Dave Bakica knows it shook them up. Marion Sheldon and Gus Jules were traveling out of town along the Alaska Highway on an all-terrain vehicle, sometime after 1 a.m. last Sunday when they passed what resembled a person standing on the side of the highway. Thinking it was a local from their small community who might be in need of a ride, they turned around and headed back, Bakica explained in an interview Tuesday, recalling his conversation with the pair. As the two lifelong Teslin residents and members of the Teslin Tlingit Council approached to within 20 feet, they noticed the figure was covered in hair but standing upright the entire time. Though the northern natural light was dusky at that time of the morning, Jules told Bakica he saw what he believed to be flesh tones hidden beneath a mat of hair. "I have no doubt they saw something and are convinced it was not a bear or anything in the ordinary," the conservation officer said. "They are convinced this was something out of the ordinary...And they are pretty shook up over it." Bakica said Jules is an experienced hunter. He described the figure as standing about 7 feet tall, but hunched over. They could see it was not a person. "I have no doubt in my mind that they believe what they saw was a sasquatch," Bakica said. "Whether it was or not, I do not know. Just because you can't prove something was there, does not mean it was not there."

WYOMING

When I was attending the Texas Bigfoot Conference in 2008, an individual who had a very unusual bigfoot sighting approached me. The man was a resident of a nearby community, but stated that he had previously lived in Colorado and wanted to know if I was interested in the incident. This is the Stewart Simpson bigfoot sighting.

Stewart Simpson was born in Ringgold, Louisiana and attended high school in Jefferson, Texas. After graduating from Jefferson High, Stewart attended the University of Texas where he graduated with two bachelor's degrees, one in computer science and one in business. Mr. Simpson is married and has one boy and two girls, and is presently a business development manager for Baker Hughes where he seeks out trends in the oil and gas industry.

Stewart Simpson

In early 1989 Stewart was living in Casper, Wyoming and working for Dresser Industries, a service company where he had 50+ people working for him. His job required him to be very aware of and responsive to client needs and requests.

On a weekday in February 1989 Stewart needed to make a business trip. He was driving his 1987 Pontiac Royale on Highway 120 from the area of Thermopolis, Wyoming towards Cody at approximately 9:00 a.m. He was 8–10 miles east of Meeteeste and slightly west of Grass Creek on a two-lane road at a slight pass with an elevation of 8,000–9,000 feet. It was a dreary day; there was light snowfall, but visibility was excellent. He estimated that it was 15 degrees outside. There was snow on the roadway and he was probably traveling 20 miles per hour when he observed a creature standing in the ditch on the left side of the road. He immediately tapped his breaks and slowed to a stop while looking out his side and rear view mirrors. Stewart saw a creature walk on two legs across the road. It had brown hair covering all of its body, was six-and-a-half to seven feet tall, its arms hung down to near its knees, and its hair was three to four inches long. The crea-

101

ture never turned to look at the vehicle, so its face was never seen. It took the creature just three steps to cross the road and go into the opposite ditch. Once on the other side of the road, the creature continued into the bush and was gone.

On reflection, Stewart said that he definitely observed a bigfoot, and he described it as being built like a big football player. He never saw any genitals or breasts. Stewart signed an affidavit.

LOCATION

The reason this sighting was placed in this section is because of the weather conditions present at the time of the sighting. It was snowing, extremely cold, and very high in altitude.

Researchers do understand that all creatures (even humans) are pushed out of prime habitat areas because of a variety of conditions. Humans can be pushed out because others want to live in a neighborhood, prices are up because of demand, and only the wealthy can live there. Animals are forced out of areas because the bigger and stronger creatures have established a region as their home and the younger and weaker need to seek out new habitat. That's what appears to have happened in the extreme environments of the Yukon and Wyoming and the deserts that are described later. There are many more examples of extreme sightings in deserts and other high altitude areas, demonstrating that bigfoot is a very adaptable creature.

HIGH HEAT LOCATIONS

In researching bigfoot in extremely hot environments I immediately thought of bigfoot researcher Richard Hucklebridge. Richard lives in Palmdale, California, a desert city located east of Los Angeles and near Edwards Air Force Base.

I originally met Richard at a bigfoot conference in Felton, California, and we immediately found we had common interests and beliefs about the creature. Richard has been a longtime researcher for the BFRO and has done outstanding work documenting bigfoot-related incidents, interviewing witnesses, and supplying the BFRO with some of the best reports available on the Internet.

One of the first conversations I had with Richard was regarding how he personally felt about some reports that came out of the Palmdale and Lancaster areas of the desert. Knowing that this area regularly gets over 110 degrees during the summer months, it was hard for me to understand how bigfoot could survive in that environment. Richard explained that, on first thought, it would seem unbelievable any large mammal could survive in that heat in a scrub brush area without large amounts of water. He clarified that if those variables were the only ones present it would be hard for him to understand bigfoot's survival. However, upon investigating the desert sightings, he found areas near the sightings where bigfoot survival could easily be understood.

Richard stated that farmers in the area have huge fields of alfalfa that they water vigorously. With the high amount of watering that was occurring, small pools of water could easily be seen amongst the crops. The first time he was in those areas he said that he was amazed at the large number of rabbits in and around the fields he visited. Richard said he's been investigating bigfoot for too many years not to believe that it has found a way to sustain life in this environment—maybe not full time, but occasionally.

The following articles were all written in 1973 and appeared in the Lancaster *Daily Ledger Gazette*. What follows is a summary of what each article contained.

April 30: A nineteen-year-old girl was returning home late at night and heard her two dogs crying. She walked around the side of her house and found the dogs injured. She started to walk towards the house and while looking towards a grassy field saw a seven-foot human-type figure stand up and start to leave the area.

May 22: Three marines reported to the sheriff's department that they saw a bigfoot. The marines' sighting was in close proximity to the nineteen-year-old girl's observation of April 30. Footprints were found near the entrance to a cave and near a reservoir.

June 1: Multiple witnesses confirmed seeing a bigfoot on the outskirts of Lancaster near Nine Mile Canyon.

July 20: A hang glider was flying on the outskirts of Lancaster and observed a bigfoot sitting in the roadway eating a raven (apparently road kill). The creature somehow heard the hang glider overhead, grabbed the dead raven, and ran into an adjacent field to make its escape.

The following is a summary of a BFRO report that came from San Bernardino County, California. The sighting was from the Mojave riverbed just outside Hesperia.

Five boys were exploring the dry riverbed in August 1997. A typical temperature for this area in August is regularly over 100 degrees. The boys heard something rustling in the thick brush down near the gravel. They got closer and saw a large, hairy, ape-like creature. They went home and told their parents.

For two consecutive days after the initial sighting, one of the boy's sisters decided to search the riverbed by horseback to see if she could spot the creature. On the second day of her search she was riding near a grove of pepper trees and cottonwoods. Her horse started to act irrationally, so she stopped and looked around the area. She stated that she saw the dark brown, hairy creature standing and staring at her for almost a minute. She left the area in the opposite direction from where the creature was standing.

Bigfoot researchers went to the riverbed after the incident and claimed they found bigfoot prints 20 inches by 11 inches, along with an area they described that had bedding material.

The girl being on horseback is a particularly interesting detail in this sighting. NABS is currently researching the connection between bigfoot sightings and people on horseback (see chapter 3, Horses).

The Bigfoot Encounters website (www.bigfootencounters.com) had an interesting entry for the summer of 1989 in Sedona, Arizona. Sedona is a beautifully situated community amid the red cliffs of the desert. It is known for its arts and crafts shops and has been associated as a spiritual healing center. Temperatures in summer months regularly are over 100 degrees and there are not many large trees or water sources.

The report states that a real estate agent was previewing a property that was still under development when several bigfoot-type creatures were observed standing in the area. It appeared to the realtor that the creatures might be guarding the area. The creatures were described as having black curly hair and interesting eyes.

It was in June 2001 on the Navajo Indian Reservation of Apache County Arizona that an individual reported to the High Desert Bigfoot Research Group that he observed a bigfoot cross the road in front of him while he was driving on a very warm summer night. He described it as dark brown, very solid in appearance, and it crossed the road and hid behind a bush near a ditch. The witness stated that he had a very good look at the biped and is convinced he saw a bigfoot.

This again is an area that regularly gets to over 100 degrees during the summer months and usually only cools to the high 70s during the nights.

SUMMARY

The physical descriptions of bigfoot offered in each of the described extreme weather locations maintain a consistent theme. Most sightings of bigfoot are 5–8 seconds in duration, sometimes under stressful conditions, and it is usually an event that the witness will never forget.

NABS would like readers to understand that we could fill an entire book with reports similar to what you've read in this chapter. Those listed here are merely to provide an overview of the harshest conditions in which bigfoot may thrive.

5 SANTA CRUZ COUNTY

There are few places in the world that have the diversity in weather, topography, and lifestyles that Santa Cruz County, California offers. If you want to work on a ranch or a farm, you head towards Watsonville. If you want to commute to cutting-edge technology you go to Scotts Valley, home to some of the biggest technology companies in the world. If you like to surf, you bring your board to the world famous Steamers Lane (Santa Cruz) where waves get to 20 feet in winter months. If your sport is mountain biking, then you ride at Nisene Marks State Park where you can meander through old growth redwoods. If you'd like to live on the Pacific Ocean, pack your bags and park the moving van on West Cliff Drive in Capitola where the views are awesome and the city is quaint. If you'd like to have snow fall in the winter months and live away from people, there are miles of open space in the Santa Cruz Mountains to build your own retreat.

Santa Cruz County is a beautiful spot and a place I called home for 10 years. We lived just below the University of California at Santa Cruz. The university sits high above the city, with gorgeous views of the coast and city. The university mascot is a slug, and that should indicate what majors are highlighted: forestry, biology, and engineering is taking hold.

The population of the county is 250,000 with 445 square miles of area to spread people out. It is located 70 miles south of San Francisco, 30 miles southwest of San Jose and sits directly on the Pacific Coast in the Monterey Bay Sanctuary. Elevations range from 0–2,600 feet, and rainfall fluctuates from 30 inches to as much as 100 inches that has been recorded in Boulder Creek. The top industries in the county are agriculture (Watsonville) where artichokes and strawberries are popular, and technology (Scotts Valley) where companies like Seagate and Borland are headquartered.

The largest cities in Santa Cruz County are Santa Cruz (57,500),

Watsonville (51,258), Scotts Valley (11,615), Capitola (9,960), and the unincorporated region of the county (133,739). Highways 1, 9, 17 and 152 crisscross the county and offer easy access to all points. It should be noted that tourist traffic on the summer weekends makes getting to area beaches difficult, as San Francisco Bay Area residents attempt to escape their heat and head to Santa Cruz beaches.

This region is known for its 14 California state parks and beaches, with the oldest being Big Basin Redwoods State Park. It's a beautiful spot with old growth redwoods, open space, waterfalls, and gorgeous hiking trails.

The one park where I've always spent a significant amount of time was Nisene Marks State Park in the city of Aptos. It is located less than one mile from the Pacific Ocean and has over 30 miles of hiking trails and 10,000 acres. There are areas of this park that are very remote and have no trails. The climate goes from almost tropical to chaparral and hot. You can walk amongst the redwoods or sweat amongst the madrone, your choice! There was logging in this area until 1920 by the Loma Prieta Logging Company. The park was purchased by the Marks family in 1950 and later donated to the state in 1963. It's difficult to see any damage caused by the logging, but there is still evidence of damage from the 6.9 magnitude 1989 Loma Prieta earthquake. The epicenter of the quake was in the middle of Nisene Marks; there are still downed trees, and creeks were re-routed. There is a map where you can walk to the area of the epicenter. Keep your eyes open, as bigfoot sightings have been reported in the park!

BIGFOOT BLIND

In early May 2008, Mike Rugg, curator of the Bigfoot Discovery Project located on Highway 9 in Felton, California, contacted North America Bigfoot Search (NABS). Mike has developed a museum of bigfoot history and memorabilia and he is in the process of developing a bigfoot "think tank and inn" with overnight accommodations, adjacent to his museum. His time is devoted to developing the new project and he has little time for investigating bigfoot-related activity. He has requested that researchers with NABS work with the

Discovery Project to investigate sightings/encounters that are filtered through the museum.

I went to Felton and met with Mike at a local restaurant. He is a very intelligent and friendly bigfoot researcher keenly aware of the politics that proliferate our arena. We discussed the museum and NABS, and the possible relationship that could develop, and agreed to initiate the partnership with NABS visiting a site in Felton where a reported bigfoot blind may be located. I would report on my findings and assist on further sightings and encounters as they filter into the museum. Mike gave me the contact information for the blind and I advised him that I would be visiting the location the following week.

FOLLOW-UP

The bigfoot blind is located at 1340 Lakeside Drive in Felton. There were residents at 1420 Lakeside drive that had reported hearing loud screams emanate from the creek area in 2005, and a museum member who lives up the street from 1420 also reported hearing loud screams in the same timeframe.

In the spring of 2007 a new resident to the Lakeside area came to the museum and reported seeing glowing red eyes eight feet off the ground near the entrance to the Granite Construction Quarry that is just north of the 1420 Lakeside Drive location. The individual stated that she saw the eyes in a forested area just off the roadway as she was making a U-turn at the closed quarry gates.

I called and spoke to the resident at 1420 Lakeside, Josh Crandall. He stated that he had reported screams coming from the creek area, but that he couldn't add more than that; it was an isolated incident and hadn't happened in a few years. He encouraged me to contact the neighbors at 1340 Lakeside who own the property where the blind is located.

After several attempts I was able to contact Kerrie Crandall, owner of 1340 Lakeside Drive, by phone and she confirmed there was a wooden shelter near the creek on her property. She said she has never heard any screams or had any unusual occurrences on the property, and had no idea how the shelter was made or who assisted in the construction. I requested permission from Kerrie to travel onto her property and view the shelter and she said it was okay.

Lakeside Drive is located off Highway 9 at the western end of Felton. It is located in the Forest Lakes Housing Tract, which is serviced by the Forest Lakes Mutual Water Company. My first stop was the water company where I contacted Matthew Faneuf. Matthew stated he was a lifelong resident of the area and had worked for the water company in this region for over 10 years. He advised that Gold Gulch Creek is the stream that runs adjacent to the homes in this area; it is a spring-fed, year-round stream that has a healthy run of steelhead trout during the winter months. The first tributary that steelhead hit as they are migrating up the San Lorenzo river is Gold Gulch Creek. The State of California is in the process of adjusting the creek flow and ensuring that the fish run is safe by building diversion systems so the trout can make the swim far up the stream. Matthew offered some interesting facts about Gold Gulch Creek. The confluence of the creek and the San Lorenzo River is where the largest single nugget of gold in the State of California was ever found, thus the name Gold Gulch Creek. This gold find was what started the gold rush into the area in the 1800s.

I told Matthew I was a bigfoot researcher and asked if he knew of any unusual circumstances or encounters in the area. He said he hadn't heard of anything, but he took my card and said that he would forward my name to anyone who reported anything similar to screams, yells, smells, or sightings.

I proceeded to 1340 Lakeside where I encountered a fairly treacherous road to the bottom of the hill near the residence. I parked and walked another 100 feet to the creek bed where I found a low-flow creek (Gold Gulch) and immediately found several large indentations made by deer. I also found the remnants of a blind that looked very similar to a structure I investigated in Del Norte County.

This blind appeared to have not been used in several years, as parts of the sides had collapsed. It was built against a five-foot-diameter fir tree that was just on the perimeter of the creek bed. The entrance to the blind faced the creek and the tree supported the back portion. Its construction was similar in design to a Native American teepee, in that all of the branches, limbs, and wood that make the

Front and side view of bigfoot blind side.

outside supports are all leaning towards a center mass area. The flooring of the blind consisted of much of the outside walls that had collapsed; nothing else was visible on the floor other than leaves, brush, and dirt.

Blind dimensions: seven feet tall at its highest exterior point; seven feet wide at its widest exterior point; five feet high at its highest interior point.

Blind construction materials: mostly broken branches and limbs native to the area.

The base of the blind had much of its wood buried into the creek bed, probably caused when the creek rose and covered the branches in gravel. Because the base of the blind was buried, the entire structure was very sturdy, wind resistant, and able to withstand direct hits from other large branches and still stay erect. The structure could offer significant odor resistance from a creature inside, depending on the level of brush and leaves added to the walls to hinder and deflect body odor.

The location of this blind was the key aspect that convinced me of its specific use. Looking south from inside the blind, downstream, provided a clear view for almost 100 yards. Looking upstream you could see almost 150 yards. Looking north and uphill was a heavily used game trail visible for almost 75 yards. Kneeling inside the blind provided nearly a 200-degree angle of view of anything coming into the area, a superb ambush position for any predator. The height and width of the blind entrance (five feet high by four feet wide) indicated that the predator utilizing this blind was not huge, probably no larger than six feet in height. A bigfoot in the seven- to eight-foot height range would need a wider entrance point to be able to quickly exit in pursuit of game without destroying the blind.

CONCLUSION

There are no mammals living in Santa Cruz County that are known to make structures similar to the blind in the photo. It is doubtful that a human made the blind, based on the design and construction. There is not enough room inside the blind for an adult human to completely lie down, yet there is ample room for a six-foot human/bigfoot to lean back against the fir tree and have a commanding view of any mammal coming into the area. The ambush location for this structure was not picked by chance.

The screams described by neighbors in this area are indicative of a bigfoot presence. The red eyes eight feet off the ground that were seen by a motorist can also be an indicator of bigfoot, and seeing a bigfoot near a quarry is something NABS has been researching for the last few years (see chapter 3). There are very, very few quarries in the area of the den, and the sighting of the eyes near a quarry is a strong signal for NABS to take the sighting seriously.

A NABS historical file search of Felton showed one report from 1932 regarding bigfoot. The narrative reads, "A boy, age 10 went outside a rural cabin and saw what he feels was a 7' tall creature brush past him. He thinks it was looking in the windows of the cabin earlier at adults." This report came from the confidential file of John Green.

Another older report from the county was reported in the "The Track Record" #125 (pg 5).

It happened in 1960 near Santa Cruz, CA in the evening. He was driving down the main highway and a large creature crossed the road in front of his car. The road had two full lanes with a shoulder on each side about as wide as each lane, so the total road was

111

quite wide. When the creature crossed the road it took no more than 3–4 steps to clear the entire area, bipedally. When it reached the other side it went up a 50–60 degree hillside and didn't even slow down. Another interesting fact was that a construction outfit that was near the same area had been noticing that something or someone had been picking up large objects, 700–800 pounds and stashing them in other areas nearby, as in hoarding them for some reason. They never found out who, or what, was doing this.

During a trip to Del Norte County in 2007, I investigated a bigfoot-and-blind sighting near the Klamath River. The blind in that sighting was very similar in construction to the blind in Felton. The same type of materials relative to the area, and the dimensions were a close size to each other. An associated note to the blind issue: the Klamath structure had an elevation of 350 feet and the elevation of the Gold Gulch blind was also 350 feet. It is my belief that this structure is probably a hunting/ambush blind used by bigfoot.

JAYSON WILLIAMS & MIKE FORAKER

Mike Rugg sent me an e-mail explaining that an individual contacted him at the museum saying he had seen a bigfoot near his house. Mike didn't have all the details, but he had heard that the person was soon going into the military and he needed to make contact or he would lose the details of the sighting. Mike asked if I would contact Jayson Williams and investigate his sighting.

I contacted Jayson at his home early on a Friday morning. He said that he would gladly talk with me and explained where he lived and what he observed.

Mr. Williams told me he is an 18-year-old married father of one living in a small hillside community just south of the city of Ben Lomond. He was raised in San Jose and Santa Cruz County, and attended Scotts Valley High School. He was being deployed for the United States Army in August and had been staying close to home in the weeks before he left.

In March 2008 Jayson and his neighbor, Mike Foraker, were outside their homes at approximately 5:00 p.m. when they heard an extremely loud scream immediately followed by a series of gunshots. They couldn't determine from which direction the scream emanated because of the way their homes sit in their valley and the manner that loud noises bounce off valley walls. Jayson said he had never heard a scream like that, and doubted it came from a human or any animal he knew.

After hearing the scream, Mike and Jayson sat on the back porch for another 30 minutes talking about the chickens that had been disappearing from the pen in Mike's backyard. The Forakers had a seven-foot-tall mesh fence encircling a 15- by 20-foot pen where they kept 10 chickens. In the weeks prior to this conversation, Mike had lost all but two of the chickens and he had no rational explanation for what or who was stealing them. The walls to the pen were high and well constructed, and nothing had pushed the fence down. There were no footprints in the hard soil surrounding the pen or in the pen. He was getting very frustrated with the thief.

It was getting dark and Mike went into his house to rest because of a bad back. Jayson stayed outside and was looking from the front of his yard towards the back of Mike's yard. It was between 7:00 and 7:30 p.m. when Jayson heard some rustling in downed logs and brush at the rear of Mike's residence. Jayson grabbed a flashlight to enhance the sunset light and took a few steps towards the

Jayson Williams next to the chicken coop.

rear yard. At this point he shined the light in the direction of the noise and saw a huge two-legged creature quickly pass across a small open area near the chicken pens. The creature had medium-tone brown hair that completely covered the body. It walked much like a human with long arms and it slightly stooped forward. Jayson was able to view this from a side angle. Jayson said he could tell that the hominid was very muscular and described it as healthy in appearance. The creature looked to be seven to seven-and-a-half feet tall, based on its appearance next to known landmarks in the area. He watched it disappear down an embankment behind Mike's residence.

The next morning Jayson went to the area where he had seen the creature and searched the entire region for footprints or other evidence; nothing was found. He explained that he had just recently seen several specials on bigfoot and footage of the creature, as well as still photos, and he felt that what he observed behind Mike's residence was a bigfoot. I asked Jayson if I could meet him at his home and investigate the location, and he agreed.

I drove to Santa Cruz County, up through Felton and up Highway 9 towards the quaint city of Ben Lomond. The turn-off for Jayson's residence is on the northern hillside off Highway 9, in a small community that is surrounded by huge Douglas fir and redwood trees. The area is lush, green, and appears to be ideal habitat for a bigfoot.

I met Jayson at his residence and he told me he had gone to the bigfoot museum in Felton and heard recorded bigfoot screams The screams on the tape matched the scream he and Mike heard just before his bigfoot sighting, so he encouraged Mike to go the museum and listen to the tape and make his own conclusions.

I wanted to interview Mike as well, so we walked over to his house. Passing the chicken coop, I could readily see that the mesh wire was very high, over seven feet in places. The ground in the area was very compacted and not conducive to footprints. I also walked down the hillside behind the coop and observed a very worn trail that meandered up from the bottom of the valley and from the area where the San Lorenzo River flows adjacent to Highway 9. Jayson and I made our way to Mike's front door and found him lying down on a sofa. He explained that he was recuperating from back surgery and was taking a needed break. Mike said he knew there were several bigfoot living in the area of his community. Many months before, a resident up the road heard a loud scream that scared him so much he bolted his doors and windows and refused to come out until daylight; he had described the sound as something that was definitely not human and couldn't be from any "normal" mammal in the woods.

Mike said that 20 years ago he owned land in the Corralitos region of southern Santa Cruz County (near Watsonville). He had made regular trips to a place above his property where there was an abandoned avocado orchard near Buzzard Lagoon Road. He had heard the same type of loud screams coming from the orchard area in the past. Mike eventually went up to the orchard during daylight hours to investigate. He found hundreds of avocado pits scattered throughout the orchard and it appeared that something had walked through the trees grabbing the fruit, eating it, and then randomly throwing the pits on the ground. He stated that he knows of no animal that would stay in a confined area like the orchard and eat the fruit and discard just the pit with such little fruit left on it. He explained that when he heard the screams, he had two wolf dogs that lived with him on the property and they were not afraid of anything or anyone, and would never back down from anything. When he was outside with the dogs and they heard the screams, the dogs simultaneously ran as fast as they could into the house and both tried to hide under a table in the corner. At this point Mike said he knew that if the dogs were scared, maybe he should be too, and went inside the house and locked the doors.

I questioned him about the screaming and gunshots that Jayson had claimed he'd heard. Mike confirmed hearing a scream that was very, very similar to the screams he had heard in Corralitos 20 years before. He said it was very odd to hear the scream in conjunction with gunshots, but shooting in the valley wasn't unusual.

I thanked both individuals for their time and reports, and encouraged them to contact me if they had any further information or they met individuals who had bigfoot sightings or encounters.

Jayson Williams signed an affidavit outlining the facts of his sighting and stating that the hominid he observed was a bigfoot.

After I completed my interviews with both subjects, I went to the end of their community and walked the entire route back to the Williams residence. I found many deer prints, lots of small mammal tracks, and a very worn dirt trail that starts at the bottom of the hill behind Jayson's house. This trail is quite long and goes through the valley behind his house (730-foot elevation) to well over 900-foot elevation. There is actually a small grove of redwood trees immediately down the hill behind Michael's chicken coop.

The location of Jayson's sighting is ideal for bigfoot. There is ample cover and food source, and water is close in proximity. The time of day is early for most sightings, but if it is related to the gunshot, it's possible the creature was fleeing from a threat.

Humans and animals utilize the trail behind Jayson's and Mike's residences extensively; I could easily see many deer and human tracks on the trail. Mike had stated that he had trapped several raccoons behind his residence and then relocated them because they were causing disturbances at his house. I spent approximately two hours walking the various hillsides in the area, and at 2:00 p.m. on a Thursday afternoon in mid June it was still relatively cool under the tree canopy. There was a large forest fire less than 15 miles from this location and there was no smoke present. The floor of the forest in this location was five inches deep in various needles, small branches, and leaves. The only location able to retain footprints would be the trail behind the Williams and Foraker homes, and I found nothing that would match a bigfoot track. I did note that there was a constant light breeze blowing from the valley floor, up the valley, and past the witnesses' houses. The direction of the wind indicates that the bigfoot probably never had an opportunity to smell the witnesses, as they were downwind from the hominid. Annual precipitation in this region is over 60 inches annually.

The report was taken more than three months after Jayson's bigfoot sighting.

JESS HAINES

One of the significant benefits of operating your own website is the information received from people throughout the world. It is amazing who looks at the site and what information arrives through e-mail. The Internet is an amazing communication tool and one that has definitely changed the scope of bigfoot research. It is our website (nabigfootsearch.com) that connected Jess Haines and North

America Bigfoot Search. He sent us an e-mail outlining some very interesting incidents that occurred on his ranch, and invited us to investigate.

I contacted Jess by phone and he invited me to his family ranch located on the far southern end of Nisene Marks State Park near the city of Watsonville. The area is pristine, with large trees, few homes, and terrific views of the Pacific Ocean. There is a very long dirt road that leads into the 136 acres that the Haines family has owned since 1894. The ranch doesn't have cattle or other animals roaming open land, just a few horses. It is mostly acreage with a few large build-ings where men work on their equipment, and there is significant open space where I saw more than 10 deer. Jess and his wife, Kendra, have lived in a variety of different locations in the past decade, including the ranch, in Watsonville, and more recently in Costa Rica.

By pure coincidence, the Haines ranch is within two miles of the avocado orchard noted in the Mike Foraker report above.

The winter of 2000 started a series of events that Jess Haines attributes to bigfoot activity. Jess and Kendra were living in a small house above his dad's residence in one of the higher elevated points of the property. Jess claims that his brother saw what appeared to be a bigfoot walking down the middle of their dirt road while Jess Haines. he passed it in a vehicle going the opposite direction. Immediately after the observation, he told everyone on the ranch that he had seen it, but what he saw did-n't sink in until he stopped his car. Jess explained that Santa Cruz County is known for its transient population and, in some places, large numbers of homeless people. If you were accustomed to driv-ing through those areas it would not be unusual to see a very ragged, dirty, unkempt person walking along a road. That's what Jess claims his brother observed on their road, except that this hominid didn't have clothes on. It was very dark outside; the subject walked much like a man and was huge and dark. The next day the boys went out to the road to see if they could find tracks. They did find giant footprints in the dust of the road, which appeared similar to human prints, but much larger.

In the summer of 2002, Jess, Kendra, and Jess's stepmother were sitting on the porch at Jess and Kendra's residence when they heard a very loud, guttural scream coming from the area above them, near the horse stable. Jess said they could all see that the horses were going crazy, running around the yard wildly, and were obviously very disturbed. After hearing four loud screams that seemed to get further and further down the road, there was a period of silence. The silence only exemplified how loud the screams were. Jess explained that he had been with the horses when mountain lions were nearby and the horses never became so agitated even when a big cat was in the region. The group retreated into their house as the sounds dissisipated and quiet came over the ranch. I interviewed Kendra after I interviewed Jess, and she said that the screams were not human and from no animal she knew.

Approximately three or four days after the screaming incident, Jess was driving his Suburban down the ranch road at approximately 11:30 p.m. He was coming from the upper area of his property and was coasting downhill when the high beams on the car illuminated what appeared to be a redwood slash pile in the small field adjacent to the roadway. He was traveling very slowly and could see that the pile was exactly the color of fresh redwood bark and it was fairly large. He wondered who would have been working and made that pile and placed it in that location, and then he realized he was the only person at the ranch during that week doing that type of work. He slowly passed the pile, still perplexed, when he realized that what he was looking at was a huge creature bent down on its knees with its arms covering its head (evidently trying to make itself look like a stump or slash pile). Jess quickly backed up, but the creature was gone.

A few days after this, Jess and Kendra were sleeping at their ranch house when they heard a thundering knock on the front door. The knock was so hard they later determined that it actually broke off the door handle. It occurred in the middle of the night (1:00–3:00 a.m.) and there were no other sounds or knocks. Both were too afraid to investigate what happened, so they just stayed in bed. Jess said that he didn't have a firearm in the house at the time; that soon changed. The next morning they found the door handle broken and lying on the ground outside the front door. There was no other evidence that Jess could find, even though he spent considerable time that morning scouting the grounds.

Two days after the door-knock incident, nearing midnight, Jess and Kendra were again in bed when they both heard wood knocking coming from the area where Jess's brother had made his sighting on their roadway. Jess and Kendra described the knocking as two-by-four boards being slammed together with great force four times. Having been to the ranch and seen the surroundings and residences in the region, I can confirm that there are no other homes anywhere near this area. Jess and Kendra both said that they were baffled by what or who could have been making the sounds. They did not believe that people had been trespassing because nobody travels that far up into the mountains in the middle of the night on a private road; they might get shot!

Jess Haines signed an affidavit.

LOCATION

One of the more convincing features of the Haines incidents is the proximity to Nisene Marks State Park and the elevation and rural nature of their property. There are historical records of bigfoot sightings in the far upper reaches of Nisene Marks. The location of the Haines ranch is exactly at the southeast corner of the park in an area where there are no trails; it is very remote. There are huge old-growth redwood trees and many deer that I observed in the area. This region gets significant precipitation. I don't believe that a transient or homeless person would wander onto the Haines property in the middle of the night 15 miles from the closest town. If Jess's brother saw something, it was probably a bigfoot.

Adjacent to the Haines property are several mountain peaks that are 2,400–2,600 feet in elevation. There are few roads in this area, but there are a couple of dirt roads that fringe the state park. I spent an entire day driving this area and talking to people of the region. I learned that this area has significant wildlife and many, many deer and wild pigs.

Corralitos is the closest town to the Haines property. With a small general store, it is more of a hamlet than a town. I learned from other witnesses and from the bigfoot museum in Felton that there have been bigfoot incidents and sightings in the hills just above Corralitos (reference the avocado orchard in Williams/Foraker report above) and this is very near the Haines property. Starting to connect the dots of the sightings, it is easy to see how all of the sightings and encounters are related.

If you look at other bigfoot reporting sites and databases you can start to see a relationship between sightings, locations, and bigfoot in this region. The BFRO site has a sighting reported on Highway 152 that crosses the Santa Cruz Mountains just outside Watsonville. This sighting is of a hominid walking along the roadway and described exactly like a bigfoot by a motorist driving over the hill. If you continue searching into Monterey County you can find more sighting and incident reports stretching into the Los Padres National Forest and wilderness area south of Monterey and east of Carmel. All of these sightings are in the mountainous areas.

One report that NABS took in this region was from Kenny Rogers. His encounter took place in southern Santa Cruz and northern Monterey County just south of Watsonville.

KENNY ROGERS

People interested in bigfoot tend to seek others with similar interests, and that's how I met Kenny Rogers. Kenny was a friend of Jess Haines, who referred me to Kenny and asked me to initiate contact. I was able to speak with Kenny by phone and later arrange a personal meeting in the area of his sighting.

Kenny Rogers is 50 years old, graduated from Watsonville High School, attended a local junior college, and at the present time teaches biology to a group of local high school students. He has spent the majority of his life living in Santa Cruz County and believes that he understands local wildlife better than most people.

The southern portion of Santa Cruz County contains a series of large agricultural fields that harbor strawberries, artichokes, and even brussels sprouts. The transition from Santa Cruz to Monterey County is almost seamless; you go from one field in one county right into another field in the next county. The only landmark that tends to signify a change is a ridgeline that runs from east to west, from the mountains to the coast, and is the only ridge for 60 miles as you travel south. It was this ridgeline where Kenny had a bigfoot encounter.

Kenny Rogers

It was late in the summer or early fall of 1977 when a friend of Kenny's was having a drinking party in the area of 170 Hayes Road, Watsonville. This parcel of land sits up against the northern ridgeline and adjacent to several large fields. There are many mature trees on the side and top of the ridge.

Kenny was working a swing shift the night of the party, so he didn't arrive at the location until very late. The group was camping near the top of the hill behind the ranch and almost everyone was asleep when he arrived. There were tents and sleeping bags littering the ground with people snoring the night away. The weather was getting cool and there was a slight amount of fog. Kenny can distinctly remember looking across the valley and seeing the old train depot two miles west from the party.

At approximately 2:30 a.m. Kenny was awake and sitting in a chair near the campfire, looking southeast towards a small saddle in the ridge. He could hear something coming down through the saddle, but couldn't quite make out what it was. The sounds indicated that it was a biped and he thought that perhaps it was a wino from the train yard, but this idea quickly faded because the footsteps were much too loud, almost thunderous, for a human. The creature kept getting closer and was in an area where another car and truck were parked. Whatever it was, it stopped and then started to shuffle backwards. The hominid stopped again and this time let out the loudest

121

yell Kenny had ever heard in his life. He was shocked that everyone didn't wake up, but maybe they were all intoxicated. Kenny described the yell as something between a pig and an ape, but extremely loud. The creature then stepped over to a small grouping of large trees and started to shake them very violently. At this point Kenny started to get concerned about everyone's welfare, as whatever was shaking the trees had to be huge and very unhappy.

After the tree shaking, everything in the area went very quiet for almost an hour. Kenny thinks he might have slipped into a light sleep, but then he was suddenly awake again, feeling as though something was stalking the perimeter of where everyone was sleeping. He said the creature continued to sneak around the group for almost an hour, making him feel very unsettled. Then it simply disappeared. Kenny fell asleep and was wakened the next morning by the bright sunlight.

The morning after the tree shaking, Kenny investigated and found all of the area at the base of the tree was completely trampled down. He is an outdoorsman and amateur tracker and he always looks for tracks of animals and predators, but he said that no hoof prints could be found anywhere near the tree. It was Kenny's feeling that whatever was shaking the tree had pads for feet, and nothing pointed, because the grass was bent over but not broken.

About a year after the party incident, the family that owned the ranch asked Kenny to housesit while they were out of town. One Saturday night at approximately midnight, the owner's small dog started to go crazy just outside the front door. The dog had been outside for the majority of the night and it was at the front door making a frantic noise and throwing itself at the door to come in. As Kenny opened the door, the dog bolted straight into the house and cowered in the corner of the front room. Kenny had a friend with him at the house, and he and his friend were simultaneously overcome by the worst possible odor imaginable. He described it as something similar to cat poop and horrible human body odor mixed together. The odor was so horrendous it made them both gag.

Kenny decided he'd grab a shotgun and he and his friend would investigate the odor. They followed the smell up into the area where they had partied the previous year and then heard a very loud screech. It sounded like something got hurt hurdling barbed wire. It

Ranch and hillside on Hayes Road.

was a very loud scream and definitely not human. The men walked back to the house and the smell was completely gone.

Several years after the encounters at the ranch, Kenny was watching the television show *In Search of...* hosted by Leonard Nimoy. The show focused on bigfoot and played a recording of a purported bigfoot screaming. Kenny said the scream on that show was exactly like what he heard the night at the party, so he believes he heard a bigfoot screaming at his friends while they were all sleeping.

A few months after housesitting, Kenny talked to his friend who lived at the house about the strange encounters. His friend told him that he remembered two specific incidents years before. The first occurred when he was a little younger and sleeping in the second floor bedroom. He explained that for some reason he awoke to his room absolutely dark, as though something was blocking the light at the second floor window. The next morning he went outside the window and found a rose bush completely crushed. He explained that there was no way a human could climb up the wall because there were no grab handles or ledges; whatever got up and looked and blocked the light in the window had to be huge.

His family also owned a dairy and occasionally slaughtered cattle, so they had a meat locker on the property with a large latch on

123

the outside of the door. In the second incident, late one night something unlatched the door on the meat locker, took a cow carcass weighing 200 pounds off its hanging hook, and pulled it outside where it took several large bites out of it, and then left it in the yard. The family was completely baffled, and never was able to figure out what could have had the dexterity to open the latch, lift but not tear a 200-pound carcass off the hook, and then take bites out of it.

Kenny Rogers signed an affidavit to all facts depicted in his statement.

LOCATION

The ranch sits at the base of a small ridgeline that extends from the coastal range to Monterey Bay. It is an ideal ridgeline to access fields of fruit in a very stealthy manner. There was sufficient cover 30 years ago to easily conceal a large creature in its movement from the mountains to the coast.

A review of various bigfoot databases indicates there have been several bigfoot sightings reported within a short distance of this location. It would appear that from everything Kenny has communicated about his personal encounters at the ranch and the owner's incidents, this spot is also in bigfoot territory.

COLETTE ALEXANDER

When I was a police officer in San Jose I wanted to live in an area far from that city and in a place where I didn't worry about urban crime; I found it in Santa Cruz County. From 1983 until 1993 I lived in the city of Santa Cruz. The city is an interesting mix of older citizens, young students from UC Santa Cruz, and a hippy culture that's stuck from the 1960s. I enjoyed the environment because there were cultural events to attend on the weekends, great restaurants, and it sat on Monterey Bay and the Pacific Ocean. Santa Cruz also has the San Lorenzo River, known for its runs of steelhead and salmon.

The San Lorenzo has its headwaters in the Santa Cruz Mountains above the city of Boulder Creek. If I dropped you with your eyes closed near the headwaters, when you opened them you'd think you were in far Northern California and probably in the coastal mountains of Del Norte or Humboldt counties. These are very remote areas, with few homes, big trees, and significant wildlife. According to California Fish and Game officials, there haven't been bears in Santa Cruz County for decades, yet wildlife biologists studying the upper reaches of Boulder Creek claim bears are living there today.

Colette Alexander.

When I was living in Santa Cruz I would routinely go fishing in the San Lorenzo for steelhead. I would usually park in an area between Santa Cruz and Felton and walk approximately one mile through the lush wilderness to get to the river. It was common knowledge that there were homeless camps in the area, and some of the homeless were known to be less than hospitable. I always carried my gun on my hip and it was obvious to anyone who approached that I was armed. The last thing I needed was a confrontation in the middle of the wilderness. In the many trips I made into this area I never saw anyone, never heard anyone, but it did feel creepy. On several trips it was very foggy, light mist, and deadly calm. I knew about bigfoot back then, but I wasn't thinking it would be in this area. Wow, I was wrong.

In 2008 I had made a trip to Happy Camp in Siskiyou County and met a fellow researcher, Linda Martin. Linda had worked in Happy Camp for years and operated a web design business. She was also the resident bigfoot researcher and routinely posted sightings from various parts of the state on her website, www.bigfootsightings.org. NABS researchers routinely scan the web for new sighting data and found a sighting reported to Linda from a Santa Cruz County incident. I contacted Bigfoot Sightings and Linda, and she placed me in contact with bigfoot witness Colette Alexander.

Linda told Colette that I'd be calling and gave her some back-

125

ground on our organization. I called Colette at her residence just outside Sacramento. She was very cordial and volunteered to give me the details of her sighting.

Colette is a 44-year-old divorced mother of an 11-year-old girl. She was raised in upstate New York, and Washington County, and eventually moved to California in 1977. She went to Cabrillo College in Aptos, a suburb of Santa Cruz and received her A.A. degree in accounting and finance and a minor in anthropology. Colette explained that she did a few different jobs in her younger years and one that she enjoyed most (because of her love of the outdoors) was being a fire fighter in the Shasta/Trinity National Forest in 1987. She explained that she did see a lot of wildlife, but never a bigfoot, and never even heard any stories about it while she was stationed there.

In June 1999 Colette was living in Santa Cruz and working as a bookkeeper. She had a roommate, Chris Kelley, and they routinely hung out together. One day both women decided to enjoy a clear, beautiful, warm Santa Cruz day by going to the San Lorenzo River to have a picnic. They packed sandwiches and headed up Highway 9 from Highway 1, approximately one mile. At the first right turn they pulled to the shoulder, parked, and exited the car to walk down the sandy beach area towards the river. They walked approximately 100 feet, laid out their blanket and enjoyed the peaceful river view and the beautiful sunny day.

The women weren't the only people in the area. There was a fisherman approximately 20 to 30 yards downriver who was intent on his hobby and did not acknowledge Colette or Chris. Colette says that they were sitting quietly for close to 20 minutes and had just started to eat their sandwiches. She was looking downriver towards the fisherman, but for some reason looked at the riverbank near the man. She saw the face of a creature inside a large cypress bush. The creature was looking directly at Colette and putting its hand towards its mouth, mimicking her gesture as she ate her sandwich. She couldn't believe what she was seeing. The creature appeared to be sitting on the ground, and its head seemed slightly higher off the ground than Colette's head. The face was not human, but not animal; she described it as "a strange cross between human and ape." The creature continued to stare at Colette as she moved her hand with the sandwich to her mouth. At one point she purposely went in super slow motion to see what the creature's reaction would be; it did exactly what Colette did. The creature even smirked at her, as though it was having fun.

Colette's description of the hominid was that it was completely covered with air except on its face; it had dark skin. The creature opened its mouth when it put its hand near that area. She could see very white, straight teeth and no fangs. The teeth looked the same as human teeth, but much larger. She was close enough (20 to 30 yards) to see the whites of the creature's eyes, but could not determine iris color. She described it as having a fairly prominent brow, but there was not a jutting chin as in some primates. She could see that the hominid had very large arms and shoulders. Bushes were covering much of the creature's body, but Colette could see that it had almost no neck. It also appeared as though it had long hair coming off the front of its head, almost as though the hair was pulled from the top over its back.

I always ask witnesses to give me their opinion of the sex and age of the creature they saw. Colette was quick with an answer; she felt it was a female. She stated that when it was mimicking her it was doing it with the grace of "someone at a tea party; it was incredibly delicate." She felt the female was "young, probably a juvenile that wasn't afraid, and was just as interested in me as I was in it."

Colette's observation lasted approximately one minute. She knew she had to lean over and tell her friend what she was observing. For just a few seconds she turned away from the creature and quickly told Chris to watch and look at what was in the bushes. Colette turned back around and the creature was gone. The fisherman was still in his position, less than 20 yards from the hominid's location and apparently unaware of what was in the nearby bush. Shortly afterwards, the women felt it was time to leave. Colette affirmed that she never felt threatened or afraid, and she felt the creature was very peaceful.

There were no unusual sounds or smells associated with this sighting.

Colette confirmed that she has seen many bigfoot films and videos and believes that what she observed on the banks of the San Lorenzo River was a bigfoot. She guarantees she did not see a homeless person or an ape. She knows that the creature she observed had distinct human behavioral qualities and it was purposely mimicking her actions. She stated that this was an incident that she will never forget. Colette signed an affidavit.

Location/ NABS Comments

The location of this incident is on the fringe of the city of Santa Cruz and 60,000 people live in the general area. The San Lorenzo River does have many remote and desolate areas just east of the sighting and is the probable home of the creature. It is amazing that the creature would place itself in the situation it was in, between the river and the roadway, hidden just by a few bushes.

I am not surprised that the creature allowed itself to be viewed by two women. Bigfoot has a history of not being afraid of women or children, and this is another example. It did not allow itself to be seen by the male fisherman. I also believe that it is extremely probable that this was a juvenile female who was interested in the behavior of a human female in the same manner we'd be interested in their behavior.

An interesting side note to this sighting is that the Bigfoot Discovery Museum is located approximately five miles east of this sighting on the same roadway in the town of Felton.

6 AMADOR COUNTY

One hundred and forty three miles east of San Francisco and 45 miles east of Sacramento sits Amador County. It has a population of 38,941 residents in 593 square miles. The county seat is the city of Jackson, with a population of 3,989. Other incorporated towns in the county include Amador City (196 residents), Ione (7,129 residents), Plymouth (980 residents), and Sutter Creek (2,303 residents).

The county is crossed by Highway 49, which has been named by *Parade* magazine as "Most Scenic Highway in California." Industries in Amador County include lumber, gold mining, clay, pottery, glass, cattle ranching, and wineries. Many of the residents commute to their day jobs in Sacramento, but there is also a significant group of retirees that call Amador County home. The area has very diverse topography that changes in elevation from 200 feet to 9,371 feet.

Fiddletown is one of the oldest unincorporated areas of the county. In the 1800s it had the largest population of Chinese immigrants anywhere in California, except for San Francisco. It was originally settled by immigrants from Missouri and was named for young men supposedly "fiddling around." The town received a name change to Oleta in 1878, and then it was changed back to Fiddletown in the early 1900s.

The Miwuk Indians lived in the Amador County region for hundreds of years. They had established over 100 villages along various rivers in the area and grew to over 20,000 members in the early 1800s. When Spaniards arrived in the 1820s they brought smallpox with them, which resulted in wiping out a vast majority of the Indian people. In 1848 the gold rush reached the region and that brought with it confrontations between the miners and the Indians, many of which were fatal.

In 1994 the United States government finally recognized the Miwuk Indians' sovereignty and gave them 40 acres for their members. In the twenty-first century the tribe has attempted to establish the Buena Vista Rancheria Casino on property that they state is theirs. Amador County has put up a huge legal fight against the casino, but it appears that the casino will be built.

Abandoned mines riddle Amador County landscape. Many of these mines were old gold and quartz excavations, and some were the deepest in the United States at the time. The Kennedy Mine was 4,600 feet deep and removed 35 million dollars of ore.

The county sits on the eastern edge of the San Joaquin Valley and extends like a finger into the high Sierra region. In the Sierras it sits between Calaveras and El Dorado Counties, with Alpine County bordering to the east; Sacramento County borders to the west. Sacramento County is predominantly a very flat region that has little cover and many people.

COUNTY STATISTICAL DATA

The following is a list of the counties and the number of bigfoot sightings/incidents taken from several databases:

County	Location	#Sightings/ Incidents	Sq. Miles
El Dorado	North of Amador	23	1,710
Alpine	East of Amador	8	738
Calaveras	South of Amador	8	1,020
Sacramento	West of Alpine	0	965
Amador		0	593

The fact that Sacramento has no bigfoot sightings or incidents does not surprise NABS. It's too flat, too hot, has too many people, and has little cover with almost no water. It is surprising that Amador County occupies space in the Sierras that is almost identical to El Dorado and Calaveras, and there have been no reported incidents in the county other than what NABS has researched.

DANIEL WALKER

A bigfoot researcher told me many years ago that I needed to be visible if I wanted witnesses to find me. I listened to that advice and put North America Bigfoot Search on the Internet with a page to report sightings and encounters. One of the many reports that filtered through our site was Daniel Walker's.

I first spoke to Daniel via phone, and then later we met at his residence just outside the old city of Fiddletown. It's a cute place with many old buildings still standing from the times of cowboys riding down the middle of the street. It is a historical landmark and is worth a trip just to see the brick-and-mortar construction. Fiddletown is located at the base of the Sierra Nevada Mountains, 35 miles west of Sacramento.

Daniel is 24 years old and attended Citrus Heights High School and later graduated from the Merchant Mariner Technical School to work the high seas, although he never pursued that. He has been living with his aunt and uncle, the Morgans, just east of Fiddletown, raising horses on a beautiful 10-acre parcel at the 2,200-foot elevation.

Daniel explained to me that he was driving his truck back from Sacramento after visiting friends. He was alone and it was approximately 11:30 p.m. with a partial moon illuminating the road. He was driving eastbound on Fiddletown Road just west of its intersection with Hale Road in Amador County. It was a clear August

Daniel Walker next to the stop sign.

night in 2007, about 75 degrees outside, and Daniel had the windows of his truck rolled up and the air conditioner on. I should state here that this region gets very hot in the summer and stays hot during the night. As you travel east into the Sierra foothills, the temperature starts to drop as you rise in elevation. When Daniel started his trip in

Sacramento it was probably near the high eighties, and as the night got later and he drove higher in elevation, the temperature dropped.

Daniel explained that as he approached the intersection with Hale Road, he started to slow down, as he saw something on two feet walking from the southern edge of the roadway. He described the hominid as walking very casually, similar to a human. It was completely covered in hair, it had long arms, and it was brown in color. The bigfoot appeared to have hair longer than the bigfoot in the Patterson–Gimlin footage. As he got closer to it, the pace of the bigfoot accelerated. It never looked directly at him, but it was at a slight angle. He could see a slight profile and could tell that it had a face more like a human than any animal—flat, with a nose, and not a snout like a dog or bear.

As the truck got closer to the intersection, Daniel watched as the bigfoot stepped down into a small creek bed on the north side of the road and quickly disappeared amongst foliage and brush. He stated that it looked very healthy and was quite thick across its shoulders, buttocks, and legs. Daniel confirmed that he has seen the Patterson–Gimlin footage and other specials on bigfoot and said that the creature he observed matched those creatures exactly, except that the one he saw was brown in color and not black.

Daniel drove directly back to his residence and found everyone asleep, so the next morning he spoke with his uncle, Norman Morgan, and explained the incident to him. Later in the day Norman and Daniel went back to the scene and Norman stood next to a stop sign in the roadway that the creature had walked by as it crossed the road. Norman brought a tape measure and Daniel pointed to the area of the pole where the height of the creature reached. The measurement was seven feet, five inches.

I drove to the area a few weeks after NABS received the report and a few items immediately struck me as interesting. First, the elevation: 2,200 feet falls within our elevation association with bigfoot. We believe that the optimum elevation to observe a bigfoot is 2,400 feet, with that range extending from 1,600–3,200 feet. As I was driving on Fiddletown Road at 10:00 a.m. on a warm morning, I saw several deer crossing the road within two miles of the intersection of Daniel's sighting.

Daniel and Norman met me at the intersection of the sighting and we reenacted the encounter. It appeared that this was not a chance encounter with this bigfoot; there is rationale as to why it was there. The area of the creek where Daniel observed the hominid retreat is filled with berry bushes. In late August these bushes are filled with berries, a food source that bigfoot relishes (see chapter 3, Associations: Berries).

I searched the entire bank of the creek bed and found a very well worn game trail that parallels the creek and the roadway as you travel eastbound. The trail turns into an old 4x4 dirt road that has one lane well worked by game. The soil in this area is like cement, very hard and not conducive to finding footprints. The southern side of the roadway has a small ridge with more brush and cover and would be an ideal place for a bigfoot to sit and ambush prey walking down the road or creek bed. Norman advised me that the creek flows for 6–7 months of the year and usually is dry in the summer, as it was on my visit.

Amador County is one of the smaller counties in this area in California. It is predominantly at the lower elevations of the Sierra range. Prior to traveling out to meet Daniel and Norman, I did a search of several bigfoot databases and found no reported sightings for the county. I thought the finding was odd because adjacent counties had many bigfoot sightings very close to the border with Amador.

After spending several hours with them, Norman took me aside and stated that Daniel was a very trustworthy person who never fabricated stories. He said Daniel's story has remained consistent since the first time he told it, and he believed that Daniel had the encounter with bigfoot.

I found Daniel and his story very credible. Daniel signed an affidavit.

LOCATION

There are not many residences in this region of Amador County that border the roadway. After 11:00 p.m. at night the amount of vehicular traffic on the roads in the county is miniscule. These are mostly blue-collar people who work hard during the day and get to bed early at night. There are the few that travel to jobs in Sacramento and take a late commute home, but they rarely arrive later than 10:00 p.m.

The exact location of this sighting, at 2,200 feet, is optimum for deer during the month of August. I personally saw over eight deer either on or near the roadway on my visit. The hills were covered in thick foliage and the vegetation was still green despite the heat of the summer. It was also remarkable that the ridge on the south side of the road came to an abrupt end at the intersection of the sighting. It is quite possible that the bigfoot was walking the ridgeline to the roadway and making its way to the berries in the creek bed (see chapter 3, Associations: Ridgewalker).

NORMAN MORGAN

Daniel Walker lives with his uncle and aunt, Norman and Susan Morgan, approximately three miles from the location where Daniel had his bigfoot sighting. The Morgans live on property on the north side of Fiddletown Road that intersects the roadway immediately adjacent to a huge berry bush (ironic). From their house they can see neighbors who have large ponds and huge 100-year-old Douglas fir trees on their properties. They routinely see bobcats,

Driveway berry bush at entrance to Morgan home.

deer, raccoons, and wild turkeys. The elevation of the property is 2,300 feet and sits atop a small ridge at the eastern end of their valley. The valley is relatively small and the Morgans are able to easily see across it. Years ago most of the bushes and small foliage were cleared away by residents.

Norman was raised in Oak Grove, California and joined the Navy close to his eighteenth birthday. He spent nine years in the service and reached the rank of Quartermaster (Navigation Specialist). He has owned a fire extinguisher company for a number of years and still makes a comfortable living in that profession. The family has three dogs that live with them on the property: a shepherd, golden retriever, and a McNab.

The inside of the Morgans' valley has a few huge trees, but is predominantly covered with grass with almost no bushes. Just outside the valley the forest line begins and is heavily covered with oak, Douglas fir, and madrone.

When I was speaking with Daniel by phone he said his uncle wanted to talk with me. At the time I thought that maybe he was going to check me out to ensure I was a professional researcher and then ask me a series of relevant questions. I was wrong.

The Morgans have lived in their valley for six years. During the winter months there is some light snow and during the summer months the temperature can get into the nineties. Norman explained that since he and his wife moved into the house they have occasionally heard *whoop*

Norman Morgan.

whoop sounds coming from the perimeter of the forest late at night, and only during the warmer summer months. He thought it was possible that this occurs more than just in the summer, but their windows are closed in the winter and they can't hear outside noises. He said it sometimes starts at 11:00 p.m. and continues until the very early morning hours, 4:00 to 5:00 a.m. The oddest part of the story is that Norman and his wife feel that the sounds are coming from the tops of the biggest trees in the area. He knows these are not owls or any

135

other common animals or birds that normally frequent the region. One time when he was on the Internet he went to a BFRO website and found a section for sounds that bigfoot can make, and what he listened to was identical to the sound he and his wife have heard at night. They have occurred on approximately four to six occasions, and the dogs are almost always inside the house and they never make any noise, an unusual behavior for dogs.

Norman is a very confident and sincere person, and I found no reason to disbelieve his story. There were certain indicators in the area of his residence that pointed toward the distinct reality that he may be hearing a bigfoot.

The Morgans live near the top of one ridge in the small valley they occupy. The ridge encircles the valley and falls off to the north and towards the American River, which is relatively close to their residence. There is ample fresh water in the area because of the huge ponds maintained on adjacent properties. Morgan's explanation that the sounds are coming from the tree lines at the ridgelines near their house makes sense. If bigfoot were a ridge-walker (as the Yurok claim) then it would make sense that they are making sounds to attract other bigfoot from the ridge tops; there would be no point in yelling for others from the bottom of a valley. Daniel's bigfoot sighting three miles from the Morgan residence indicates that the hominid does frequent the area, and may be attracted to the isolation, water, and food sources available in the immediate area.

LOCATION

I spent several hours driving and hiking the area around the Morgan residence. I also reviewed United States Geological Survey maps of the region. Reviewing a 10-mile radius from the property and the sighting location, I found that this region was one of the few optimal places for a bigfoot encounter. If we take into consideration elevation, water sources, ridgelines, food sources, and lack of people, the area of the Morgan property is an ideal location for bigfoot encounters and sightings.

7 TRINITY COUNTY

If you want to live in a rural area with beautiful scenery and few people, Trinity County may be the place. Trinity was one of California's original counties in 1850. It's the size of Vermont (3,200 square miles) yet only has 13,000 people. It sits between Humboldt County on the coast and Shasta County to the east. It ranges in elevation from 600 to 8,000 feet, and gets as much as 50 inches of rain per year. The rain accounts for the county having 29 square miles of surface water that includes lakes, streams, and rivers.

The county seat is Weaverville, with 3,500 people. As late as 1999, Weaverville didn't have one brand-name store or fast-food restaurant. Times have now changed. They have most of the major restaurants, drug stores, and markets. The lumber mill, one of the major industries, is still in the middle of town. The economy of the area has always been slow, making a living difficult, but the riches of the environment are vast.

Just outside of Weaverville is Trinity Lake. It is the third largest lake in California, with 147 miles of shoreline, and has crystal clear water. The lake was one of the first places in the state where the bald eagle started its comeback. It is now home to a large contingent of eagles and you can routinely see them diving into the lake for fish. The lake is famous for good fishing, water skiing, and kayaking.

The Trinity Alps Wilderness Area sits at the northern edge of Trinity County. It is famous for its beautiful rocky peaks, bear, alpine lakes, and bigfoot. The area is vast; with 517,000 acres and 550 miles of maintained trails, it is the second-largest wilderness area in California. It was formerly known as the Salmon Trinity Alps because of the proximity of the Salmon River. The wilderness area was doubled in size under the California Wilderness Act in 1984. It has 55 lakes and streams with an elevation range from 2,000 to 8,000 feet. The Trinity Alps have some of the most beautiful alpine lakes and scenery available anywhere in North America. The upper lakes do not usually open until mid July because of a high snow pack.

There are three major highways that cross the county: 3, 36, and 299. The main towns are Weaverville, Hayfork, Hyampom, and Lewiston.

HAYFORK

Hayfork is located in Trinity County, California in the northwestern portion of the state. Trinity County only has 17,000 people and Hayfork has 2,500 of those. Hayfork was established in 1851 and was originally named Kingsbury. In a span of three years, the small town changed its name three times, eventually staying with Hayfork because of the amount of hay the surrounding valley produces. The town sits at 2,300 feet in elevation and does get a moderate amount of snowfall each year. There is one main street (Highways 299 and 3) that boasts three diners, a Mexican restaurant, two markets, two auto repair shops, a gas station, regional USFS office, and Hayfork Elementary and High School. There is a Trinity County Sheriff sub station at the end of town, along with a small county library. The Trinity County Fair opens every Labor Day weekend in Hayfork and draws a significant crowd from Northern California.

On the northern side of Hayfork and just up the hill sits Ewing Reservoir. It is the local water supply and recreation spot. It has good trout fishing in the summer and offers a great place to hike and see the countryside. On the western edge of town is the county's fairgrounds and Hayfork Airport. The airport has one runway that is paved asphalt for its 4,100 feet.

Hayfork has not been without controversy. In the mid 1800s the Wintu Indians, with 150 in their tribe, had settled in a small meadow approximately seven miles south on Hayfork Creek. There was an altercation between some sect of the Wintu and a local rancher, J.R. Anderson. Anderson ended up dead and the Wintus were blamed. The Hayfork sheriff mounted a posse and rode out looking for the Natives. Late at night the tribe was found. The posse waited until morning and at this point the story gets vague. Something happened and almost all of the 150 in the tribe were slaughtered; three children were spared. The murderers weren't even in the tribal party killed. It was a horrific event that scarred the region for years.

There are two communities that are neighbors of Hayfork. Twenty-six miles northwest is Hyampom, an even smaller town than Hayfork; and still smaller yet, Wildwood is 15 miles south.

Hyampom is a two-store town that takes a narrow road to reach its location. There is a nasty cliff that you have to follow to finally reach this beautiful spot. There is no law enforcement presence in this town, and it has some degree of notoriety for being a bit rowdy. It does have a beautiful valley and surrounding mountains. In the late 1800s and early 1900s Hyampom was a thriving lumber town with its own mill, and once you arrive you will quickly see how its thick forests could maintain a profitable timber industry. In more recent times Hyampom has become a home for ranchers and some farming.

Wildwood has a small country store that boasts the town's name, and that is the only way you know you've arrived. It is a beautiful drive out to the store as you follow a small creek and gorgeous valleys. At 3,500 feet of elevation, Wildwood is a fairly busy place during the summer months, as fishermen and hunters flock to the location because of its isolation, good outdoor sports, and beauty. The easiest method to reach Wildwood is to take Highway 36 west from Red Bluff. It is also accessible by traveling east on Highway 36 from the coastal city of Rio Dell. Either direction will offer you great mountain vistas and a guarantee of seeing deer, bears, and other mammals indigenous to the area. From Hayfork you drive the Wildwood Road and follow Hayfork Creek for 15 miles until you reach a small housing community of Wildwood, or until you see the store.

Hayfork, Hyampom, and Wildwood all sit inside the Shasta-Trinity National Forest, approximately seven miles west of the very isolated and desolate Chanchelulla Wilderness Area. The southern area of Wildwood is split between the Hayfork and Yolla Bolly Ranger District.

As the baby boomers in California have started to retire, they have looked for areas of the state to extend their dollars and live comfortably. There are many retirees from throughout California who have moved to Trinity County because of the inexpensive housing and access to the outdoors. In my travels throughout this region I continuously saw 60- to 70-year-old couples, and a distinct absence of younger people. There aren't many jobs in the county except government, which is the largest employer at over 40 percent. The age of the residents actually played well into my research of bigfoot, as there is a broader base of knowledge that has stayed in the community.

BACKGROUND

I came to Hayfork after conducting an exhaustive Internet search of all relevant bigfoot activity in Northern California. The three towns offered a mixture of activity and a consistent time span of sightings.

The catalyst that prompted me to devote significant time to this region was Greg Fork. He had investigated three separate bigfoot incidents for the United States Forest Service in the Wildwood area. All of these are documented later in this chapter.

I didn't have any specific names to contact when I arrived, so I knew I would have to canvass the area for locals willing to talk.My first stop was the local grocer. I had been advised that the Wildwood Store kept a listing of all the bigfoot activity in this area. Apparently they used to have a map showing all of the sightings and incidents that occurred in the area, but unfortunately no one knows what has become of it.

DICK MURRAY

I drove into Hayfork and stopped at the first market I found. I had just made a one-hour drive from Weaverville in a blinding snowstorm. The trip would normally take 25 minutes in good weather, but the snow and ice slowed my Jeep to a crawl. I wanted to grab some food and drink at the store, show my goodwill, and then contact the manager to see if I could dig up bigfoot witnesses.

I found the store assistant manager, Mike Wycoff. He was interested in my research and did what he could to lead me in the right direction for contacts. He told me I should first try to locate the retired fire chief, Bob Young, as he knew the town very well and usually knew all the local gossip. He also told me to go out to the fairgrounds and contact the fair manager, Ron Hall. Mike said that Ron was well connected and could fill in what Bob missed. Just as I was ready to thank him and walk out, he made one final recommendation. He said that I should step across the street and talk with a retired Forest Service employee, Dick Murray. He explained that Dick knew what was happening in the woods and in the Forest Service around Hayfork, and he was a very intelligent guy. I thanked Mike for the help and went to find Dick Murray.

Dick has a small but professional office in one of the newest buildings in Hayfork. He was alone when I walked in and greeted him. He is a short haired, clean-cut, well-spoken man who looked me right in the eye. I explained the reason for my contact and went into a short background of North America Bigfoot Search.com. I told him that we were interested in his personal background and any knowledge he had of bigfoot.

Dick said that he came up to Hayfork from Long Beach in 1966 where he was a financial consultant and mortgage banker. He has a bachelor's degree in electrical engineering, and a master's degree in math from Michigan State University. He wanted to get away from traffic and congestion and felt Hayfork offered what he needed. He applied for a job as district engineer with the Forest Service in their engineering department, with the idea he would be able to work in the field, but his job actually had him in the office a lot with his main role of engineering mountain roads—many throughout the Weitchpec and Orleans area. He explained that the bureaucracy forced the engineers into the metropolitan areas like Redding, so he decided after 12 years to leave the Forest Service. Dick's final year with the government was 1977.

Dick explained he didn't have any firsthand accounts of bigfoot,

141

as he never heard or saw the creature, but he did have a colleague in the Forest Service who had seen it. He said that it was a day and a meeting he would not forget.

In the summer of 1974, Dick was working out of the USFS Hayfork office. It was near 4:00 p.m. one day when a good friend and fellow Forest Service employee, Glen Cecil, came into his office, all excited and wide eyed. Dick said that there were other Forest Service employees sitting in his office when Glen entered the room. Glen was a no-nonsense employee, very competent, and knew the forests and animals very well. He was quite respected as a senior member of their team and nobody ever questioned his competence.

Cecil was a fire prevention officer and on the day of this meeting he had been checking on firebreaks up Dubakella Ridge. This was an area east of Wildwood in a desolate portion of the district. He was specifically up Dubakella ensuring that shrubs and other smaller foliage were clear of the fuel break lines in case of a fire in the region. His visit to Dick's office wasn't a normal occurrence because their offices were in different places, but he had specifically made the drive to Dick's location to tell him about his sighting.

Cecil looked at everyone gathered around Dick's desk and they all went quiet for a second. He said, "I think I saw bigfoot today." Everyone in the room started joking that he wasn't serious. Cecil looked at Dick with a complete stone face and stated, "I'm serious, and the thing is human, it walked like a human and it's huge. I know what I saw." Cecil was not the type of employee to fabricate or exaggerate anything; if he said that he saw bigfoot, then Dick believed him. He went onto explain he was working up on Dubakella alone when he saw the huge creature a moderate distance away. He was shocked, but reiterated he knew exactly what it was—bigfoot.

Cecil retired after 30 years with the Forest Service. Dick later hired him as a plumber for some of his land properties, and they always stayed in close contact. He died eight years ago and his wife passed soon after that. He had one son who is an English teacher in China.

Near the end of my conversation with Dick, he looked at me and stated, "If anyone else had made the claim of seeing bigfoot I would have forever questioned it. Cecil knew the woods as well as anyone

I've known. When he made the claim of seeing bigfoot, I know that he must have seen it. He wouldn't lie and he wouldn't have mistaken it for another animal."

LOCATION

I listened intently as Dick Murray was explaining Glen Cecil's sighting, and when he stated that the location of the sighting was Dubakella Ridge, I nearly fell off my chair. Dubakella Ridge is directly adjacent to Mud Springs and the location where Greg Fork, a Forest Service officer, investigated a purported stalking and harassment of four campers in the middle of the night by a bigfoot (see this report later in this chapter). The size of the Hayfork Ranger District is enormous. What are the chances that two confirmed bigfoot sightings, both involving Forest Service employees, would occur at the same location? It's mind-boggling.

Deer crossing road.

This location is within five miles of two other bigfoot incidents also investigated by Greg Fork, and the Wildwood Store is at the center of all of this activity.

The elevation in the Dubakella Ridge/Mud Springs area runs from 3,200 to 3,800 feet. It is heavy timber with very few visitors at the site. It's an eight-hour drive from the metropolitan Bay Area (San Jose/San Francisco) to get to the location. You need a four-wheel-drive vehicle to reach the spot if the roads are wet or snowy.

On one of the drives I made into the Wildwood area in January just after a heavy snow, I was amazed at the number of deer I saw on one single drive—10 in less than 20 miles. The photo is of a doe crossing the roadway at 3,500 feet in heavy snow. There is ample big game in this region to support a family of bigfoot.

KATHY SCHRAEDER

One morning I was having breakfast in Irene's, the lone café in Hayfork. It's a place where all the locals meet early in the morning and exchange gossip about local events. As would roll my way, I was the local gossip for a few days. The waitress came to my table and saw some of my materials I was reviewing and asked if I was the bigfoot man. I said that I was and we exchanged a few stories. The retired fire chief of Hayfork, Bob Young, was with me and he validated my credentials for the waitress. She told me there was somebody I needed to talk to who lived on the outskirts of town— Kathy Schraeder, a special education teacher at Hayfork School.

Directions in this part of California can be a little sketchy. There aren't a lot of street signs, addresses, or gas stations to ask for directions. There's a lot of "turn right at the big oak, go 3/10th and make a left at little 13." If you want to ensure you get to where you need to be, you need to ask a lot of questions in a variety of ways. With the help of Bob and the waitress, I was able to discern where Ms. Schraeder lived.

I was told that the Schraeder residence was seven miles from Hayfork, across a creek and up a hill, around a bend, and look for a bus stop. Okay. It had just snowed the night before and there was snow accumulation of three feet on the ground. It was getting a little mushy and the ground was quite muddy when I turned off the main highway onto the Schraeder dirt road. I made my way up several small hills while I paralleled a very steep embankment that would have had me fall into an adjacent creek—a bit nervy. The road was slippery, even in my lifted four-wheel-drive jeep. After approximately one-half mile I reached a beautiful large house and adjacent barn. The area around the residence was well maintained. I walked a short distance down a paved walkway to the front door and knocked, and Kathy answered.

I knew before meeting Kathy that she was home from school, sick with the flu. I'm not one to push my luck with getting sick, but sometimes in bigfoot research you have to make a contact when it presents itself. Kathy was gracious enough to meet, even with her condition. After a few minutes of small talk I could see her eyes start

Schraeder residence.

to brighten and the energy start to return; she looked interested in our topic.

Kathy attended the University of Montana for three years and transferred to Sonoma State where she obtained her bachelor's degree and teaching credential. She was looking for a property in the Hayfork area and 28 years ago (fall 1980) found the location where she and her family presently live. She has one 32-year-old son who is a hunter and outdoorsman and lives at the house with Kathy, her husband, and her sister, Diane Sprow.

Kathy said that she bought the property because of its remote location and its setting, just below a ridgeline, just above a creek, and its beautiful meadow behind the house. When she bought the place there was only a small hunting cabin on the property that three of them lived in; it was a little cramped. Months after they moved in, a gentleman visited who claimed to be a bigfoot researcher. She didn't remember the name or the organization, but distinctly remembered the incident. It was in the summer and they were barbecuing and invited the guest to stay for dinner. He told a variety of stories and incidents he had investigated, and the reason for his visit. The people that Kathy had bought the property from had reported bigfoot activity in the past and the researcher was there doing follow-up work. Nobody can remember what type of activity was reported, but they all remember the visit.

In the summer of 1982 Kathy, her husband, and son were trying to sleep in the small cabin when they heard the strangest sound they ever heard in the mountains. It was a warm night and they had the windows open, and near midnight they heard a high-pitched, very, very loud scream come from the top of the ridge past the meadow behind their cabin. The yell occurred multiple times over three consecutive nights. By the third night she said it was obvious that the incident was unsettling to her husband, who was starting to get concerned about what could make such a loud noise. Kathy obtained a voice-recording device on the fourth day to record the scream, but as is almost always the case, the screams stopped.

On Thanksgiving 2007 Kathy and her sister Diane were on their deck enjoying the brisk autumn day when they saw their dogs start to go crazy on their back hillside. They were barking in unison and they appeared agitated and nervous, unusual for her dogs. Just a few seconds after the barking started they also heard a very large branch break, as though something had just stepped on it. They both became quite concerned and decided to walk up the hillside and determine what was annoying the dogs. Kathy knew there were bears, deer, and raccoons on her property, and there was probably even a mountain lion somewhere nearby, but the sound of a large branch breaking was very unusual.

The women started their climb up the hillside when they were immediately overwhelmed with a "death" smell. Kathy said it was the worst smelling, most horrific odor she had ever encountered, similar to something that was rotten or dead. They continued up the hill to where the dogs had congregated where they could immediately see the carcass of a dead deer. It struck both women that this had been no ordinary kill, quite the opposite. The deer was in a small clearing on a thick bed of fall leaves, and it was lying on its back with its legs spread apart. She stated that the legs were not broken but just spread on either side of the carcass. It appeared that the deer had been partially skinned with the lining evenly pulled to each

Hillside behind Schraeder home.

side of the body. The ribs were exposed and spread apart in the mid section to expose the inner cavity. There were no organs inside, or on the ground adjacent to the body. There wasn't even any blood anywhere near the body of the deer. It appeared that the organs had been cleanly removed while all of the meat of the deer was still intact. The deer appeared to have been killed quite recently, as rigor mortis had not yet settled in and there was no odor of decomposition coming from it. They searched for bullet holes or other causes of death, but nothing was obvious. I asked if there was a possibility that its neck was broken, and she stated it was possible. There were no chew or bite marks anywhere on the body. They did look for tracks in the area, but the large amount of leaves on the ground eliminated the possibility of seeing the tracks of what killed the deer.

As the women were examining the dead deer, their dogs were running further up the hillside towards the horrendous odor. Something up the hill smelled putrid. They made their way up the hill towards the dogs and found them huddled around and pawing at a pile of scat. Kathy described it as a large, roundish pile, almost like a cow patty in initial appearance. It was brown and contained acorns, grass and hair, but the consistency was quite different than scat from any cow, horse, or bear. Kathy was really baffled as to what would have made it.

Kathy and Diane took a sample of the scat, bagged it, and placed it in their freezer. They also dragged the deer back to their residence and photographed it. Kathy said that she couldn't reconcile what would have killed the deer, opened it up like it was, and not taken the meat. She took digital photos of the deer, but never downloaded them onto her computer. Weeks later her husband needed space on the camera memory and erased the deer photos. Recently when cleaning out her freezer, Kathy saw the bagged scat and decided it was time to throw it out. In hindsight she regretted that all of the evidence was destroyed, but she stated that her memory of that day would live forever.

LOCATION

The Schraeder residence sits at 2,400 feet (NABS has associated the 2,400-foot elevation as the center point for many bigfoot sightings in Northern California. See *The Hoopa Project*). This is another example of this elevation point playing a critical role in a bigfoot incident. The actual physical location of this property is ideal for a bigfoot appearance. In the last five miles on the drive to the residence I saw 10 deer near the roadway; it was amazing the number that was easily visible. Salt Creek sits at the base of their property and was running full when I was there. Behind the property is a large meadow with a ridgeline that runs for 15 miles from one end of the valley to the other. This ridgeline is responsible for several bigfoot incidents chronicled in this book. A review of U.S. Geological Survey Maps indicates there are several year-round springs in the mountains behind this property, all of them reachable with a moderate two- to four-mile hike from the Schraeder residence.

An interesting part of this entire property is the history of this location for bigfoot incidents. The incidents are sporadic, but have occurred regularly over a 30-year period. It was quite beneficial to have Kathy involved in the most recent incident. She is educated, interested, and conscientious, all of which lends itself to a factual reconstruction of the evidence and circumstances surrounding the incidents.

LINDA HILDERBRAND

One of the first stops I made in Hayfork was at the Trinity County fairgrounds where I met the manager, Jerry Fuller. When I asked Jerry if he knew someone who'd had bigfoot encounters, he told me he had a friend, Ron, who was a plumber who had some unusual events on his property down Highway 3. Jerry immediately called his friend, and from the brief conversation it was clear that Ron wanted to meet. Jerry got off the phone and gave me directions to Ron Hall's home. I said that I would try to get there that same day.

My day was very full and I never did get to the Hall residence that first day, but the following day I found myself at another residence within two miles of the Hall house. It was an odd coincidence that I was meeting with another family on the same side of Highway 3 just further down the road from the Halls. I was excited to be an in area with several bigfoot encounters. I knew I was in a region where bigfoot thrived, so I was "all eyes."

At the conclusion of my first meeting I made my way down Highway 3 and found the turnoff for Mill Gulch toward Ron's home. It had snowed heavily the previous three days and the Mill Gulch Road had not been plowed. I drove my Jeep 100 yards down the road when I came to a small bridge over a large creek. The snow was approximately three feet deep and wet. I successfully made the bridge and started to drive uphill towards the residence. The road started to get very muddy with a mixture of solid ice and mud. The driving got a bit nervy, but I slowly made my way up three small hills until I came to a house. There was a small road up the driveway to the front of it, and I decided that it was probably prudent to stop here rather than risk total gridlock by trying to go further up.

I stopped the Jeep and jumped into three feet of fresh snow. I walked the 10 feet to the front porch and was immediately met by

149

three barking dogs and a nice woman. She greeted me warmly and said, "I know that you aren't here by chance; you must want to see me!" I told her that I was looking for the Ron Hall residence. She stated that it was further up the hill, but that she was his sister-in-law and asked if she could help.

I introduced myself and said I had been told there had been a bigfoot encounter somewhere near Hall's house. She smiled and said that it was on the hill behind us, and then invited me inside.

She introduced herself as Linda Hilderbrand, 63 years old. She said that she and her husband, Bob, had lived on the property for over 30 years. When they first moved here, there was a just small cabin on the property, which they eventually tore down, and then they built the house in which we were sitting. They accomplished all of this while Bob worked for the United States Forest Service as a graser.

Linda said that in the time frame of 1978 to 1980, in the summer months, when she and her husband were out on the front deck they heard very loud screams coming from the hill just in front of their house. There was a series of several high-pitched screams, louder than any human could possibly produce, obviously coming from the top of the hillside less than 300 yards from their residence. Linda's husband opened the door to their house and called for their dogs to go outside. He wanted the dogs to chase away whatever was making the sounds. They had one large German Shepherd and three Labs that would chase anything: mountain lions, bears, deer. They were tough and not afraid of anything—well, almost anything. Linda watched as the dogs went to the door, sniffed, and then refused to leave the immediate area of the door or deck. Her husband was very upset and also a little confused as to why the dogs wouldn't go get whatever it was. It was obvious that the dogs were very scared.

Bob decided to go up the hill himself. It was starting to get dark, but there was just enough light to see a little. Bob got halfway up the hill and both he and Linda could hear branches breaking where the creature was moving up the hill, as though it was going back behind the house. Bob got concerned and decided to return to the house and get his rifle. He retrieved the gun and told Linda he was walking over to Ron's so they could both go looking for the creature.

Linda stated that as Bob was walking down the driveway she could hear a bipedal creature walk down the hillside paralleling him

View of hill in front of Hilderbrand home.

at a distance where it couldn't be seen. Bob got to the bottom of the driveway and turned to go uphill to Ron's residence. The creature had now crossed the roadway and was on the other side of the creek, uphill from Bob and parallel to him, out of sight. Linda could tell it was a two-legged creature because of the walk, the sound of the thump the creature made as it hit the ground, breaking of branches, and the way it stalked her husband. She further stated that the most uncomfortable part of this incident was that they could tell this creature was intelligent by the way it moved, walked only when Bob walked, and its ability to stay just out of sight.

Bob eventually got to Ron's. Ron also got his rifle, and Linda explained how the men stood near the end of their road and just looked and listened for an extended period of time. It appeared that the creature never moved the entire time they stood there. They both felt that they were being watched as they were staring into the dark forest. When it got extremely dark, both men decided to go back to their homes and call it a night.

Everything stayed relatively normal at the Hilderbrand residence until the winter of 1982. Linda explained that it was near 7:00

Looking from Hilderbrand porch, down the driway towards the creek and the other hill.

p.m. one evening when they heard the same scream they had heard three times two years earlier. She said it was very discomfiting. They could tell that it was coming from the creek area near the bottom of their driveway. It was just one loud scream this time, and the Halls also heard it. Linda reiterated that the scream was much, much louder than any human could ever make. It had a relatively high pitch, but they could tell it was coming from a very large creature.

Linda said that a bigfoot organization came by her residence after both of her incidents and told her that the behavior the creature exhibited, the screams, the stalking behavior, coupled with its ability to stay out of sight, all indicated that it was a bigfoot.

I never did have the opportunity to meet with Ron Hall.

LOCATION

The elevation of the Hilderbrand residence is 2,800 feet. It sits about

Deer under tree outside Hayfork.

a quarter mile from Highway 3, out of sight and in a remote location. They are approximately two miles from another residence that has had bigfoot encounters—the Schraeders. The ridgeline in front of the Hilderbrand's house runs parallel to Highway 3 and extends the entire distance until it reaches the Schraeders, where it terminates at a valley.

A review of a United States topographic map shows that the Chanchelulla Wilderness Area is less than seven miles from these residences. Remember, designated U.S. wilderness areas are truly remote. They do not allow any motor vehicles into the region. If you want to get deep inside the area, you have to walk or ride a horse. There are several springs within two miles of the Schraeders' home and there is a year-round creek fronting Highway 3. The entire valley that parallels Highway 3 is the home of herds of cattle, which means that hay is in year-round supply to any nearby creature. Behind the residence there is nothing but dense forest, and there isn't another town for miles.

One of the most interesting aspects of my day visiting both residences was the seven-mile ride from Hayfork. I actually stopped the Jeep four different times to photograph deer that were taking a mid day nap under adjacent pine trees. I have never seen so many deer in such a short distance in the middle of winter. It's obvious that there's significant food source in this region for a bigfoot to survive.

BOB YOUNG

When I first arrived in Hayfork I had been advised to seek out the retired fire chief, Bob Young, because he had lived in the city for decades and knew almost everyone in town. He was described as a nice guy who was easy to talk to and usually willing to help on almost any cause.

Bob was assisting the local tow company on some of their jobs and I made three separate trips to the auto yard to locate him, but he was always out on the road. I eventually found him two miles out of town on an emergency call with the tow truck. I arrived in the front yard of a residence that bordered Highway 299. A large propane truck had slipped off the driveway and was leaning precariously on a hillside. It was half on the driveway and half on the hillside; it did-n't look pretty. Bob was standing at the highway calling for a second tow when I arrived.

I introduced myself and explained my purpose for being in the area, and asked if he could help us. He was very pleasant and con-genial. We set a meeting for 9:00 a.m. the following morning at a local café.

I arrived the next morning and Bob and I chatted for over an hour. He told me that he would gladly help me locate anyone in town that I was interested in meeting. I asked him if there were any bigfoot inci-dents that he personally investigated or played a role in, and he smiled slightly and said that the most famous incident in the valley that he personally knew about occurred seven miles out Wildwood Road. He stated that he had been a reserve deputy sheriff in Trinity County from 1972 to 1982, and sometime between 1972 and 1975 in the winter months the sheriff's office received a call of suspicious circumstances at the residence of Jim Crosswhite, down Wildwood Road. Jim owned a large mining claim in a desolate location. He lived alone with his two huge German Shepherd dogs. The dogs were similar to junkyard dogs: rough, tough, loud, aggressive, and not afraid of anything. Jim didn't have a phone, so had made his way to someone's house to call the sheriff and ask for help.

The sheriff's office in Weaverville contacted the resident sergeant in Hayfork, Sam Jackson. Sam had been on the job for many years and had essentially seen almost anything and everything that happens in a small town. When he received the call, he phoned Bob and said he would pick him up and they would ride out to the Crosswhite residence together. Details about the incident were a little sketchy, but it had something to do with one dog that had disappeared.

There was a little less than a foot of snow on the ground. They arrived at the Crosswhite residence just as it was getting dark. Jim answered the door, and Bob remembers him being very upset and distraught at the same time. They were invited in and they saw one of the dogs cowering in the corner, shaking. This was unusual because the dog was always vicious and tough.

Crosswhite explained that early that morning (3:00 to 4:00 a.m.) he awakened to hear someone walking around the perimeter of his yard. He could hear things being kicked over and the distinct sound of someone on two legs walking and prowling the area surrounding his cabin. He stated that he grabbed his gun and waited several minutes to see what would happen. He thought he heard the person start to leave the area by walking into the forest behind the cabin, but he wasn't sure. At this point Crosswhite sent the dogs out.

Bob Young.

Jim said that he heard the dogs run towards the back of his property and he heard a lot of barking, yelping, and then nothing. He decided to wait until morning to venture out into his yard and see what had happened. Two hours later, at first light, Crosswhite walked into his yard and found one dog cowering next to his back door. The dog looked to have been rolling in the dirt but had no apparent injuries. It was acting completely out of character and had no interest in leaving the area around Jim's feet. The second dog could not be found. Crosswhite found huge, human-looking footprints in the snow around his yard and leaving the area, walking towards the rear of his claim. He became very nervous and irate. He waited several hours to see if his other dog would return, but it didn't. He had to find a neighbor who had a phone so he could call the deputies. This all took a lot of time.

Bob and his sergeant searched the yard and the surrounding area and found the tracks that Crossswhite had seen. Bob described them as huge—four times the size of a normal human footprint—but looking very similar to human footprints. The prints could clearly be seen in the snow and soft ground. Some had been indented deeply and they could easily see identifiers in the prints. Bob had read some information about bigfoot, but really never believed anything about the creature until that very moment that he saw the tracks. He stated it was extremely obvious to him that "no animal made the tracks, no way." Bob had been around the woods his entire life and had always seen bear, deer, mountain lion, and other large-mammal tracks, but never ones like these.

Sergeant Jackson was interested in following the tracks, but it was cold, getting dark, and he was past the end of his shift. Bob wanted to track the creature, but he was only a reserve deputy and the sergeant was in control of the scene. Jim Crosswhite was very irate that something that big and aggressive had abducted his dog and nothing was going to be done about it. Bob admits that he was a little apprehensive about going into the woods with something that huge lurking around, but he thought it would be an interesting search. The sergeant and Bob left after spending about an hour at the location and drove back to their homes.

Bob said that of all the stories he has heard and all the people who claimed to have seen bigfoot, this was the one occasion where he is convinced that a bigfoot did make an appearance and abducted Crosswhite's dog. Months after the incident, Bob saw Jim in town and asked if the dog ever came back. He said it was unreal, but no, the dog never came home and his other dog was never the same. It never had the same aggression or passion to protect his property. Bob believes that the characteristics of the Crosswhite incident match bigfoot behavior that has been documented in this county for years.

LOCATION

Take Wildwood Road south from Highway 299 approximately seven miles. You make an eastern turn at South Fork over a bridge and drive a short distance to the claim location. The Crosswhite residence has been torn down, but remnants of the mine still remain. Hayfork Creek is within several hundred yards of the residence, and there was nothing else in this area at the time of the incident. The elevation of the property is 2,700 feet.

The Crosswhite claim was in a fairly desolate spot in the 1970s. Approximately two miles east of the claim is the Chanchelulla Wilderness Area and that is very remote and desolate. There is one USFS dirt road (31N04) that starts at South Fork and makes its way to the perimeter of the wilderness area, where it ends. This road would essentially cross the path of anything leaving the area of the Crosswhite residence. At the time of the incident, the area was not classified as a wilderness area. It received its designation in 1984, and now has 8,200 acres.

Chanchelulla Wilderness Area is known for having very few visitors. The slopes of the hills are as much as 70 degrees, there are no lakes or dams, and few accessible creeks or streams. It is loaded with deer because of its isolation and lack of hunters. Different wilderness Internet sites make claims that if there were some place where you'd like to be left alone, Chanchelulla would be your place. The wilderness gets approximately 50 inches of precipitation annually.

The name for the region that now encompasses the wilderness area was originally coined after a Wintu Native American phrase. *Chan* meant sun and *chelulla* meant black. The current name is a compilation of Native American and local names. Chanchelulla Peak has great importance to the Wintu Indians native to the area. It is a location very sacred to their tribe.

An interesting historical element to the bigfoot sighting on the Crosswhite claim is that the Wintu Indians massacred by a Hayfork posse in the 1800s lived very close to this location. This is another example of bigfoot and Native Americans living side by side.

MEL HESTER

I spent many days driving the area around Hayfork and Wildwood, and the region south of Hyampom. During those days I met many people who had traveled throughout the region, and several had told me about Mel Hester. They had all stated that if anyone knew anything about what was happening in Hyampom, it was Mel. They described him as a retiree who had lived in the area for over 40 years and always stayed close to home. He is the retired area manager for the United States Forest Service in the Hyampom region.

I was able to contact Mel by phone and we had a lengthy conversation about his background, Hyampom, and what he was currently doing with his life. We agreed to meet in early March 2008 when he would spend the day showing me the area and explaining the strange occurrences in the valley.

I drove into Hyampom from Hayfork on a very winding road that took 40 minutes of solid driving. It meanders through beautiful mountains and valleys that appear ripe for bigfoot activity. The road drops into the Hyampom valley where the Trinity River cuts through its middle, and you can view many small ranches. The one structure in the valley that appears out of place is the airport. It's in perfect condition, great asphalt, nice lights, new fence, and not a plane on the ground. Odd.

I made my way from paved road to dirt, crossed a small creek, and arrived at Mel's residence. He had three horses on 17 acres of flat ranch land. Mel is a big guy, and he greeted me warmly and invited me inside for breakfast. We sat and discussed what we'd do for the day and he asked a few questions, testing my knowledge about bigfoot, the mountains, and wildlife.

Prior to us leaving, Mel explained that there really isn't any law enforcement in Hyampom, and the government types tend to stick together for self-preservation. It was that way in the early 1970s, and it's still that way today. Mel said that in November 1970 he got a call from a resident deputy sheriff who asked for his assistance on a prowler call at a local ranch. It was a location where kids stayed for a week at a time learning about wildlife and farms; it was also an ongoing school project. The deputy picked him up at his house and they drove out to the scene late that night.

They arrived at the ranch to find five kids, 15 to17 years old, with one counselor who was 19. The witnesses explained that they heard something walking around their camp, so they went out to investigate. Two of the kids were walking down a small trail when they saw a huge two-legged creature with hair all over its body cross the path just in front of them. The kids said that the creature walked on two legs just like a human would walk. Mel and the deputy looked at each other and headed off into the night to see what they could find. They found 12 tracks of varying quality, 17 inches long and five to eight inches in width, with a huge stride. They looked very similar to human footprints. The creature's prints were almost in a straight line, without deviation step to step. They went back to the kids, who were still scared and shaking. Both Mel and the deputy had no doubts that the kids saw a bigfoot, and that they were not fabricating. The eleva-tion of the ranch was approximately 1,200 feet.

Mel Hester with his horse.

Prior to getting into my jeep, we walked around Mel's ranch and I met his horses and dogs. As I was walking the area, I made a men-tal note of how serene it was and how easy it would be for a bigfoot to be sitting on an adjacent hillside watching us that very moment. The region directly around Hyampom has big trees with many meadows. It's a perfect place to sit on the ground and have a vast view of the terrain.

We got into my jeep and Mel directed me up into the mountains on his adjacent road. He explained that Big Bar Road is closed in the winter because of snow, and he wasn't sure how far we'd make it until we'd have to stop. Many of the locals go up the road to the point where it's closed and then walk the roadway in the snow for exercise. On this day the road was blocked by snow at 3,800 feet and there was a car parked at the side of the shoulder.

Mel explained that in March 1971, at almost the identical time of the year and same location, he had several people report to him that there were odd-looking tracks going down the middle of the Big Bar Road just after it closed. They said the tracks looked huge, even a bit human. Mel made his way up another 600 feet in elevation until he was near 4,400 feet, and at that point he could definitely see something odd. The tracks appeared human, but were huge, maybe 20 inches long. The stride in the prints was nearly six feet and they were not affected by grade or soil. The tracks were in almost a perfectly straight line without deviation from side to side with each step, as there would be in human tracks. It had snowed four inches the night before and the tracks were very easily identifiable in the fresh powder. Mel was able to follow the tracks along the side of the road for almost half a mile, until they went off the embankment and down a steep hillside.

On the way down Big Bar Road, Mel asked me to pull to the side of the road and look out to my left. He told me that the highest mountain out that direction is Ray's Peak, and that locals have been seeing strange lights on the peak for years. Many people called them orbs and said they moved at low altitude near the treetops. In December 1991, Curly Lingemann (now deceased) went to Mel and asked him to go up to that area with him. Curly explained the lights he had seen, and he wanted to show Mel the location and ask if there were any roads into the area. They made their way to where we were now and Mel told Curly there were no roads anywhere in that area. Curly and Mel got out of their vehicle and scouted the region looking for more lights. The road was closed behind them and they walked a short distance in the snow. They found huge, human-looking footprints coming down the road, but they were obviously not human, as they were enormous. They had a giant stride and were nearly in a straight line. Both men thought it was very odd that they went up Big Bar to look at the lights and they found bigfoot tracks.

In the fall of 1975 Mel was partnered with fellow USFS employee Tom Hutchinson and they were burning slash on Kerlin Ridge. Mel and Tom were good friends and worked together often.

Tracks on snow-covered roadway

Mel said they always saw game when they were working, but they were never allowed to hunt while on duty, as it was against Forest Service policy for them to carry firearms. After a long day of work, the two started back down the ridge, driving into the valley. They were at approximately 1,200 feet when they both saw a giant ball of white light high above them hover over the area of Pelletreu Ridge. The light didn't move for two or three minutes; it just hovered. It was nearly dark by this point, but the object was so bright that the land and trees underneath the light were lit up like it was daytime. It wasn't a spotlight, but more diffused. Both men remained fixated on the light until it moved off in a westerly direction at a fairly high speed. Mel had another one of his employees, Cecil, in the area where the orb was traveling and he got him on the radio. "What the hell was that?" Cecil asked as the orb went by him.

Mel said that the orb was at approximately 3,400 feet (a little higher than the tree line), made no sounds, and emitted no lights other than what was pointing towards the ground.

This is an interesting aerial sighting for a variety of reasons. Both of these men were trained government employees and on duty. Both knew the landscape and aerial phenomena of the region.

Neither could imagine what the orb was doing other than attempting to watch something on the ground. Mel said there were no roads or residences anywhere in the vicinity where the light was shining on the ground, and it seemed like the orb had a specific task; it appeared to be searching for something. Mel made many statements during our day together that aerial sightings of unknown craft in this valley were not unusual occurrences. He said locals often see weird stuff in the sky, and some associate the aerial sightings with bigfoot sightings. When many members of the community make an aerial sighting, it is not uncommon to also hear about a bigfoot sighting or bigfoot tracks soon after. Mel was almost making an association between bigfoot and UFOs.

As our day was coming to a close, we were driving along the Trinity River at the far end of the valley. Mel explained that the road leaves the river and goes into the mountains not far from our location, and the road doesn't get back near the river again until it nears Willow Creek/Salyer. He said that the area where the road is off the river is prime bigfoot area; it's a region where there is no vehicular traffic in the winter because of the snow and downed trees.

Mel explained that Hyampom sits at 1,240 feet in elevation and it gets very hot in the summer. In his 30 years in Hyampom, Mel doesn't think that bigfoot spends much time in the area anymore. He said that it probably skirts the Trinity River and stays close to the water source, but doesn't go into the surrounding mountains much. This area regularly gets over 100 degrees in the summer, and is a very dangerous place to get lost in.

JEANNIE LEWIS

In 1966, Highway 299 was a winding road that led from the Pacific Ocean easterly to Weaverville. It crossed the coastal mountain range just above Arcata and McKinleyville, and it sliced between some of Northern California's most famous wilderness areas. The region surrounding Highway 299 is also home to a famous creature that hadn't yet been acknowledged by the world—bigfoot. The famous Patterson–Gimlin film of bigfoot walking through the creek bed in Bluff Creek of Northern California wasn't allegedly filmed until October 20, 1967. People living in the rural parts of Humboldt and Trinity counties had heard the stories about a large, hairy creature making a sporadic appearance, but the national press hadn't yet caught the story.

Jeannie was a young girl in 1966, but she has distinct memories of growing up in Northern California and living in the small valley town of Corning. The summers in Corning are very hot; the San Joaquin valley can get over 110 degrees in the summer. There are few areas to escape the heat, so when Jeannie's mom said that the family had to drive to the coast to pick up their father, she knew it would be a welcome change of scenery and weather. The drive from Corning goes north on Interstate 5 to Redding, and then west to Weaverville, which is the county seat for Trinity County. This area is very rural and with little commerce. They followed the Trinity River west to Willow Creek, and then traversed the coastal range and down into the coastal towns of Arcata, Mckinleyville, and Eureka, where they picked up Jeannie's dad, who had just finished a week of long-haul truck driving.

It was still daylight when the family left the coast and headed up Highway 299 towards home. The roads in 1966 were not great, but they were adequate. During the daylight hours, loggers hauling timber from the coastal forests mainly used Highway 299. By the early evening hours most of the truckers had gone home and there were just locals and some commuters on the road.

It started to get dark and the family was somewhere between Salyer and Weaverville on an isolated and winding stretch of Highway 299 adjacent to the Trinity River. There was very little traffic on the roadway, and Jeannie's brothers and sisters started to fall asleep in the car. Jeannie was the only one who was wide-awake, listening to her mom in the right front passenger seat and her dad driving.

Jeannie remembers her father suddenly braking as they were rounding a turn in the road. She looked out the front window of the vehicle and saw a huge creature standing on two legs in their lane of traffic. It was covered in hair, had a human-type face, and was very tall and large; it immediately took two huge strides on two feet to get out of their way on the road. Jeannie's dad swerved the vehicle out of their lane to avoid a collision. Jeannie got a good look at the creature;

it looked nothing like a bear and moved on two feet very swiftly like a human would move. Her dad continued eastbound on Highway 299 until they eventually reached their house in Corning.

Since this incident, Jeannie and her dad have talked about the sighting. They have also seen bigfoot specials on television and seen photos of bigfoot in magazines and television specials. Jeannie can now state that the creature she saw on the roadway on Highway 299 was a bigfoot. Her dad doesn't feel comfortable talking about the sighting, but he does agree that this is what he observed.

Jeannie lives in San Francisco, has a master's degree in education, and lives with her husband and son.

She signed an affidavit.

SHIRLEY FORK

Shirley Fork is a mountain person. She was raised in a small city outside of Sacramento, California. She attended high school in the county made famous for its jumping frog competition, Calaveras County. Shirley wanted to be a teacher and attended Sacramento State College. Her plans got derailed and she ended up getting married and moving to Willow Creek.

Shirley is in her seventies now and still lives in a quaint house just on the outskirts of Willow Creek, the same house that her husband and family occupied in their younger years. The house has a great view out the back of prime bigfoot country. Shirley said that after they moved to Willow Creek they started to hear stories about bigfoot and its history in the community. Her husband owned the Union 76 gas station in Willow Creek. Her four boys were always working or hanging out at the station when they were younger, and all heard stories about bigfoot. Shirley's husband always wanted to see the creature, but never really had the time to go out into the woods and search because he was always working. The flip side to working all the time was that he often heard things from locals that others didn't, because his station was the center of the town's gossip. Between Al Hodgson's store and Fork's gas station, bigfoot talk was fairly constant in Willow Creek.

In June 1979, Shirley was 42 and had to make a trip to Medford, Oregon to visit relatives. Shirley's son Greg, then 23, was going to drive her and her nine-month-old grandson to Medford. They wanted to leave early in the morning so they could spend a majority of the afternoon with the family. It was just the break of daylight when Greg loaded the supplies into the family car, and with the grandson in the back seat and Shirley in the front seat, they started on their journey.

They decided to drive eastbound on Highway 299 from Willow Creek along the Trinity River. The route is gorgeous, as it follows the cliffs and also gets down near the river. This section of highway was named the Trinity Scenic Byway by the California highway system because of its pure beauty. Heading east on the two-lane highway there is literally nothing but deep forests to the north, as the Trinity Alps Wilderness Area occupies this region. The highway makes its way through several small hamlets—Burnt Ranch, Salyer, Junction City, and more—and finally reaches Weaverville.

Shirley Fork.

Highway 299 at 6:00 a.m. has very little vehicular traffic. I have personally been parked on the road at noon during the week and didn't see any vehicles for 30 minutes. Even though it's the only highway to cross the mountains from Eureka to Weaverville, there just isn't a lot of traffic. During fishing season there are sportsmen that use the highway as an easy access point to the Trinity River, as it has world-class salmon and steelhead action. During the late spring the sport turns from fishing to kayaking and rafting, as the river produces some great rapids for thrill seekers.

As the Forks made their way east, the sun was just starting to rise in front of them and life was starting to move in the valley. They were headed to an area east of Big Bar called Pigeon Point by locals. It's a location on the river known for huge, flat rocks along the banks and deep forests on the opposite shore. As they rounded the

Trinity River.

bend on Highway 299 Shirley got a good view of the opposite bank, and she saw something that she couldn't believe. Less than 100 feet away she observed a huge, hair-covered biped kneeling down and bent over, and it appeared to be taking a drink from the river. The creature either heard or saw the car and stood up on two feet. The creature was huge, over seven feet tall, very dark in color, almost black. It turned slightly towards the car and she could see that the biped was a female with large breasts. As it turned she could also see that the creature had a lighter color in front than on its sides, and it appeared that it either didn't have hair on its breasts or the hair was very faint in color and thickness. After the creature briefly turned towards the car, it immediately turned again and started walking towards the forest 20 to 30 feet away.

As Shirley was watching the creature, she was telling Greg to stop the car. He looked for a place to safely pull to the shoulder, but when he was able to turn around, the creature had made the forest line and wasn't visible. Shirley has seen several bigfoot specials on television and lived very near the bigfoot museum in Willow Creek, and knows positively that she saw a female bigfoot.

I also interviewed Greg about this incident, and he confirmed that his mom had been very excited and absolutely sure about what she observed. Greg is now a retired law enforcement officer for the United States Forest Service and states that he believes his mom saw bigfoot. He said that her immediate reaction to what she saw and her abrupt instruction to him to stop the car and pull over was not her normal demeanor. "She was definitely excited." It should be mentioned that in his days with the Forest Service, Greg interviewed several people who claimed to have seen bigfoot. You can refer to his statement and interview under "Greg Fork" in this chapter.

Shirley said she has relived that observation in her memory thousands of times and the bigfoot she saw looks like the classic bigfoot in the Patterson–Gimlin film from Bluff Creek. She also felt bad about seeing it because her husband was interested in the creature and had always wanted to see one, but never had the opportunity. Once Shirley and Greg arrived in Medford she called her husband and told him they arrived safely and also advised him what she had seen. He was stunned.

Fast forward several decades and Shirley said that she continued to read about many people who claimed to have seen the creature, and that many sightings were made in the area where she observed it. She stated that she knew Al Hodgson well from his days of owning the variety store in town, and also knew he was the curator of the bigfoot museum. She said that it was only after decades of keeping this information to herself that she felt comfortable enough to talk with Al about her experience. It was after talking to Al that she heard about more people who claimed to have seen bigfoot in and around the Trinity River valley.

LOCATION

The Trinity Alps Wilderness Area just to the north of Shirley Fork's sighting has always yielded many quality bigfoot sightings and incidents. It is extremely remote, gets few vacationers, and has the habitat ideal for the creature's survival. This book contains an additional sighting that was actually made on Highway 299 further east of the Fork sighting by two hunters (see Bollman/Kibble below). The Fork sighting is also in the exact area where Jeannie Lewis had her sighting many years earlier. I'm positive that if we erected signs on Highway 299 instructing drivers to contact North America Bigfoot Search if any bigfoot are sighted on this roadway, we would be surprised at the number of reports.

This specific area has many deer, bears, and other mammals that live very comfortably because of the abundant food sources available. Berries, flowers, acorns, mushrooms, and other food grows throughout the area and are plentiful in the late spring and early summer. The area does get extremely hot during the later summer days, and the opportunities to observe a bigfoot are limited once July, August, and September roll into the valley. The bigfoot move into the cooler, higher elevations or towards the coast during these times, or stay strictly nocturnal to avoid the heat.

FORENSIC SKETCH

Shirley had a trip scheduled for the week Harvey Pratt was going to be in town, but at the last minute it was cancelled. She invited us to her home where we met in her living room. Shirley is a very warm and sweet woman who made us all feel very comfortable. She told Harvey that the sighting occurred 30 years ago, but she still remembers it like it was only a few days ago.

Harvey started the process as he usually does; he ran through a list of questions to refresh the memory of the witness and to get a framework of what he was going to draw. One item that Shirley described, and reiterated, was the size and length of the breasts she saw on the creature. She explained that as she saw it stand up from the river, the breasts drooped all the way to the waist, and there was no doubt in her mind that it was a very mature female, six to seven feet tall and weighing several hundred pounds.

Harvey started to sketch and Shirley described the way the hominid walked as an "amble" or "shuffle" as it made its way to the forest line. She said that it was hard to know the exact color other than it was very dark. Harvey continued to talk to Shirley and to draw, and occasionally she would sneak a peek at the sketch and tell Harvey, "The breasts hung lower than that." Harvey would make the breasts longer and color the creature to make it darker. Shirley did not remember the creature having much facial hair, but its head sat directly on the shoulders and it did not appear to have a neck. Shirley was doing a fantastic job and Harvey was still asking questions and refreshing that 30-year memory.

Near the end of the process, Shirley was still sneaking peeks at the sketch and still saying to Harvey, "Make the breasts hang lower." She also made a comment that the outline of the creature appeared dark and the main body had a lighter color, possibly a tint of grey. Harvey showed Shirley the sketch and she stared at it quietly for over a minute. She stated that it was a remarkable job and she was amazed that Harvey was able to draw as much information out of her as he did. She was equally stunned by the artistic creation that she was holding in her hands.

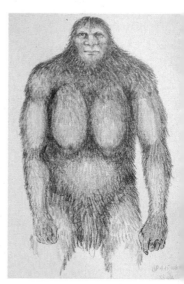

This drawing is important in the overall understanding of bigfoot, age progression, and the consequences of the creature's aging. Shirley's notation of the breasts drooping as low as they were indicates either an older creature, or possibly a creature that was pregnant or nursing. Shirley did not state that the hominid had a huge stomach, and in fact called the chest muscular. The probability is that this is an older female bigfoot where gravity has started to have its effect.

Sketch of female bigfoot seen by Shirley Fork.

The "shuffling" comment from Shirley was also an important aspect to this sighting and could be a significant part of this bigfoot's aging. Many of the bigfoot sightings describe a creature that moves like lightening and takes huge steps. Maybe an aging bigfoot changes its stride and shuffles along, as older age, injury, and maybe joint disease take hold. For a witness to specifically comment on the way a creature moved away from them as a "shuffle," indicates that it must have been significant.

CHRIS BOLLMAN & JEFF KIBBLE

Through a combination of contacts I was made aware of two hunters who had witnessed a bigfoot on Highway 299 somewhere between Weaverville and Willow Creek. I was given Chris Bollman's phone number and told that he worked in Vallejo, California. I had also been told that Chris was with Jeff Kibble when the sighting occurred and that Jeff lived in Concord.

Chris was raised in bigfoot country—Eureka, California. He graduated from Eureka High School and attended the College of the Redwoods. He had a fairly strict lifestyle while living in Eureka, as

Jeff Kibble.

his dad was a deputy sheriff for the Humboldt County Sheriff's Office. Chris is married, has three boys, and is employed in the lumber industry.

Jeff was raised in Ventura County until he was a sophomore in high school when his parents moved to the Walnut Creek area. Jeff is married and has a son and daughter. He is a project manager for a major telecommunications company.

I contacted Chris by phone at his business and we set a meeting for his lunch hour. I arrived and was immediately struck by the fact that Chris looked rugged, just like a hunter. He was a very friendly person who was eager to talk to me about his sighting. He told me at the onset of the meeting that he doesn't talk about this often, and that he was happy to hear that someone was actively looking into the bigfoot issue.

A few days after meeting Chris, I met with Jeff. He is an easygoing and polite guy who was easy to question. I could quickly see how Jeff and Chris became friends. Both men told essentially the same story with few minor differences. The minor differences make sense because of the time lag between the sighting and the reporting. If the statements were mirror images of each other it would indicate a rehearsal of the story. That wasn't the case and this only added to the credibility of each story.

Chris said that the story starts with him and Jeff Kibble deciding in October 1989 to travel to the area of the Trinity wilderness for deer hunting. Both guys had many years of hunting experience and had been friends for a very long time. They made the drive up Interstate 5 to Trinity Lake where they made their camp. It was October 17 and their favorite baseball team, the San Francisco Giants, was playing in the World Series. The guys decided to hunt, and traveled west on Highway 299 towards Willow Creek from Weaverville. They drove past Junction City for several miles and decided that since they hadn't seen any game anywhere, to make a U-turn and head back in an easterly direction and possibly drive into the wilderness.

Chris Bollman.

Chris was driving Jeff's new 1990 Ford Bronco with Jeff as a front seat passenger. They were just west of Junction City traveling eastbound when they came around a turn in the highway and saw something in the roadway—an image that will stay with them the rest of their lives. Approximately 100 yards in front of them, on the northern shoulder of the roadway, they saw a creature standing on two feet, with chestnut-colored hair completely covering its body. It had arms that hung near its knees and it was approximately seven feet tall. Chris immediately started to slow the Bronco, and they watched the creature take four long and quick strides to completely clear both lanes and head towards the Trinity River. Chris gunned the engine and quickly accelerated to the location where the creature disappeared towards the river. He stopped the Bronco in the middle of the roadway and both guys jumped out and ran to the embankment with their guns. They saw nothing.

The area of the Trinity River where Chris and Jeff saw the creature escape is a wide section that has shallow water. Both men stated that there wasn't much cover on their side, and it was almost 150 feet from the point where they saw the creature leave the roadway until it could reach the cover of the forest on the opposite bank.

Chris was interviewed at his business in Vallejo and I met with Jeff at a park near his residence. Both men signed affidavits, and both told stories that were nearly identical and extremely credible. They each said they could not believe the creature could cover the amount of ground and water necessary to get to the opposite bank in the amount of time it had from when they saw it. The water was moving fairly fast, but they didn't see any murkiness where something might have crossed. They searched the area of the bank near where they were standing, but couldn't find anything. There were a few large trees that something could have climbed, but they doubt it could have completely hidden in the trees available.

Chris and Jeff said that they've seen many photos and videos of bigfoot over the years, and they are positive that they observed the creature. The bigfoot ran across the road in a very swift manner and in such a way they could tell that it was a biped and not a bear. They guessed that the creature weighed between 600 and 700 pounds and thought it was very well built and quite healthy. They each confirmed that this sighting occurred at 11:00 a.m. and there was very little vehicular traffic on the highway. Both guys made a point of saying that the entire time they were out of their truck looking for the creature by the river they never saw another vehicle travel past them in either direction. They both also made a point of telling me that the location where they initially saw the bigfoot (northern shoulder of the roadway) was a spot where there was a cliff immediately behind the creature. They guessed that the cliff was over 200 feet high, and they doubted that any creature could have leaped that distance without serious injury or death. Both were asked if there was any roadkill in that area and both said they didn't think so.

When I was concluding my interview with Chris, he asked if I knew what major event occurred on October 17, 1989. I paused for a few seconds and then remembered that it was the date of the largest earthquake in the San Francisco Bay Area history (since the huge 1909 earthquake), the Loma Prieta quake. Chris said that this is why he will never forget the day he saw bigfoot, "This is a date I will never forget, as it's the date that my in-laws' residence was heavily damaged because of the quake."

172

LOCATION

There are several places in Northern California where the sighting of a bigfoot would be a major surprise, but not here. Highway 299 in this area runs on the southern perimeter of the Trinity Alps Wilderness Area. The Alps have a history of bigfoot sightings, and this specific piece of roadway also has a history of motorists seeing bigfoot. Just as a coincidence—or maybe not—just east of this sighting location is the Bigfoot Campground which sits alongside the river. Shirley Forks was on this same roadway, also traveling eastbound but west of this location, when she saw a female bigfoot leaning over and drinking from the river. There are a multitude of others who have reported sightings from the road over the years. On the southern side of the road and river are 15 miles of nothing but heavy forest. Further south is the Yolla Bolly Wilderness Area, and this is another region with a long history of bigfoot sightings.

The time of the sighting (11: 00 a.m.) is slightly unusual. Many bigfoot sightings are either during darkness, or very early in the morning, or just before dusk. The time of the year (autumn) indicates that California had shorter daylight hours, but each man stated that it was a fairly warm day, not unusual for this part of the state. One aspect of this date that cannot be overlooked is the earthquake that occurred approximately six hours after the sighting. Science has validated that animals act differently many hours prior to an earthquake; they aren't sure why. It is possible that this bigfoot was moving during daylight hours in some odd response to the impending quake.

The color of this bigfoot (chestnut) is different, but not unheard of by bigfoot researchers. This color is often associated with sightings near the redwoods. Whenever I hear a bigfoot report where the color is anything other than black, it rings with credibility. People not associated with bigfoot research usually only hear of bigfoot with black hair. It is not widely publicized that bigfoot has every hair color that humans have.

The fact that the bigfoot simply disappeared after it left the shoulder of the road is puzzling. Chris told me that this is one aspect of his sighting that has bothered him for over 18 years. I have personally heard of confirmed sightings where bigfoot has been recorded running 30 miles per hour. Using basic mathematics, at 30 mph the bigfoot could have covered 45 feet per second, or 225 feet in five seconds, easily traveling the distance needed to run around the eastern corner of its location and disappear upstream while staying on the same side of the river. This is purely conjecture regarding its disappearance.

The size of the bigfoot indicates it was a mature creature. Neither witness made any mention of breasts, thus the assumption is that it was male.

Hunters usually make excellent witnesses with regard to seeing wildlife. Chris was raised in Northern California and has seen dozens of bears during his life, and has also seen videos and photos of bigfoot. Jeff has the same background, and his degree of certainty of what he witnessed is absolute. They specifically mentioned that the creature they witnessed moved on two feet with fluidity, ease, and speed. They also said that its stride was huge, as it took the creature only four steps to completely cross 24 feet of roadway, or six feet per stride. They each tried to cross the roadway with the same number of strides and it took them almost triple the number of steps.

I have no doubt that Jeff and Chris witnessed a mature bigfoot cross Highway 299 in late morning in October 1989.

DOUG MORTENSON

Loggers should be high on the list of individuals who have bigfoot sightings, but few come forward. Understanding that it would be against their interests for bigfoot to exist may account for the lack of reported sightings. If bigfoot was proven by conventional scientists to exist, then it might follow that additional protection acts would evolve, more open space be allocated, and fewer logging operations approved. It may be best for the loggers to keep their eyes open and their mouths shut. When I heard from Al Hodgson that a former logger had a sighting and would be willing to talk, I jumped at the opportunity to interview him.

Al explained that he had heard from Doug Mortenson years ago and that he was a very private person who didn't talk a lot about his

Logged hillside, Humboldt County, California.

sighting. He said he would contact Doug, explain my role, and ask if I could conduct an interview. Days later I got a call from Al saying that Doug appeared to have mellowed with age and would be willing to talk. I called him and set up a meeting at his house in Trinidad.

The community of Trinidad sits predominantly on the west side of U.S. 101 north of Eureka. It is an idyllic town on the ocean with a few restaurants and shops with a view of giant rocks and small islands just off the coast. In 2007 there were reports from local community members that they had seen several bigfoot in the ocean swimming around the rocks. It is generally felt that the bigfoot were attempting to accumulate or eat eggs from the birds that nest on the rocks and islands. This story was widely reported through many news agencies and bigfoot sites, but I have never personally met anyone who actually had a sighting in this area.

Doug lives in a more remote area of Trinidad surrounded by giant redwoods. His home is in a beautiful setting, ideal for bigfoot. The afternoon of my visit came with gorgeous weather and lots of sun.

We sat outside in a garden area surrounded by a well-manicured yard and lawn. Doug asked me a few questions about my background and the number of sighting reports I've documented and then started in on his event.

In the early 1970s he was a logger working in a region east of Willow Creek along the South Fork Trinity River near Friday Ridge Road. This was a time when logging throughout Humboldt County was declining and jobs were fewer and farther apart. It was a clear fall day at 10:00 a.m. when Doug was working near the bottom of a hill where the trees had been dropped and the limbs had been removed. His job was called a "choker setter." A D9 CAT (tractor) would drop a thick steel cable down the hill to Doug and he would wrap the cable around the tree. Once the cable was secured to the tractor, it would pull the tree to the top of the hill and to a place where they would be loaded onto a truck for the trip to the mill. This was not a very economical way to move the trees, but under some circumstances it was the only method. There was sometimes a 30-minute wait from the time the tractor started its pull until it came back. During that wait, Doug explained, he would sometimes sit quietly and just watch the beautiful landscape, as he did during this event.

The crew had done several trips up and down the hillside with trees, and Doug had just finished attaching one large tree to the tractor as it started its pull up the hill. The entire basin he was in was still filled with downed trees and he was alone when the tractor started to crest the hill. He knew he had at least another 30 minutes until the tractor could be heard again, so he found a tree, sat down, and was starting to rest when he heard something odd off to his left further down towards the ravine. Doug quietly got up and started to move cautiously from log top to log top towards the noise. He got within 100 feet of the noise and saw a creature bent over doing something on a downed tree. The animal appeared similar to a bear because he could just see the backside of something bent over and black hair. Doug feels that the creature must have heard or smelled him, because after just a few seconds it stood up and ran downhill on two legs, similar to a human but much faster than any person could ever run. It never turned to face Doug; it just stood up and bolted.

Later in the day, Doug approached another logger about what he had observed and told him he had seen a bear running super fast

176

down the hill on two feet. He described the creature to the logger and awaited a response. The logger laughed and stated, "You didn't see a bear; you saw a bigfoot."

Since the time of his sighting Doug has viewed numerous bigfoot specials, footage, and news clips. He now believes that the creature he observed that day above Friday Ridge Road was a bigfoot. The creature was bulky and stout like a bear, but had a body structure more like a human, or a bigfoot. He reiterated that the speed of the bigfoot as it ran down the hillside was very, very fast considering all of the obstacles that it had in its path.

Doug signed an affidavit.

LOCATION

I have spoken to many people who have had bigfoot encounters and sightings in the area where Doug Mortenson had his. It sits in an ideal spot near a river (Trinity), near a wilderness area (Trinity Alps), and miles and miles of open space between Salyer and Hyampom. There are many people in this immediate area who have heard bigfoot sounds. These incidents go back 50 years, and sightings go back an equal amount of time. The elevation in this area is 1,000 feet, but peaks in the region go well over 4,000 feet.

Knowing that many birds in this area nest in the upper reaches of many Douglas fir, pine, and redwoods, it is possible that the bigfoot was foraging through a downed nest for eggs. It's also claimed by some that bigfoot sometimes live in the canopies of trees. If this bigfoot did live in the canopy, then it might have been retrieving food or personal items. Who knows? It seems likely that small birds or eggs were the target. Doug said that he never went up to the tree the bigfoot was leaning over to see what was of interest. He was a little too shook up over the incident to wander farther in that direction, fearing there may have been another creature in the brush and foliage.

GREG FORK

In my many years of bigfoot research and investigations I never had a United States Forest Service police officer step forward and talk candidly about bigfoot, until Greg Fork. It was Greg's mother, Shirley Fork (story above; Greg was driving when Shirley spotted a bigfoot adjacent to the Trinity River in June 1979), who first told me about a series of incidents that occurred in the Yolla Bolly Ranger District, and that her son had investigated many of these alleged bigfoot encounters and had come to some interesting conclusions. She encouraged me to call him.

I first spoke to Greg on the phone and advised who I was and how I received his number. I could tell from the beginning that Greg was very protective of his privacy and his profession. He said that he'd first like to talk to his mother before he spoke with me. No problem. Several minutes later Greg called me back, apologized for the inconvenience, and stated that he just wanted to ensure that I was a legitimate researcher.

Greg Fork.

Greg started off by stating that he was a fire protection officer for the U.S. Forest Service. That job is a sworn law enforcement position, and Greg has specialized training in many types of investigations. He often investigated other Forest Service employees for various crimes, but his primary focus was arson. He has the ability to look into almost any issue in a United States forest. Another part of Greg's job was to conduct general patrol and prevent forest fires through proactive policing. There were many times when he was patrolling the area around Hayfork and the Yolla Bolly District that he heard calls from the highway patrol, Trinity County Sheriff, or California Department of Forestry of an emergency in progress. The Trinity County Sheriff Department is so restricted in its funding that it has limited patrol functions, and the highway patrol has limited units available for rural areas; therefore,

Greg was sometimes the only law enforcement unit available to respond to some calls. Greg just recently retired from the Forest Service and was still in the process of organizing his paperwork, but many of the bigfoot stories were fresh in his memory.

I asked him if he had ever heard of a policy from the Forest Service about not talking to the public about bigfoot, and he said he knows there is an informal policy to that effect. He said that he personally called their staff biologist several times when bigfoot was suspected of being involved, and she responded on many occasions. He felt that the biologist was open to the idea of bigfoot being a real creature and was attempting to work the incident from a scientific perspective. I asked him if he was ever told that he couldn't investigate a bigfoot-related incident, or was it was ever implied to him to steer clear of the topic. He replied that he investigated Forest Service employees on criminal matters, so nobody told him what he could and could not investigate.

Greg said that ever since his mother had seen the bigfoot while he was driving their car he has been interested in the creature. He had always kept one ear open for any news inside his organization regarding bigfoot, and he always tried to be the first on scene to any bigfoot-related activity.

It should be made clear that Greg grew up in Willow Creek, the purported bigfoot capital of the world, yet he never saw a bigfoot while growing up. He graduated from Hoopa High School in 1974, and he heard many stories about bigfoot while attending Hoopa High and living in the Willow Creek area. There was one famous bigfoot story that was always being circulated around the school.

Supposedly in the late 1940s the tribe members knew that a bigfoot lived up Supply Creek and everyone left it alone. In fact, many of the members use to bring it food and generally help it survive. At this point the specifics about how the creature met its death are a little vague. The creature was either shot or died of natural causes. The tribe decided to bury it along the Trinity River across from where the Hoopa Airport is today. They supposedly dug a deep pit and buried the creature in the gravel. Kids talked about the story at school, but elders would say nothing.

Greg also explained that his father owned the only gas station in Willow Creek for many years, and his dad had always believed in bigfoot and usually kept an ear open for bigfoot activity. However, it wasn't until June 1979 when Greg was driving his mom towards Weaverville that bigfoot really started to sink into his mind. The location of Shirley Fork's sighting is 25 miles due north from the area that Greg described as having bigfoot encounters.

Tom Willhoite

Greg joined the Forest Service in 1977 and worked with local fire crews. He said that you had to work your way up through the ranks to get the pristine jobs, and working on a fire crew was one of those starting positions. One of the people he first worked with was Tom Willhoite. Tom was a seasonal firefighter with Greg, and they were good friends. Tom spent a great deal of time in the Bluff Creek area looking for bigfoot and attempting to study the creature. He encouraged me to try to contact Tom, because he felt he knew much more about bigfoot than almost anyone, but unfortunately he had no idea where Tom was now living.

Greg did mention one interesting aspect about Tom Willhoite that didn't make sense to him: Tom drove a NASA truck into the Bluff Creek area and around Willow Creek. Why he was driving the truck at that time, nobody knew. I was eventually able to contact Tom Willhoite regarding this portion of Greg's story. He confirmed that he was driving a NASA truck at that time, but the reason had nothing to do with bigfoot.

NASA Freedom of Information Act Request

You may be wondering how and why NASA would be involved with bigfoot. When John Green and I had a long conversation about the Patterson–Gimlin film, he explained that he had taken a first-generation copy of the film to Stanford Research Institute (SRI) in Menlo Park, California (five miles from NASA's Ames Research Center at

180

Moffett Field) and requested their assistance in enhancing the footage. They took the copy and stated they would get back to him, but they never did. SRI and NASA have a long history of working together on major projects; both have very intelligent scientists working for them, and Ames has huge budgets to bury secret projects.

In a document search for a relationship between NASA and bigfoot, I did find one article that definitely caught my attention. Fred Bradshaw wrote a lengthy latter to Ray Crowe about the issue of "To Shoot or Not To Shoot a Bigfoot" in "The Track Record" #103, (pg. 17) January 2001. Fred was a deputy sheriff in Oregon and had reported on strange incidents he had while camping. One incident occurred in the area of Grays Harbor County, Oregon in August 1985. They were camping, heard noises, investigated, and saw a bigfoot hiding behind a tree. He said that he immediately knew it was much too large to bring down with rifles they were carrying. The campers were scared, and immediately packed up their gear and left the area. The following is where the story is picked up.

> We came back the next day and found an 18″ track near the camp and another 14″ track in the camp. We followed the tracks into the woods but it was so brushy they ended soon. I measured the tree from where we were sitting at the fire and it was 120′ to where the head was 10 feet tall. I guessed its weight at 600-800 pounds, maybe more.

> I contacted other people who came in to check the area. Three from California and one from Washington and they found the same as I did. One fellow works for NASA, one was a forest ranger, and one was an ex-ranger, and the other was working for NASA in California. They have dealt with Sasquatch in California and had seen them before. They found four different sets of tracks.

That's evidence that NASA had traveled to Oregon to investigate bigfoot, and they claimed they had dealt with them in California. The one answer that continues to elude me is to the question: what would NASA want with bigfoot? NABS sent a Freedom of Information Request to NASA regarding bigfoot activity, and their return stated that they knew of no such investigations.

HORSE LINTO

Greg said that he has spent his life combing the woods in and around Willow Creek. He says that he has seen many unusual occurrences, but one of the most unusual was in the Horse Linto area (immediately south of the Hoopa reservation, between it and Willow Creek). It was in 1984 or 1985 when Greg and his younger brother were hunting just south of Tish Tang. Greg remembers that all of a sudden the woods went dead quiet, not a sound, a very odd occurrence and something he had never experienced before. They walked into a small clearing where all of the small bushes and trees had been torn down and purposely destroyed; the ground was even disturbed. In the middle of the clearing was a dead bull cow, approximately 600 pounds. It had an obvious broken neck that was extremely distorted. Although it appeared as though the cow fell to its death, there was no cliff or mountain nearby, so it could not have fallen. The creature had obviously gotten into a life and death struggle with something and lost. They looked for an additional cause of death, but could find nothing. It seemed that it was attacked and killed by a huge creature—maybe bigfoot? Greg never determined what killed the bull and never went back into that area to hunt again.

YOLLA BOLLY INCIDENTS

The first Forest Service bigfoot-related incident Greg investigated occurred in September 2004. A group of hunters came into the Yolla Bolly office to report they had been harassed by a creature that was screaming and stalking their campsite. This activity continued throughout the night and they were too afraid to attempt to sleep. They staggered into the office the next morning explaining the activity to the rangers and asking if anything like this had ever happened before. Two of the four hunters were so concerned that they left the area and to Greg's knowledge have never come back.

Greg had been on a day off when the hunters reported the incident. When he returned to work the next day, the rangers told him about the report and he decided to investigate. The incident occurred near Dubakella Creek approximately one mile from Mud Springs. The hunters had decided to camp in an area that was not a formal camp, but was near a water source. Greg advised that there hadn't been anyone camping at this location in years, and it didn't have any of the improvements normally found at a dedicated camp. Greg wanted to ensure that the hunters hadn't been partying too hard and that they had level heads, so he went to their camp where he met two of the hunters. One was near 30 years of age and the other in his early fifties; one was from Redding and the other from the Bay Area. They stated that they hadn't had anything happen the previous night, but the night before had been horrific.

The hunters explained that at approximately 11:30 p.m. they were sitting quietly near their fire when they started to hear bipedal walking on the perimeter of their camp. They tried to see who was there, but they couldn't see through the darkness with their weak flashlights. The walking and stalking continued until midnight when the loudest scream they have ever heard occurred. It was an unusual scream, a guttural-type sound that didn't sound like anything human. The screams went on until near 3:00 a.m. and the stalking on the perimeter of the camp continued during the screaming. Greg said the two hunters seemed very believable, were totally sober, and appeared very concerned about the incident.

The hunters told him that their partners left because they were genuinely afraid for their lives. They said that there was nothing in the world that is rational and human—knowing there were four hunters fully armed with high-powered rifles all pointing in its direction—that would harass them that way. They felt that if the creature had the nerve to harass them under those conditions, they would never go to sleep until daylight, and at that point they would pack their bags and get out of the area, never to return.

Greg thanked the hunters for their report and asked them to point him towards the area of the sounds. He spent several hours in the vicinity looking for tracks and other evidence of the creature, but found nothing. The ground in this area is very hard-packed soil in which it would be nearly impossible to leave any type of track, even from a truck.

At the culmination of his investigation Greg believed the

hunters had been possibly harassed by a bigfoot. Their description of the sounds and yells they heard didn't match any known mammal in the area, and he couldn't determine what else might have caused the disturbance.

From September 2004 until August 2005 Greg spent the majority of his time patrolling and investigating arson fires in the same Yolla Bolly area. There hadn't been anything unusual happen until he heard a radio broadcast of an accident in late August 2005. He was patrolling a region just outside Hayfork when he heard a call from the California Highway Patrol requesting assistance. The patrol stated that they didn't have any units to respond to a reported call of a motorcycle down on Highway 36 near "three towers" (three large transformer towers), approximately five miles west of the Wildwood Store. This location is approximately two miles from Mud Springs.

Greg advised dispatch that he was only a few miles from the location and he would respond in his forestry truck. He arrived at the scene in just under five minutes and found a fully dressed Harley Davidson motorcycle lying in the middle of the highway. The cycle had sustained major damage and it was obvious that it wasn't driveable. The operator of the cycle was strutting around the scene in full leathers and helmet, and appeared to be very upset and nervous at the same time. He also appeared to be clenching his right fist as though he was either holding something or about ready to hit someone.

Greg parked his truck partially in the road to act as a barrier at the scene and then slowly exited his truck to approach the operator. The cyclist was a white male and stated that he wasn't seriously injured, just bruised up, and didn't need medical attention. Greg asked the guy what had happened, and the cyclist stated that he was traveling the speed limit and came around a turn in the road. Just as he was coming out of the turn he saw a huge bigfoot standing directly in the middle of the road. The cyclist described the bigfoot as nearly seven feet tall, standing on two feet, reddish in color, completely covered in hair or fur, and very well built, very sturdy. The cyclist had nowhere to go and had to run directly into the creature. He got knocked down, his cycle knocked down the creature, and in the process of falling he grabbed a clump of hair from the creature. That's what was in his clenched fist. The cyclist stated that after he fell he looked up to see the creature stand back up on two legs and run off the side of the road, apparently uninjured.

Greg examined that hair in the cyclist's hand and found it to be over six inches long, reddish in color, with the appearance and consistency of pubic hair. Greg said he has lived in the woods his entire life and knows the type of hair and fur that animals have at various times of the year. He stated that the hair that was in the cyclist's hand was not from a bear, as bears do not have a coat with hair that long or that color during that time of the year. Greg asked the cyclist for a sample of the hair and he was told "no way."

An investigation was conducted at the scene and Greg could not find any hair on the pavement or cycle, and there was no blood visible. A search was also made for tracks on the adjacent hillside and none were found. There were no witnesses to the crash. The cyclist was picked up by a friend and the motorcycle was later hauled from the scene by another acquaintance.

Greg explained that the motorcyclist was absolutely adamant that he had hit a bigfoot and seemed to be quite hyped up about it. The claim by the cyclist, and Greg's knowledge that bears don't have hair like that in the cyclist's hand, made Greg believe that the motorcyclist actually hit a bigfoot. It may also be quite a coincidence that the accident occurred only three miles from where the hunters were harassed by a creature near Mud Springs. Greg's conclusion was that the motorcyclist hit a bigfoot and the creature left the scene uninjured. The California Highway Patrol did not respond to the incident because the motorcyclist did not claim an injury.

BIGFOOT SHOOTING

The summer of 2006 started as another mundane fire season for Greg. It had your typical arson fires, and some scattered fires started by thunderstorms and rowdy campers. It was late in September when the year started to get interesting.

Greg was working one day when a 16-year-old boy and his sister came into the regional office. The boy was hysterical as his sister walked to the counter and placed on it an expended 30-06 brass casing. The sister stated that her brother had just been hunting and shot something that he needed to tell the rangers about.

The boy started the story by saying that his sister had dropped him at the top of Knob Peak before sunlight that morning (located four miles east of the Wildwood Store). He was going to slowly walk down from the peak looking for bucks and then have his sister pick him up at the bottom later in the afternoon. It was about 11:00 a.m. and he was walking parallel to the hillside in a chaparral area when he heard something uphill from his location. He turned slightly to look up at the brush to see what was making the noise. Initially he saw a dark object in the bushes and thought it might be a bear. He stood still and turned more towards that direction. At this point the thing in the bush stood up, spread the bushes with its hands and made itself visible. At a distance of approximately 50 feet the kid said he saw something standing on two feet, with hair over its entire body, and a face that had human quality, but it was huge, over seven feet tall. The creature appeared to be coming at the kid, so he turned more and shot the creature in the chest. The creature let out a gut-wrenching scream and simultaneously ran up into the hillside and disappeared.

The kid said that the thing he shot was not human and probably was a bigfoot. He was concerned because the creature had facial qualities that appeared human, but it had hair covering its entire body. The kid actually drew a picture of the creature for Greg, and it confirmed that the creature did appear to have human traits. (Greg searched for the drawing in his records, but couldn't find it.) The sister felt the story was strange and wanted to report it to the authorities in case a creature was laying somewhere in the bush in agony.

Greg notified his supervisors and a team met him and the kid at the scene of the incident. The supervisor made a decision to dispatch local fire crews to the top of the peak. The fire crew walked the peak from top to bottom searching for a body, blood, hair, or any other evidence of a creature or a shooting, but nothing was found. After an exhaustive search, a long debriefing with the kid, and questioning the sister, it was Greg's opinion that the kid shot a bigfoot. Greg was so concerned about this incident that he called the Redding Highway Patrol office to contact the local California Fish and Game staff and advise them of the incident. The response by the warden was that his agency didn't want anything to do with a shooting of a bear. Greg responded back that the shooting victim didn't resemble a bear, didn't run like a bear, and didn't scream like a bear. The warden again affirmed that his agency didn't want anything to do with the investigation.

This incident occurred approximately eight miles from the Mud Springs incident and in close proximity to the motorcycle crash.

Greg believes that the incidents he investigated all had bigfoot involved. He says that all the witnesses were too positive of what they saw, and that the creature witnessed couldn't have been mistaken for a bear, elk, or other mammal.

Greg signed an affidavit to all of the above incidents and stated that he would continue to look for evidence of the creature.

CONCLUSION

The incidents Greg Fork described are important for one primary reason. He was a law enforcement officer for the United States Forest Service with the primary task of investigations. The Forest Service trusted Greg, his judgment, and expertise, and asked him for his expert opinion on a daily basis. In Greg's affidavit he described a variety of incidents where claims were made that bigfoot was involved, all inside an eight-mile radius. It is interesting to note that directly across the wilderness area from Hayfork is the exact spot where Greg was driving his mom over 30 years ago when she spotted bigfoot. I found Greg to have impeccable credentials and to be a man of high integrity.

BIGFOOT KILLINGS

I found the incident of the young hunter shooting the bigfoot quite credible. The sister and the hunter were so concerned that the creature might have been human that they reported the incident to Greg, even though charges could have been filed if a body had been found.

There have been several credible stories over the years about individuals shooting bigfoot. Although there has never been a bigfoot body found, there have been hunters who claimed that they shot and killed the creature. The major concern by hunters after a shooting was that the creature looked human and they were uniformly afraid they would be charged with homicide. I am including two stories that were recorded in "The Track Record" about hunters killing bigfoot. They come from a variety of sources, and they obviously haven't been confirmed. You be the judge of the veracity of their stories.

"The Track Record" #56 (pg. 8–10) published a story from "Bugs" about killing two bigfoot in central California (see news clippings reproduced on following pages). This story received national publicity. If Bugs didn't kill the bigfoot then he spent a lot of time studying bigfoot reports, because his description of the creature is very accurate. The concerns he expressed about coming forward with his killing were real; if the hominid did come back with a DNA profile as *Homo sapiens,* they could have been charged with murder.

"The Track Record" #52, included this report (below) by Ted from Carson City, California.

This region does have bigfoot sightings on a fairly regular basis and is not far from the area covered in our four-county mapping efforts. The nature of this shooting is disturbing and the shooters seem callous. It is against the law in California to hunt at night, so these hunters were not hunting ethically even at the start of the trip. The idea of shooting something you cannot identify is against every hunter's code of ethics.

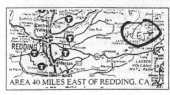

* Ted from Carson City shot, in 1956, a Bigfoot between the eyes with a 30-.30. It was deer season and he was at Oak Rut, about 40 miles east of Redding, CA. It was 10:30-11 PM, and no deer had been sighted. In an open field in manzanita, 50-60 yards away he noted eyes 10-12 inches apart that appeared gray. Richard, my companion, noticed that it kept looking at us, and decided to go ahead and plug it (he was 18 at the time), "let's see what it is...the woods came alive with an incredibly loud sound and our hair stood on end. It went backwards and we held a spotlight on it. All we could find later was tufts of hair, and the ground was torn up, and a thick and tough manzanita branch was broken off." There was no blood found, and he didn't keep any of the hair, but wishes he had. A Dreamland call in.

Ted from Carson City clipping from "The Track Record" #52..

* "Bugs" called into the Art Bell show Tuesday, Apr. 16th. He said in a southern drawl, that he and two other men, ex Marine Vietnam snipers, had shot and killed two Bigfeet in central California somewhere, and had taken ten Polaroid photos. In December of 1973 the three had been varmint hunting at night...a good time to get bobcats and coyotes for the bounty. It was night and they caught eyes in their lights at a distance of 100 yards, which at first they thought was a deer. "Bird Dog" was in back of the pick-up with the twin lights, 500 thousand candlepower quartz spotlights...they had been hunting critters as professionals for some time. Bugs put his scope to the "deer" and immediately saw and recognized his target, at first hunched over...and it stood up when they stopped the truck, as a Bigfoot...he automatically fired his .243 with 125 grain bullets (heavy load, but still only good for varmints). At the

WEATHERBY

same instant Bird Dog fired his .300 Weatherby, 280 grain bullet (would have brought down an elephant). The thing went over, then got back up, stood on its hind feet, and started to run across the field. "We fired again at it...then he jumped up again and we fired a third time and it lunged and hit a fence, but got down

WEATHERBY .300 RIFLE WITH SCOPE

8

LYLE VANN APRIL 17, 1989 LADY, N'RN AZ. CAMOUFLAGED IN TREES. NOSE IS IN CENTER OF PHOTO, SHE RAISED HANDS UP TO CHIN WHILE VANN WAS WATCHING.

into a riverbed and was gone." It was 2:30 AM and the thing was too big to go looking for at night, so they drove around until light. At 7:30-8 AM they picked up a heavy blood trail and followed it along the river for a half mile, and up into a ravine where they came to a dense thicket of wild plum trees, about 50 X 150 feet long. Dense enough you couldn't see into it even though there were no leaves. The blood trail led directly into it.

Who was going in? They flipped a coin. "I never was very lucky!" Bugs crawled in fifty feet on his stomach, and a creature came at him, crawling on all fours. "The sound was a scream like nothing I ever heard, as she kept coming at me. The hair on the back of my head went up. It sounded nothing like the Bigfoot record that was played earlier on the radio. I had a .44 magnum hand gun with 240 grain shells with an overcharge, and I fired.. She was knocked back three or four feet...and again she came at me, and again I fired, and again a third time, all three shots being head shots. This time she stood up, and I fired a fourth time, and at the same time both of my companions could see her as she stood up to 5 or six feet in the brush and they both fired at her also."

"That time she went down and I could see that she wasn't breathing...and we crawled in. The dead male laid about ten feet away. I noticed they both had organs like humans. The three of us drug the bodies out of the brush...The bodies looked like they were human, except they were covered with brownish red hair...there was no clothing. The nose was like a humans though the mouth was ape-like (no lips). The eyes were half human, half ape, and they had large, protruding, foreheads, with a short neck in front. The overall face was half ape and half human. From the back-side there was no neck at all...it was all muscle. The male stood about 7 1/2 feet at about 350 pounds, and the female at 7 foot and about 300 pounds.

"What're we gonna' do with 'em'," they asked? They came to the conclusion that they might go to prison for killing them, and decided to stay quiet. This is probably true...Some DA out to make a name for himself, win or lose, would have everything to gain with a murder trial...lots of publicity. They took ten Polaroid photos and divided them. Bugs lost his when his house burned down. He's also gotten rid of his extensive gun collection and doesn't own a weapon anymore. A year and a half ago, Bugs finally told his wife the story. When they heard me (Ray Crowe) on the radio, she talked him into contacting us...(Art Bell show) I had made a plea earlier for artifacts, hair or what-ever, never dreaming this fax would come in.

"I happen to think it's true...I lean towards believing it," Art Bell said. He had received hundreds of faxes, and they seemed to be running 80-90% belief in the story. Others called in.... "Shooting at something you haven't identified isn't a very responsible sportsman...it gives sportsman a bad name." (Bug's called in...he knew definitely what it was. Ray...who says you have to be a "Good Sportsman" anyway?). Bugs called and said, "we knew what we were shooting at, using two 500,000 candlepower quartz spotlights. We always knew what we were shooting at...these spotlights

BIGFOOT DOLL FOR SALE

9

"Bugs" clippings from "The Track Record" #56 (pg. 8–10).

Continued on page 190.

ALIEN GRAY DOLL
FOR SALE

would light up the area for 300 yards, we used 9 to 15 power scopes...there were a lot of cattle in the area." Noting that human eyes don't reflect light. Bugs went on to tell of coyote pelts being in the $50-100 range, bobcats at $750, and on a good night they could make $400-500.

Fred, a Kansas Police Officer, said the bullet loads were not available commercially, and that momentum would carry the female forward after he shot her in the brush. Why were they carrying a 300 Weatherby (very powerful, violent recoil), for bear? He doubted the story. Mike in Las Vegas...My partner and I were on patrol, "the government listens to you nightly, just waiting for something like this to happen. The caller Bugs gave too much information about himself, and I'd bet my paycheck that as he spoke the government people are zeroing in on him...once found, we'll never know the truth abut his story." Art Bell said "I think you're probably right (Bugs was Vietnam vet, pelt sales records, house burned, a number on his fax will tell where it was posted from...what else did I miss?). Art Bell show, call in from western America. 800-618-8255.

From Bugs..."I'm afraid I'm going to jail...I can't take that chance now." Nothing has been heard from him since... think he might have been scared off. Or maybe, as one caller suggested, go to the site, mark it with fluorescent spray paint, give Art a map, and Bugs is uninvolved. Did that happen? Don't know...I haven't heard anything...maybe he's in jail...maybe the thing's been dug up. A month now, and no word. Maybe it was a hoax, or he couldn't get his buddies to consent to digging the dead Bigfeet up.

Peter Byrne sent a fax to Art Bell that he read over the air; in four parts: (1) We will guarantee you full legal protection with us covering all costs (2) You can choose your own attorney, contact us for a retainer (3) We will cover all expenses in any physical recovery (4) A monetary reward, for an exceptional and unique contribution to science. Very generous, I think

A fax back to Bell from Bugs. "Please. I'm scared, please do not repeat the interview...I'm sorry I sent you the first fax...based on what he'd heard, he didn't think it'd be such a good idea to continue with what he'd been doing. Calls coming in, if not in the vein of a hoax, were not encouraging. There is no immunity for murder or manslaughter...a defense fund...his buddies must agree

"Bugs" clipping, page 3, from "The Track Record" #56.

* Had another call from someone in the TV industry, and they said that when an Oregon State Police officer in Bend Oregon was interviewed, he commented on four hunters being killed in the Bend area in the 1970's. The rifle had been found twisted out of shape. Also there was a report from the Forest Service of a Sasquatch footprint of large size had been found in a lava bed near Bend, the lava had been fresh at the time. The Forest Service people said also that a large-breasted female had

FRANCOSIS DE LOYS, F.G.S., PRIMATE PH
1920, NEAR TARRA RIVER, VENEZUELA.
PHOTO FROM 10 FT. AWAY, COARSE, LOI
GRAYISH BROWN HAIR, NO TAIL, 112#, 5
1-3/4 IN TALL, 32 TEETH (NEW WORLD IS
FEMALE HAS LARGE CLITORIS, TYPICAL
SPIDER MONKEY, *ATELES*

been seen in the Bend area. The TV person had talked to someone that had been in the intelligence community for 20 years. He had read a report that said there was a satellite photo of a Bigfoot walking across a field in Washington State (no confirmation on that from other responsible sources...think a hoax). Also, while searching for missile sites in abandoned caves and tunnels, one group had large boulders thrown at them and found giant bones while searching the caves, where they think the Bigfoot lives. A drilling rig in Canada 2-3 years ago ran into flesh about a mile down. In the 1920's or 30's, a photo was taken of a Sasquatch in Canada that had been shot and killed (Loy photo from SA? Don't know if any of this stuff is good or not, but just something readers outta' know about - Ray). Also there are reports of a Pterodactyl, the large "extinct" Jurassic flying lizard of dinosaur times, once common in North America,

Clipping from "The Track Record" #54 (pg. 8)

The following incident happened in October 1978 near Yankton, Oregon, and was reported in "The Track Record" #57 (pg. 11, para. 9) in 1996.

> Jim Hewkin was a volunteer investigator/researcher for the Western Bigfoot Society. He reported that near Yankton (south of Mt. St. Helens), a bigfoot was shot between the eyes four times (that in itself seems highly improbable) and then it rolled off the road. The shooters didn't go down the road to chase it, but the next day investigators (not sure from where), came in and the creature's body was gone. Don't know if it was only wounded or a possible mate took the body away. There were tracks of a possible family group, male, female, and a child.

I haven't seen any additional details to the story listed above. I'm not sure how anything can be shot four times between the eyes.

"The Track Record" #54 (pg. 8) reported several incidents (clipping shown on opposite page). The description of a bigfoot walking across a field is almost identical to the photo that Greg said he saw, except it was supposedly taken in the Bluff Creek region. While there is no confirmation on the story, there is some validation, as Greg's story still holds true, and everything he has stated to me that I have been able to check out has been true.

JOHN LEWIS

Stories about bigfoot have come to me from all parts of North America, and even from other parts of the world. Just when you think you've heard every feasible bigfoot story possible, wow, there is a new twist. As long as you can keep an open mind, a great imagination, and have some knowledge of the outdoors, you'll enjoy this story.

A reader from our website (nabigfootsearch.com) forwarded us a summary of this event. There were enough details that I was able to contact the author, John Lewis. After a few e-mails he started to feel comfortable enough to open up and relate one of the most fascinating stories we have heard emanating from northwest California. This may be unbelievable to some, but I have confirmed many of the historical facts quoted by John and he did sign an affidavit. Here is John Lewis' story.

John is a 51-year-old merchant marine living in San Francisco. He has spent his life working on the waters of the Pacific, from the Yukon to San Francisco; he has seen a lot. He grew up in Santa Cruz and moved to Hawaii when he was young. He says that it seems he was always living around water and that it steered his life decisions. He has been a merchant marine for over 30 years. Presently he is the captain of a boat in San Francisco bay and has a beautiful office outside Pier 39. He is married and has one five-year-old son and an 18-year-old stepson.

John explained that he lived a fairly normal life without many unusual occurrences. He has always been interested in scientific odd-

John Lewis.

ities, and one summer night in 1993 he was watching a television show with his mom, Esther, about the Patterson–Gimlin film footage of a bigfoot in Bluff Creek in 1967. They both felt that it was a bit unusual to see a female bigfoot with breasts walking across a remote creek bed being filmed by two cowboys. John knew that his mom had lived in remote areas of Washington State and had heard there were bigfoot sightings in those areas. He asked Esther if she had ever seen a bigfoot while in Washington; she said no, but immediately followed that up with the statement that her father had told her a story about bigfoot nearly 70 years earlier. John says that he sat upright in his chair, shocked. This was big news. He says that his mom never joked about these types of issues, and she had a stone-cold face when she made the statement. He knew she was telling the truth.

Esther stated that her father (Theodore Spagopoulos) worked for the Southern Pacific Railroad as a line worker during the early 1900s and was employed there for 20 years. The family lived in the valley area of Northern California (Redding/Red Bluff/Corning) and her

192

father would leave the family for two-month stretches to work. She can remember that her dad told her this story sometime between 1925 and 1928 while she was in elementary school. She and the rest of the kids had made fun of their dad because his story seemed contrary to what they learned in school, and they routinely made fun of him when he brought it up. It wasn't until seeing the Patterson–Gimlin footage that Esther started to put the pieces of the puzzle together. She related the following details.

In the late 1800s, Southern Pacific Railroad was frantically trying to cover the State of California with a rail line that linked southern California with Sacramento and then push it north to cross into Oregon. The building of the line was dirty, lonely work that went into some of the most desolate regions of the state. The northern line from Sacramento to Oregon cut through what is now Redding, Red Bluff, Dunsmuir, Weed, and points north. These were areas that didn't have population masses and there weren't good roads. Many of the towns that existed in the early 1900s were actually thriving based solely on the railroad. Much of the land the railroad crossed was in close proximity to Native American settlements, and there was sometimes disdain for the railroad by the residents. There were rarely confrontations, and most of the conflicts facing railroad workers came from grizzly bears living in the far reaches of the forests and drunken co-workers blowing off a week's worth of work.

The workers building the new railroad were called line workers. They would report to a local boss in a main camp that was in a fairly remote area of the state for a two-month assignment clearing the forest and building the line. They were placed in a two-man team for a week at a time. They would leave their main camp on a Monday morning and hike out to their remote site where they would build their own tent, set up fires, and lay out their food for the week. From their camp they would take a short walk to their line location and start to clear trees and brush that was pre-marked by a surveyor. As the week progressed, the team would start to work more remotely from their camp, usually with each employee working in an opposite direction from each other so they could each be responsible for the amount of their own progress. By week's end this strategy usually resulted in employees working quite a distance from each other. They would agree on a specific time to meet back at their camp, close shop for the weekend, and start the hike back to their main camp.

On weekends the workers would congregate at the bar in the middle of camp and sometimes fraternize with local girls who arrived to meet the men. It was very rare that relatives made the trip to the main camp, as it was usually not an easy trek and sometimes the weather in these locations didn't accommodate travel plans. At the conclusion of the weekend the workers would again meet up with their partners, get their axes, shovels, saws, and food, and make their hike back to their line site. All of the line employees were on a two-month rotation, a long time away from families and friends.

In 1914, Southern Pacific decided they needed a line between Eureka and major cities south. They put logistics in place and started building the line in 1915. This line was to go through some very desolate, wet, and heavily wooded regions of Northern California, and some of the locations known for bigfoot sightings. Communication through this region of the state wasn't well developed and this area wasn't heavily populated. There were established Native American populations, but they didn't communicate much with railroad workers.

The Southern Pacific line was built adjacent to the South Fork Trinity River and next to the small logging town of Hyampom. The line went through portions of Humboldt and Trinity counties before reaching Hyampom. Winters in this area can be very cold, with some snow, lots of rain, and are very depressing because of a lack of sun. The forests are very dense with large, old-growth redwoods, Douglas fir, and pine, depending on the exact locations, elevations, and distance from the coast.

After one two-month work stint, Theodore came home and was quite distraught about what had occurred at their camp. He told all of his kids that he and his partner had come back to the main camp after a week and found that one of his co-workers on another team had disappeared. The missing worker's partner had stated the worker had gone missing late in the week and never returned to their camp after a day of work. He said that he made a brief search of the work area and didn't find anything, and thought his partner had quit and walked off. The main camp said that nobody had returned, so the foreman formed a search team. They spent the weekend scouring the hills near the work site and found nothing.

The following week, work resumed as usual and line workers went back out and pushed deeper and deeper into the woods. Three, and then four, weeks passed from the original disappearance when one two-man team on their way back to their site found the missing line worker naked and delirious in a pit near one of their work sites. They made a small litter and carried him to camp. Theodore was on the trail and saw the line worker and heard his statement. The worker was partially alert and could speak, and said a female ape kidnapped him and held him in the pit against his will and he was forced to have sex with it. The ape licked his hands and feet until they were raw so that he couldn't walk and escape. Theodore stated that he saw the worker and could see that his hands and feet had no skin on them at all, and he could easily see red, raw, bleeding flesh on both hands and feet. The workers carried the employee back to the main camp where limited medical aid was available. He was suffering from severe wounds and hypothermia, and died late that night.

Esther was in elementary school when her dad told this story to her and her siblings who were older and had more education. The kids laughed at him and said the story was impossible. They had all learned in school that gorillas, apes, and monkeys do not live in the forests of North America; it just couldn't be true. Her dad had a bad temper when people disagreed with him on a point that he knew was fact. She said he became irate and stated, "This is not a story; this is what actually happened." Esther was convinced that her dad completely believed the story. Theodore died in 1961.

After John and Esther watched the bigfoot special, Esther told John she understood how her father and the abducted line worker mistook bigfoot for an ape/gorilla. She felt the story was believable because of the appearance of the bigfoot, and because it was photographed (Patterson–Gimlin) in the same geographical location as where Southern Pacific had been building their line.

The Patterson–Gimlin film was made in Bluff Creek, approximately 30 air miles from the southern fork of the Trinity River. The area between the Patterson–Gimlin film site and the Trinity River fringes the Hoopa Indian Reservation and a dedicated wilderness area. This region has many, many bigfoot sightings going back to the early 1900s. The term "bigfoot" was coined in the 1958 and used in newspapers in the 1960s because of the large human-like tracks that were found in the Six Rivers National Forests around Bluff Creek.

Esther died in 2006, but John is adamant about the story and the facts. I have interviewed him several times about various facets of the story and he has always remained consistent. The exact location or year of the event cannot be determined, but based on the timeline we know that Theodore was working somewhere on the South Fork Line.

A review of bigfoot sightings along the South Fork Trinity River and Hyampom shows a consistent pattern. There are years when sightings haven't been reported, but there is a long and frequent pattern that stretches over 70 years of sightings and incidents.

FOLLOW-UP

I traveled to Sacramento to visit the California State Railroad Museum. The museum does maintain a record of some railroad employees from the early 1900s. The staff was quite interested in the project and volunteered to search their records for any correlating information on the employees. Unfortunately the records at the museum do not include labor employees from the early 1900s. No records were available.

When I first started to investigate John Lewis's rendition of abduction and rape, I was quite apprehensive about many aspects of the story. I wrote the summary and sat on the story for many months while I was accumulating other bigfoot sightings for the book. It was only after I had finished reading every copy of "The Track Record" NABS had purchased that I realized that pits in the middle of the forest had been connected with bigfoot for many years. It was at this point that the Lewis story started to take on a different level of credibility and believability to our team. Below is a list of "The Track Record" stories about pits and bigfoot.

"The Track Record" #9 (pg. 1).

Other slides had shots of old pits that had weathered and were coated with moss. Another fresher pit was dished out, like a meteorite crater, the large rocks rimming the pit.

Probably the most dramatic pit shots were those where Glenn Thomas had reported Bigfoot activity in October of 1967. Rocks were stacked, one seemingly about to topple. One of the shots showed discolored rock, fresher than the rest. Jim visited the area this year, a month or so ago, with John Green and another friend, noticing that something had been digging in the same hole, as it was even deeper than before.

"The Track Record" #36 (pg. 4).

Seaside, Oregon, April 1994

A large built up rock pit was photographed, over the side of quarry. Not a place you would expect humans to erect a pit. Also, photos of tracks left recently were exhibited.

Note: Seaside has a long history of bigfoot sightings. NABS has made statements in the past that we believe there are associations between quarries and bigfoot that we do not understand. To find a pit adjacent to a quarry, with tracks, is truly unusual.

8 SISKIYOU COUNTY

There is probably no other county in the State of California that's as remote and as sparsely populated as Siskiyou County. It is located at the farthest edge of Northern California and shares its northern border with the State of Oregon, its western border with Del Norte County, the eastern border with Modoc County, and the southern borders with Shasta, Humboldt, and Trinity counties. If you have a hunch that this county is big, you would be correct. Siskiyou is the fifth largest county in California at 6,300 square miles.

Siskiyou County claims to be the most ecologically diverse county in California. They say on their website that they don't know where the word "Siskiyou" originated, but it is probably related to a Native American word. The major industries in the county are agriculture, mining, and construction, with tourism adding to the list.

The Klamath River runs across two-thirds of the county and is a recreational stronghold for fishing, rafting, and swimming during the summer months. There is a historical disagreement over water, as the Klamath tribes have fought for a high water flow from the Iron Gate Reservoir to keep salmon and steelhead healthy, while farmers and ranchers have claimed they need the water for crop survival. The water feud has never been settled and is a yearly fight.

One oddity to Siskiyou County is that there is not a city in the county with a population over 10,000. The largest is Yreka (7,275 population) and it's the county seat. The next largest towns are Mt. Shasta (3,840 population), Dunsmuir (1,910 population), and the unincorporated region accounts for 23,700 people. The people in the county are spread thin and there aren't a lot of retail centers, motels, hotels, and restaurants once you leave Yreka. You need to bring your own necessities wherever you are headed in this county. Law enforcement is also a little thin in this region. They do not have a big tax base and they can't afford a huge sheriff's office. You will see a resident California Highway Patrolman (CHP) occasionally on

Highway 96, and you will see CHP regulars on Interstate 5, the two major highways in the county.

Another oddity is the number of wilderness areas Siskiyou County has inside its borders. The Red Buttes, Mt. Shasta, and Marble Mountain Wilderness Areas are totally inside the county border. It shares the Trinity Alps and Siskiyou Wilderness Areas with other counties. Remember, wilderness areas are heavily restricted: no cars, no bikes and you can only use horses or mules to help you into the backcountry. The amount of rural open space in this county is mind-boggling.

The largest wilderness area inside Siskiyou County is the Marble Mountain Wilderness Area, with 89 lakes stocked with trout inside its 242,000 acres. It was first assigned a primitive area status in 1931 and then reclassified as a wilderness area in 1953. The area has many peaks that range from 7,000 feet to Boulder Peak at 8,299 feet.

The open space areas of Siskiyou County contain large numbers of bears, deer, elk, eagles, osprey, wild turkey, quail, and fishers. If you look real hard and you are very quiet, you might also see a bigfoot! There have been many bigfoot sightings inside the Marble Mountain Wilderness Area, and there is famous film footage shot by a family group while they were on vacation. They found a small shelter built beneath trees where the branches were broken by being twisted and not cut. As they were examining the shelter, one of the kids spotted a bigfoot several hundred yards away walking a ridgeline towards them. The father shot footage of the hominid as it slowly made its way down the ridge in the general direction of the people and the shelter. It appeared from the gestures the bigfoot made that it was slightly disturbed that the people were near its home. The creature eventually disappeared on the backside of the ridge. This footage has its supporters and its non-supporters, but it was shot in the Marbles.

HISTORICAL BIGFOOT PERSPECTIVE

During my research of the four county regions I spent hundreds of hours hunting through libraries, universities, and small bookshops, and scanning newspapers looking for older articles about each county. As a researcher you must read almost all odd-sighting documents thoroughly to see what words are used to describe strange sightings; you cannot look for "bigfoot" as that term was only coined in 1958. Many early sightings of bigfoot-type creatures throughout the United States describe the creature as a "wild man." That is one of the main phrases I look for when browsing documents from the 1800s and early 1900s.

One of the most intriguing pamphlets I've ever found is "The Hermit of Siskiyou." This was printed by the Crescent City News in 1896 and authored by L.W. Musick.

The copies are scattered throughout rural counties and are light in color, quite worn, and equally hard to read. The following story from a Del Norte County correspondent, writing from Happy Camp, Siskiyou County, California on January 2, 1886, was found inside the document and it is relevant to bigfoot research.

"The Hermit of Siskiyou" pamphlet.

I do not remember to have seen any reference to Wild Man which haunts this part of the country, so I shall allude to him briefly. Not a great while since Mr. Jack Dover, one of our most trustworthy citizens, while hunting saw an object standing one hundred and fifty yards from him picking berries or tender shoots from the bush. The thing was of gigantic size—about seven feet high—with a bulldog head, short ears and long hair; it was also furnished with a beard, and was free from hair on such parts of its body as is common among men. Its voice was shrill, or soprano and very human, like that of a woman in great fear.

Mr. Dover could not see its footprints as it walked on hard soil. He aimed his gun at the animal, or whatever it is, several times, but because it was so human would not shoot.

The range of curiosity is between the Marble Mountain and the vicinity of Happy Camp. A number of people have seen it and all agree in their descriptions except some make it taller than others. It is apparently herbivores and makes winter quarters in some of the caves of Marble Mountain.

The article is one of the first coming from the Marbles where they talk about a bigfoot-type creature. The descriptions vary according to who wrote the story and who told the story, but there are many specifics about the article that are fascinating. Bigfoot in Northern California is routinely thought to be seven feet tall or taller. I have personally taken several sighting reports where bigfoot is standing and eating berries or actually picking berries off a bush. The statement of the witness that the creature looked too human to shoot is a major point of validity that this person saw a bigfoot.

Since the Patterson–Gimlin film footage in October 1967, many people have only associated the appearance of the bigfoot with the hominid on the footage. That creature had a very hairy face and hair over almost all of its body. In this specific region of California this is not the normal appearance of bigfoot that we've documented. Ninety percent of the sketches Harvey Pratt has completed for witnesses I have confirmed show a bigfoot with significantly less facial hair than the bigfoot associated with Patterson–Gimlin. This is not a fact that is routinely discussed in bigfoot circles.

North America Bigfoot Search specifically accelerated Harvey's return to Northern California so that nobody could claim that witnesses were influenced by the release of our first book, *The Hoopa Project,* and the associated sketches in it.

Witnesses for this book have also gone against the grain of general public perception of what bigfoot looks like. They have consistently described a hominid that looks much more human than the Patterson–Gimlin creature, and the sketches all appear somewhat similar, even though the witnesses don't know each other, live hundreds of miles apart from each other, and none had an opportunity to view the sketches published in the first book. The consistency of

NOTE 4. A Del Norte Record correspondent, writing from Happy Camp, Siskiyou county, Jan. 2, 1886, discourses as follows:

"I do not remember to have seen any reference to the 'Wild Man' which haunts this part of the country, so I shall allude to him briefly. Not a great while since, Mr. Jack Dover, one of our most trustworty citizens, while hunting saw an object standing one hundred and fifty yards from him picking berries or tender shoots from the bushes. The thing was of gigantic size—about seven feet high—with a bull-dog head, short ears and long hair; it was also furnished with a beard, and was free from hiar on such parts of its body as is common among men. Its voice was shrill, or soprano, and very human, like that of a woman in great fear. Mr. Dover could not see its foot-prints as it walked on hard soil. He aimed his gun at the animal, or whatever it is, several times, but because it was so human would not shoot. The range of the curiosity is between Marble mountain and the vicinity of Happy Camp. A number of people have seen it and all agree in their descriptions except that some make it taller than others. It is apparently herbiverous and makes winter quarters in some of the caves of Marble mountain."

The article from "The Hermit of Siskiyou" on the wild man of Marble Mountain.

the witness descriptions validates the likelihood that bigfoot exists and probably thrives in northwest California.

There is another historical article about bigfoot associated with Siskiyou County. Thompson Creek is located in the Seiad Valley just east of Happy Camp. Chinese workers in the 1860s were utilized for a variety of labor work in the Happy Camp region, including building paths for the railroad and mining. This quote is from "The Track Record" #52 (pg. 10).

The article by Marie Stumpf goes on to mention that there was a sighting on Thompson Creek, a tributary of the Klamath River in the 1860's. A group of Chinese workers were building a ditch to

202

carry water to a hydraulic mine were so frightened when they sighted a Bigfoot creature that they refused to return to work.

I'm not sure where Ms. Stumpf obtained her information, but the location fits well within the historical bigfoot perspective of Siskiyou County.

Ray Crowe had a section in his newsletters that he called "Letters to the Editor." The following letter appeared in the March 1994 edition of "The Track Record" #35 (pg. 13).

Billy Guffin, March 15, 1994. In the summer of 1992 I was trying to find a trail in the Siskiyou area. About 8am I decided to take the main road back down to the Klamath River (at Happy Camp). Within a block of the summit there was a road curve and after the curve there was a straightaway for about 3 blocks. I glimpsed a black furry animal walking down the road that was only about three strides from going around another curve. I sped down as fast as I could, but it was gone. It had a straight-up gait on two legs, and when it turned his head around, did not turn its entire body. I had a short glimpse of his face, and it was devoid of facial hair in some spots and it made me think it was indeed a Sasquatch.

The following summer, 1993, having found the trail I was originally looking for, I camped in the same place. That evening when I went to bed I was listening to my radio. When I turned it off to go to sleep, something hit my tent like a piece of wood or rock. Since this was an almost level open space, there was no way something could have rolled down and hit the tent. Something had to throw a small rock or piece of wood, or maybe hit it with a paw or fist. It startled me as I was almost asleep. I yelled, "go away and leave me alone". I never heard another sound the rest of the night.

Billy's description of his sighting is another example of the human quality of bigfoot in this region.

The following article was taken from "The Track Record" #53 (pg. 12, para. 3) and exemplifies that many people have sightings and most have no idea they can be reported, or to whom to report.

I read an Article of Bigfoot in Tonic newspaper. I live in the Klamath Nat. Forest on the Scott River (CA). It's been over 25 years but I saw two Bigfeet at separate locations—1967 on the Klamath River my ex-husband and I watched one for about 15 minutes thru' field glasses. It was much as the "standard" description, except no pointed head, and had curly reddish brown hair. I could draw you a picture still after all this time, it's paved on my memory. Second time was on the ramp onto I-5 at night, full moon, winter of 1969. this huge fellow, over 7 foot, had long wavy silver-gray hair and caught in my head lights. Looked almost human in the face despite brow ridge and other anatomical differences. The eyes impressed me so, I thought if I only stopped I could communicate. I didn't stop. There are still sightings in the area occasionally. Barbie S., California. 12/9/95

DARRELL WHITEAKER

In the spring of 2008 I made an overnight stop in Hoopa to contact a few witnesses and to have a decent bed to sleep. As I was loading my 4x4 in the morning, a gentleman stopped by and started an interesting conversation.

At 9:00 a.m. the parking lot in the motel was empty except for my rig. Darrel Whiteaker is someone I had seen at this spot before, and it appeared that he worked for the Hoopa tribe. We started talking about the weather, fishing, hunting, and the conversation slowly started to move towards bigfoot. He asked me if I was interested in hearing about some unusual incidents he had encountered while hunting. I told him I was always interested to hear about anything unusual in the woods.

Darrell is a 62-year-old, semi-retired retail manager who has lived the last 10 years in Willow Creek. He said that he had previously lived in Ukiah, but always enjoyed the outdoors and loved the environment offered by the north coast. He explained that he is divorced and has a son who is a police officer and doesn't believe anything about bigfoot. Darrell is a Navy veteran and he spent time stationed in Guam and Vietnam.

During my conversation with Darrell he told me that some people had given him some verbal challenges when he talked about bigfoot and the possibility of it existing in the forests of Northern California. He stated that he had hunted for most of his adult life and enjoyed the personal challenges that are encountered while in the woods. He knew the sounds, wildlife, and dangers that the forest possesses, so when something is out of place, he notices it. He explained that to be a good hunter you need to have keen senses about yourself at all times, and if those senses are working to their optimum they can save your life and also give you an early warning signal that something is wrong.

Darrell Whiteaker.

As we were talking, I could tell that Darrell was a very smart and careful person. He obviously knew the woods well and was accustomed to spending considerable time there. He stated that of the hundreds of times he had been by himself deep in the forests, only once did things seem out of place.

It was approximately 10 years before, in October 1997, when Darrell decided to go deer hunting. He traveled from Willow Creek, through Hoopa, Weitchpec, Orleans, and then branched off and headed for the Forks of Salmon. He took a trail north from the highway just up from the Forks and headed towards the Marble Mountain Wilderness Area. He was approximately four miles into the hike when he came upon a valley that was about 10 acres in size. It had large trees on the perimeter, but was fairly clear in the middle. Up until this point in the morning Darrel said that it felt like a normal morning hunt; small game would run across the trail, wind was blowing in the trees, and there was a smell of the fresh outdoors.

Darrell had a strange sense that something was very wrong with that valley. He felt that he was being watched, but couldn't quite explain why. He also said that in 50 years of hunting, hiking, and being in the forest, this was the first time that he was ever in a forest situation where he couldn't hear anything. No leaves rustling, no animal sounds, complete absolute silence. This gave him one of the most uncomfortable feelings he had ever had in his life. For the 30 minutes that he walked the valley, Darrell never saw any wildlife— no birds, small game, large mammals, nothing. Since he was on the fringe of one of the most wild wilderness areas in the State of California, this was a very odd scenario.

After several minutes in this valley, Darrell decided he should leave. He hadn't seen any game, no people, there were no sounds, wind, and it was all too odd and too uncomfortable to stay any longer. He walked back to his vehicle and never went back to that spot.

A short note about the Marble Mountains; they have a long history of unusual stories and sightings. "The Hermit of Siskiyou," noted earlier in the chapter, is the first unusual documented case. There is also an account from a noted bigfoot researcher who observed a bigfoot while on an expedition in the Marbles. The sighting left the researcher so shaken that this person has refused to talk about it in an open forum. There are many, many more stories from this region.

Before Darrell left he asked if I wanted to hear one more story. Absolutely, I told him. He stated that he had a great friend who was also a hunter and went to Onion Lake above Bluff Creek. He thought this was in 1993 during hunting season, September to November. The hunter took his toughest bear dog with him, even though he was going after deer. Darrell described this dog as the most vicious, toughest dog that he had ever seen that wasn't afraid of anything or anyone.

Darrell's friend drove down into Onion Lake, parked his vehicle and hopped out with his dog. At that immediate moment the attitude of the dog instantly changed from frisky and alive to whimpering at the owner's feet. His friend couldn't believe what he was observing and decided to take a short walk towards the lake. It took only a few steps for Darrell's friend to realize that something was abnormal about Onion Lake. There was no wind, no sound, no wildlife; it was as though the sky opened up and sucked all of the sound from the small valley. The dog was still cowering between Darrell's friend's legs even as he attempted to walk. At this point it was obvious to the friend that there was something scary and wrong about this lake, and he immediately went back to his vehicle with the dog and left the area.

LOCATIONS

Onion Lake: This area has a history of bigfoot sightings, tracks, screams; almost everything associated with bigfoot has been reported to have occured at this location. I have personally spent considerable time at the lake and have hiked the outside perimeter. I have been there on days where it was very unpleasant and too quiet. I was there with another researcher and we spent an entire day in a small valley just east of the lake experiencing the same phenomena that Darrell's friend experienced. We each attempted to find a rational reason for what we were experiencing, but couldn't understand it. You could look to the tops of the 200-foot-tall pine and Douglas fir trees and see that they were blowing in the wind near their tops, yet there was no sound down on the floor; it didn't make any sense. It was also extremely odd that we didn't see any small ground mammals, squirrels, rabbits, nothing.

Onion Lake is very close to Blue and Bluff Creeks and the Blue Mountains. These have been historical hot spots for bigfoot activity. It is also just up the hill from Laird Meadow where researchers have found bigfoot foot and hand prints. This lake is really in the middle of all activity for this portion of the Six Rivers National Forest.

Marble Mountains: I have personally spent considerable time in the region around the Forks of Salmon. The beauty of the Salmon River is enough to hold anyone's interest. It has notoriety for being a spot for competitive kayakers to run the rapids. The water has a beautiful aqua tone and it's easy just to sit and stare. The landscape in this region is brutal. The Marbles have tall mountains with steep canyons that make hiking difficult. You couple the brutality of hiking with the extremes in weather and you get a location that can make even camping in the wilderness a very challenging ordeal. The challenge of life here may be one reason that bigfoot thrives in its

207

interior. It takes almost eight hours to travel to this wilderness area from any major metropolitan area of California. You have to really want to be here to get here after a winding eight-hour drive.

I found both stories very credible.

LARS LARSON

In May 2008 I was in Happy Camp to make local contacts when I stepped into a local mining supply store. When in an area where I have few bigfoot leads, I would often visit locations outdoor people frequent. A mining supply store has contacts with some people who rarely come to an urban environment, and they are an enticing prospect to me. Three women ran the front desk at the store and were very polite and helpful, even though I wasn't buying supplies. I advised them what I was researching and asked if they knew of any locals who had encounters.

One of the women made a joke of bigfoot, but another politely took me aside and said that I should speak with a local miner, Lars Larson. She told me that many years ago Lars had discovered several bigfoot prints near his mining site, and he had made the trip back to town to get plaster of Paris to cast the prints. She advised that Lars was a very crusty old man who didn't really like talking to anyone, but it might be worth my time to attempt contact. The lady pulled out a copy of a map and gave me directions to his house.

The name, Lars Larson, had actually already been mentioned to me by Tara Hauki (see her report below), who lives up the road from him. She had said she would ask Lars if he would talk with me, so I hoped he would be willing to meet.

Lars lives down a lonely stretch of dirt road on the outskirts of Happy Camp. It is a beautiful location at the base of a mountain somewhat near the airport (a location where people have reported bigfoot encounters). I made the drive to his house and was greeted warmly. I explained to Lars who I was and what I was researching, and he invited me to pull up a lawn chair in his front yard. It was a beautiful spring day with a great view from his yard as it overlooks Indian Creek, which runs through Happy Camp.

Lars said he had heard about me from Tara, and that he'd learned that I had lived a long time in Hoopa, which made me a good person in his opinion. He said that he would tell me his story.

Lars was born in Van Nuys in Southern California and spent the first 18 years of his life there, until moving to Napomo (near San Luis Obispo) where he sold hay and lived a somewhat rural existence. He always enjoyed mining, but wasn't really efficient at it until he attended a mining seminar in Happy Camp in the mid 1980s. The seminar taught him how to reduce his time panning, which he really appreciated. He decided that Happy Camp was a place filled with miners and people willing to help each other, so he moved there in 1987.

Lars was looking for a good place to mine and decided to go 10 miles up Indian Creek from the center of Happy Camp and place his tent and supplies near where Bald Hornet Creek flows into Indian Creek. He was there for two months and was doing pretty well looking for gold. He had found several quarter-ounce nuggets and was making a decent living. He was living off the land, often catching trout out of Indian Creek for his dinner, and eating herbs and berries from the surrounding area to sustain his nutritional needs. The elevation of his mining site was approximately 2,200 feet.

Lars Larson.

During his two months on Indian Creek, Lars saw four or five black and cinnamon bears of varying size. He often saw a pack of dogs, and an unusual sighting was a group of five different bucks that frequented his area. On July 31, 1989 there was an unusual summer rain that soaked the area. The creek rose slightly and he had to take extra care around the mining site. Four days after the rain the weather turned hot. Lars remembered working a lengthy day and, being very thirsty, he walked up the creek slightly and bent down to take a drink from the cold, fresh water. At that exact moment he saw several large, human-like footprints coming and going from the creek.

This photo of Indian Creek was taken near the center of Happy Camp in mid May. The location where Lars found prints was 10 miles upstream from this area. The creek would have been much smaller, slower, and not as deep at his spot.

Lars said he had heard about big-foot over the years, but had never spent any time thinking about it and never contemplated if it was real or a hoax. In the two months on his site he had never had another visitor, and never even saw anyone in the immediate vicinity. He didn't immediately know what to make of the prints he was looking at. They looked like giant, human prints, as they were in gravel and weren't absolutely recognizable. He explained that the prints would have had to be made just after it had rained four days earlier. The prints made Lars very inquisitive and he started to backtrack from the creek and towards the woods to see how far he could track the visitor. His efforts didn't take him far, just to a point near the edge of the bank of the creek where there was clay soil and a slight embankment. This is where he located his best impression. It appeared that the creature was walking downhill towards the creek and had slipped slightly. The toes made a good impression but the heel and ball of the foot made only slight impression, as it appears that the weight of the creature was on the toes. I asked Lars if he could make an impression in the clay similar to what the creature made, and he said, "No way." If he attempted to make any impression it would only slightly dent the clay, nowhere near as deep as the impression that was left.

Lars was so interested in preserving the tracks that he stopped mining for the day and made the journey back to Happy Camp to buy the materials to cast the print—five pounds of plaster of Paris. The cast he made was the one print that appeared clear enough to get a good impression. He said he had never made a cast before, so it might appear a little rough. He felt that the cast had caused his personal credibility to suffer in Happy Camp, so he had loaned it to a local restaurant where it was on display.

I asked him if he kept any firearms or ammunition on his mining site or campsite; he said no. He said that several years after making the cast, he was told that two other locals had been just upstream from his location on Indian Creek in an area near Preston Peak (a location where NABS made a bigfoot expedition in September 2008). A mother and son who were mushroom hunting had supposedly observed a bigfoot in that area sometime in the last few years.

I asked Lars if we could drive to the restaurant where his cast was located, and he said that we absolutely could. Then he asked if it would help NABS if he gave the cast to me, and I said that we would love to have it in our collection. He reiterated that he would be happy if it left Happy Camp and he didn't have any association with it any longer. He also told me that several years ago there had been visitors in town claiming to be professional bigfoot researchers, and they told him they didn't believe his cast was real; they stated it was a hoax. This made him very upset, as he finds it incredibly unbelievable that anyone would have gone to that area of Indian Creek at that time of the year and placed prints along the creek. It wasn't especially close to his camp, and, as well, it was unusual for him to even be in that area, let alone to find the prints. He said that the researchers that came to Happy Camp (Tom Biscardi) upset a lot of people, and many wouldn't talk to anyone associated with bigfoot for quite awhile.

We drove to the restaurant where we met the proprietor, Alexis, who said the cast stirred a lot of local attention and she would miss it when it was gone. Back at his residence, I took a photo of Lars with the cast, and he again thanked me for taking his story seriously and making a good home for his cast.

LOCATION

Indian Creek Road runs parallel with Indian Creek from the center of Happy Camp for approximately 20 miles north until it reaches the Oregon border. Approximately halfway to Oregon, the south fork of Indian Creek flows out from Preston Peak and the center point for the region of the Siskiyou Wilderness Area. It is this exact point of the wilderness area that is only seven miles wide, its narrowest point. Directly west of Preston Peak is the region of Patrick Creek, an area with documented bigfoot sightings. Another interesting aspect of Indian Creek is its location between two wilderness areas, Siskiyou and Red Buttes. These would be two of the ideal locations where bigfoot could live unmolested for a majority of the year.

The specific location where Lars found the print has many conditions that make it an ideal spot for a summer bigfoot encounter. Lars had stated that there are berry bushes in the immediate area, there are fish in the river, and the elevation is 2,200 feet. In my first book we showed a direct correlation between elevation and sightings/encounters of bigfoot. The ideal elevation for a bigfoot sighting in this region of Northern California is 2,400 feet; 2,200 feet is too close to overlook its importance.

A close examination of the cast shows four distinct toes with a fifth possibly visible near the big toe. The toes are clearly visible and it's obvious they made a deep impression where the cast was made. The ball and heel area of the foot are not well defined, and it does appear that the creature was walking downhill and was placing considerable force on its toes in a downhill trajectory. The cast is still held by North America Bigfoot Search.

TARA HAUKI

It's fascinating the way bigfoot eyewitnesses find each other. There have been several instances where one witness I interview tells me about another person they are communicating with who also had a sighting. This is how I met Tara. Travis Cover (see chapter 9) is a truck driver from Brookings, Oregon who had made contact with Tara through a "My Space" account and they had traded experiences over the Internet. Travis eventually made his way to Happy Camp, met Tara, and said that he felt she was credible and wanted me to meet her. I eventually made e-mail contact with her and we set a date and time to meet when I'd be in Siskiyou County.

There aren't many towns in California more remote than Happy

Camp. It is located on Highway 96 approximately 15 miles from the Oregon border. During winter months there isn't a direct route to and from Happy Camp from Oregon. The only way to access the town is by traveling from the west via Somes Bar, and from the east via Highway 96 off Interstate 5 for a 90-minute, twisting ride along the Klamath River.

When I arrived at Tara's residence I was immediately struck with its unique location. It is between the base of a mountain and Indian Creek, a creek that runs year round. As I exited my vehicle, another major bigfoot element hit me—her entire back yard is filled with berry bushes. I briefly scanned the surrounding area and couldn't see any others.

Tara greeted me, extending an invitation to sit in her beautiful backyard on a warm spring day. She told me that her first bigfoot sighting was right in that yard. She explained that her bigfoot journey actually started several years before.

Tara Hauki.

Tara was born in Fresno, California and attended high school in Coalinga, California. Coalinga is known for having one of the largest cow pens in the world. This is a town where ranchers bring their herds to sell and trade; the yard is huge. Before she finished high school, Tara and her family moved and she attended Placer High School. She moved again, this time north to Weed, California at the foot of Mount Shasta. She briefly attended College of the Siskiyou. Her father was living in Applegate, California and had very bad health. He had several bouts of cancer until it got so debilitating that he took his own life. In the early 2000s Tara got a small inheritance and decided to live off the money and enjoy the great outdoors. At one point she lived adjacent to the Marble Mountains near the Salmon River, a spot she described as beautiful. In the summer of 2003 Tara went camping at Fish Lake just below Bluff Creek. She was there for two weeks and experienced some unusual events.

She said that one night when she went to sleep in her tent around

10:00 p.m. she started to hear whistling coming from opposite side of the lake. The whistling was loud and almost sounded like communication back and forth. She could tell it wasn't human-type whistling because it was much too loud. Much later, twigs and small branches hitting her tent awakened her. There was no wind that night, and the branches didn't come at her all at once. It was almost as though something was trying to get her attention. Tara felt scared of what was trying to disrupt her sleep, so she just stayed still until it stopped and she was able to go back to sleep. She awoke the next day and tried to look for traces of what had bothered her, but the soil and ground in the area of her tent was much too hard for anything to make a track.

In 2004 Tara moved into the residence where she now lives in Happy Camp. Her property had been vacant for many years prior to her moving there. On a warm night in July 2005, Tara was in her house and had all of her windows open. It was near 11:00 p.m. when she heard something outside her front door area between her yard and the shoulder of the roadway across the street. She went to her front door and 50 feet away she observed a bigfoot standing on two feet adjacent to a blackberry bush; it appeared to be eating berries. She said that she doesn't remember making any noise, but for some reason the creature turned to look at her and then immediately started to run. Tara described the running as similar to the way a person would run, except the arms hung straight down to its sides. It appeared skinny and was very tall, maybe ten feet, but proportional to a 30-year-old man. The creature did not make a sound and there were no smells associated with it. The next day Tara went to the front of her house and looked for tracks. She found one footprint, 18 inches by seven and three-quarters inches.

For the purposes of this report, I will call the front of Tara's residence the area facing the front street. The rear of her house faces Indian Creek. There is a huge hedge of berry bushes that are between her house and the creek. She took me down a trail along the left side of her yard and showed me a very interesting aspect to her property. Almost in the middle of her yard at the base of a large tree is a thermal spring. It doesn't matter the time of the year or weather conditions, there is always water coming from the spring, and the water is almost always near 70 degrees. Around the time that Tara saw the bigfoot near her front yard, she made a trip to the back of

Tara next to berry bushes.

her property and found a bed of horsetail ferns adjacent to the spring. As I was standing next to the spring I could easily understand that a bed made in the middle of the high grass could go undetected for weeks. There is a healthy game trail that leads directly to the spring, and its source is only 30 feet from Indian Creek. Tara told me that in the area of the mud near the spring and adjacent to the bed of ferns, she observed one large toe print that looked similar to a human toe, but much too large.

After Tara had her bigfoot sighting and found a print, several bigfoot researchers approached her and asked if they could install thermal imaging cameras on her property. Tara, being the helpful person that she is, allowed the installation of the cameras. She showed me on the property where the cameras were installed, the same locations where I would have set them. They were operable for many months, and to the best of her knowledge they never caught a bigfoot on film. They did see a mountain lion, bears, and other small mammals. There was one interesting event that Tara did mention.

During the July to October 2005 timeframe that the cameras

215

were operational, Tara had a local female teenager at her house hanging out one night. The girl was watching the cameras and Tara was doing chores when she heard her friend scream. The girl claimed that she first saw a bear enter the camera frame, then she briefly saw a bigfoot, and then they both quickly jumped off screen. Tara said that she told the researchers about the incident, but she also stated she was concerned that nobody ever reviewed the hours of videotape taken at her house.

In the time frame of late July to early August 2006, Tara heard a loud noise coming from her next-door neighbor's yard area. Her neighbor has a side-yard light that is always on at night, and as she looked toward the light, between the light and her residence she saw a bigfoot violently shaking an apple tree. As she watched, the bigfoot stopped the shaking and then walked across the area of her yard adjacent to the berry bushes, where she estimated its height at nine to ten feet. She could tell it was a bigfoot because it was tall, walking on two feet, dark in color, but she really could not tell much more than that because of how dark it really was that night.

Capped cave entrance.

Tara said she has done a lot of thinking about why bigfoot visits her yard, and she has come to a few conclusions. One major issue is that a hill comes to an end across the street from her house and her yard would be a natural path for a bigfoot wanting to get to the creek. Her residence is also the only property that doesn't have a fence completely surrounding the yard, and there is an access point to her rear yard between her house and the neighbor. Possibly the most important reasons the creature returns may be the spring and the availability of a multitude of berry bushes—water and food in one spot.

Approximately a month after seeing the bigfoot shake the tree,

Tara was asleep on her couch in her front room. It was a warm night and the windows were all opened. She believes it was near midnight when she awoke gagging because of a horrible stench in the air. She described the stench as so bad that she almost was ready to vomit. As she started to rise from her couch she looked directly at the window that looks towards her backyard and she clearly saw a bigfoot looking in the window. She described the creature as having dark-colored skin, two red-colored eyes, having little hair directly on its face, and appearing much more human than animal. She said that as they looked at each other the creature cocked its head to the side with "a motherly endearing look." After less than a minute, the creature ran from the area and wasn't seen again. Tara believes that the creature she saw at the window and the one in the front and rear yards are all the same bigfoot.

Tara said she believes there was something about the cameras going up in her yard that affected the visitations by the creature. She has not had a visit in over 18 months.

There are many locals who know of Tara's sightings, and word spread throughout the community. At one point an individual told Tara about their Native American grandmother who, when she was young and just after giving birth to this person's father, was kidnapped by a bigfoot and held for two years. The creature kept bringing her raw meat to eat, something she never enjoyed. The woman finally escaped and made it back to her home, but then died shortly after. Tara said that she believed the story. I asked her if she could attempt to track down the individual who told her the story, and she stated that she would.

Near the end of my day with Tara she took me to an area on the backside of the Happy Camp Airport, less than half a mile from her house. The elevation is 1,200 feet at the airport and it is the home of the Cal Fire Heli-attack Fire Center. I had been to this airport before regarding reports of bigfoot sightings and prints on the far side of the runway. Tara explained there is a long history of sightings in this immediate area, and part of the explanation is that young bigfoot like the hum of the engines on the helicopters. She also said that there was a cave that she wanted to show me that has been there over 100 years.

We walked to the back of the runway and I was immediately drawn to the region because of the swampy conditions. Bigfoot likes to stay near water; it's a nutritional source and an ambush location for other prey. We walked around the swamps and stopped at a location where a cave was visible. It had a five-foot diameter opening and there was water draining from its entrance. Tara said she was surprised to see that the cave now had bars on it prohibiting entry, since as little as a year before anyone could enter it. She knew of several boys who had entered and gone deep into the cave and were in water up to their waists. She had no idea how far the cave went into the mountain and no idea what made it. It does appear to have been a manmade cave based on the structure and location.

After spending nearly an hour at the back of the airport I could clearly understand why a bigfoot would have visited this area. It had water, shelter, an area to ambush game, and a location to amuse itself—the airport.

Tara signed an affidavit detailing all of the incidents that have happened to her regarding bigfoot.

LOCATION

Across the street from Tara's residence is where a ridge comes to an end. It appears as though bigfoot makes his entry into her yard at this point, because at the opposite end of her yard is Indian Creek and the town of Happy Camp. It would be unlikely that it walked through the middle of town.

I reviewed a topographic map of Happy Camp and the region north to the Oregon border and found some fascinating facts. Directly behind the Happy Camp Airport and directly up the hill across from Tara's residence is nothing but open forest. If you walked far enough, you would eventually end up in either the Siskiyou or the Red Buttes wilderness areas. There is a small gap directly north where you wouldn't hit either wilderness, but that gap is narrow in comparison to the size of each wilderness area.

Tara's sightings, Lars Larson finding a footprint, and the residence of Kirk Stewart in Del Norte County (see chapter 9), all fit in a very neat triangle. Kirk's story describes a sighting he and a friend had of two bigfoot in the Siskiyou wilderness. Kirk's sighting is in general proximity to everything that occurred at Happy Camp. Even though the incidents are in different counties, the overall proximity cannot be ignored.

This book covers a vast area of Northern California and a detailed map of the region would be too large to include here. North America Bigfoot Search (NABS) produced a detailed four-county map of Northern California with over 350 details and locations of bigfoot sightings and encounters. This map can be purchased through NABS website. The map provides readers with a much deeper appreciation of the continuity of the sightings and their proximity.

It would appear from the dates of Tara's incidents that the creature arrives in Happy Camp during the mid summer months, likely simultaneously with the availability of berries and fruits.

Indian Creek runs behind Tara's residence and does contain trout. The spring and the fall bring steelhead and salmon into the tributary, as they utilize the creek as a spawning ground. Bigfoot would not need to come all the way into Happy Camp to catch trout from the creek; that could be accomplished much higher in the mountains away from humanity and would have minimal risks of detection. The obvious attraction to Tara and her yard were the berries and the fresh spring water.

Tara told me that she didn't have any animals at the time of the bigfoot sightings. When I visited her in May 2008, she had two dogs that were very loud and challenging. There are many historical accounts indicating that bigfoot does not like dogs. Tara made a statement that she felt bigfoot stopped visiting since the cameras went up, drawing the conclusion that the cameras made them uncomfortable. I believe that since the cameras have come down, Tara got the dogs. Bigfoot will avoid a yard where there are dogs.

9 DEL NORTE COUNTY

Of the 58 counties in the State of California, Del Norte sits in the farthest northwest corner of the state, sharing one border with the Pacific Ocean, one with the State of Oregon, and others with Siskiyou and Humboldt Counties. It's one of the greenest counties in the state, and that's predominantly because it gets 66 inches of rainfall annually. The beautiful Smith River recreation area sits in the middle of the county and its 300,000 acres take up a significant portion of the Del Norte interior.

The people of a county are the real story in any area of California. Del Norte claims that 17 percent of the citizens are of German ancestry and 8 percent claim to be Native Americans. The Native Americans are split between the Yurok along the Klamath River and the Tolowa from the Smith River region.

The real economic development of the Del Norte started in the 1850s with a gold rush near Myrtle Creek. Copper and placer mining started to replace gold in the 1860s and continued in the mountains east of the Smith River (above Gasquet) well into the 1900s. The other industry that was doing well in the county in the 1800s was logging. In 1853 the first mill was established in Crescent City and that's when the area started to grow.

The wild river status of the Smith River allows huge salmon and steelhead runs to rush the river in the fall and winter months. The largest steelhead in the state can be found on the Smith, with fish over 25 pounds not uncommon. The aqua waters of the Smith coupled with a backdrop of giant redwoods makes for a memorable vacation spot.

The counties along the north coast of California have historically had unemployment issues, mainly caused by the shutdown of the logging business and the lack of new industries moving into the region. The one part of state government that has continued to grow in the last 20 years is the California State Department of Corrections (CDC). The building of Pelican Bay State Prison just outside of Crescent City made the CDC the largest employer in the county, and also caused a major spike in housing prices as employees needed a place to live. Pelican Bay is a notorious prison, as it's the one location where the corrections department sends the most violent gang members and prisoners that have assaulted correctional officers. You have to be a pretty tough person to work an environment like Pelican Bay.

There has been one major natural disaster in Crescent City in recent history and that was a tsunami on March 28, 1964. Earlier on that day a giant 8.8 quake hit Anchorage, Alaska, which caused an undersea surge and subsequent tsunami. The ocean surge in Crescent City killed 11 people and destroyed 150 businesses and homes. The city still practices tsunami drills and has a disaster plan for emergencies of that magnitude.

HISTORICAL BIGFOOT PERSPECTIVE

During the many weeks I have spent in the county, I have been told by a variety of Native Americans and local Smith River and Crescent City residents to look into the history surrounding the French Quarter area of the county and bigfoot. The French Quarter is the region around Gasquet and the hills to the south, French Hill. People stated that there were always rumors of unexplained deaths to residents and gold and placer miners in that area, and many felt they were tied to bigfoot. People were not only killed, but many disappeared and were never seen again. The abduction of a person is more of a crime of the twentieth century than the 1800s. Miners and residents just disappearing was not a common occurrence in the 1800s. Transporting someone against their will was not an easy task in that era; there were limited places to hide someone who was abducted, and chances were the locals knew where all those places were. But in the French region many people who disappeared were not ever found.

The pamphlet, "The Hermit of Siskiyou," by L.W. Musick has several interesting stories about the northwest region of California, and a few directly apply to bigfoot.

On page 78 under "note 2," Musick talks about a series of 18 deaths in the French Hill area in 1895. There were 18 deaths and where he has been honored with various offices of worth and confidence, and where his name was ever synonymous with that of strict integrity.

Assuming that the Lost Cabin story is not entirely a myth, it is doubtful if any other region could establish a more authoritative claim for its location than is presented in the foregoing account, for the westerly slope of the Siskiyou range, included principally in Del Norte county

NOTE 2. Relative to the prolonged and unaccounted-for absence of a miner from his cabin on French Hill, (subsequently found dead) the Crescent City News, dated Jan. 25th, '95 says:

"It is said that of the eighteen disappearances around the French Hill country only one of the bodies has been found."

The preceding quotation would answer as a text upon which to found many speculations with respect to the mysterious disappearances from the region indicated. Such occurrences have become so common, in fact, as to elicit, locally, only the stereotyped allusion: "Another French Hill victim." Many strange stories have obtained with reference to the misfortunes that have, or, peradventure, might have, befallen those who were thus seemingly spirited away, and yet no definite solution of the great problem has been reached. They simply become merged in obscurity, but whether of violence, of accident, or of self-volition, is a matter solely of conjecture; however, a sombre cloud of reproach has settled upon the shaggy brow of old French Hill; and, alas! the vaults of her native treasury are less inviting to the miner and prospector than if the spectre of the mysterious foe stalked not her lonely caverns.

French Hill deaths story, from
"The Hermit of Siskiyou."

only one body was ever recovered. This was an extremely odd occurrence for the 1800s, although the region around the disappearances was quite remote. The idea that the deaths were related to theft is not really feasible, since many of the mines were only mining placer and not worth killing. If they were straight robberies, there would be no reason to take the bodies, since forensics at that time couldn't place a firearm with a bullet or blood splatter with clothing. Nobody knew what DNA was in the 1800s, so blood evidence didn't mean anything. The entire idea that 18 disappearances occurred in the relatively small region of French Hill was baffling, until you start to think about bigfoot. If there are no survivors, there are no witnesses, and there is no suspect.

In my two years researching the Del Norte region, the area of French Hill has the most fascinating stories related to bigfoot. Del Germain, Aaron Carroll, Kirk Stewart, Travis Cover, and James Renae have all had bigfoot sightings or incidents in that immediate area (see reports below). Aaron Carroll actually had a sighting at an old mine, and I personally investigated large bigfoot-type prints in an area just above the Tyson Mine. This is a hotspot for bigfoot sightings and incidents even today.

If you drew a straight line from French Hill directly south across the Siskiyou Wilderness Area, you would be at Bluff Creek. Remember, Bluff Creek is probably the most famous location in the history of bigfoot sightings as the Patterson–Gimlin film was shot there in 1967. The entire Siskiyou Wilderness Area has a history of bigfoot sightings and incidents.

From my research of the historical background of bigfoot it would appear that the creature had a questionable past, possibly mixed with violence and acts of abduction and sexual assault. Refer to the Humboldt County historical bigfoot section and the section on Al Hodgson (chapter 10) for accounts on violence and sexual attacks.

ROGER HUNTINGTON

In early May 2007 I received a report on our website (www.nabigfootsearch.com) of a bigfoot encounter just across the California border into Oregon. I was headed there the next week, so I called the reporting party and spoke directly with Roger Huntington. He was well spoken and stated that he'd like to meet me in Crescent City the following week.

It was a Tuesday morning when I met Roger at my favorite meeting spot in Crescent City, McDonald's. Roger brought photos with him and immediately laid them on the table in front of me. They were photos of what appeared to be a human footprint in sand, but it only had four well-proportioned toes. I had heard of and seen suspected bigfoot prints with three, four, and five toes, but I had never met someone who claimed they had actually discovered a four-toed print.

Roger Huntington with photo of print.

Roger is a very clean-cut young man and speaks with authority and conviction. He started by saying that he is a long-haul trucker and regularly drives the route from his home in Cave Junction, Oregon to Medford. Cave Junction is just across the California border in northwest California. He explained that he was born in Cave Junction, is 26 years old, and is married.

In November 2006 Roger and his wife, Sheila, were driving back from a short vacation to Grants Pass, Oregon. It was cold, but he was getting tired of driving and asked Sheila if they could stop for awhile to allow him to do some prospecting, something Roger liked to do in his spare time. Sheila said she needed a break as well, so they stopped at a spot near Eight Dollar Mountain just outside Cave Junction. They were approximately 20 miles from the California border and maybe 400 feet across a bridge that ran over a small creek. As he stopped the truck and opened his door, his feisty bulldog pug jumped out and ran directly to the hillside adjacent to the small creek. The dog was was barking the entire time it was running, and it seemed to be targeting something specific.

After he exited the truck, Roger heard some movement in the

large bushes adjacent to the creek and in the same area where his dog ran. After only a few seconds his dog came running from the bushes and Roger could see rocks being lobbed from the bush in the direction of the escaping dog. The rocks were large, nearly ten pounds each. Roger said that he couldn't believe what he was witnessing. There were no vehicles in the area, it was winter and cold out, so there would obviously be no swimmers in the area. He kept thinking to himself, "Who could possibly be throwing the rocks?" The first rock barely missed the dog and the second hit the ground and rolled partially into the dog.

Four to five rocks were thrown a distance of approximately 30 feet, and then bushes in the area started to move, then nothing. Everything went quiet and Sheila was getting upset and wanted to leave immediately, but Roger wanted to understand what their dog chased and what was throwing the rocks. He walked down the creek in the direction of the bushes to see what evidence he could find.

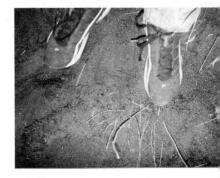

Four-toed print between shoes.

Roger walked approximately 300 feet past the point where rocks had been thrown when he found a medium-sized, human-looking footprint in the sand near the creek. The foot was smaller than Roger's size 12 boots, but was highly unusual because it had only four toes. He found three total prints in regions where prints would be discernible. He never smelled any odors and heard no other sounds. The approximate elevation of this location was 2,000 feet and it was fairly remote.

Roger reiterated that he had not seen anyone for many miles as they drove into this site. He also stated that he had not seen any other footprints in the area, as recent rains had probably washed them out. The four-toe prints that he did find looked as though they had just been made. The ridges were crisp and there was no debris inside the print cavity. The print was pushed into the sand deeply as though something heavy had made it.

Roger signed an affidavit stating he had no idea what could

have made the print, and that he did not know the prints were there before he arrived.

LOCATION

This specific spot of Roger's finding fits nicely with our elevation theory (bigfoot tends to stay near the 2,400 foot elevation in this region of California). Another belief is that bigfoot normally stays near water, and the print was found next to a creek. This entire area of Northern California and Southern Oregon is a hot spot for bigfoot activity, and it has been for decades. The behavior the creature exhibited is consistent with a bigfoot being angry. The creatures have repeatedly shown that they do not like dogs, and when pushed they will respond with anger. There are reports of dogs being killed. A small bulldog or pug would not be a threat to even a small bigfoot, thus there wasn't significant violence.

The size of the print discovered does not concern me. There must be many adolescent bigfoot roaming the forests in order for bipeds to have an active breeding group. The size of the print indicates that this was an adolescent or younger bigfoot that probably got scared because of the dog. It threw the rocks to scare the dog and humans away while it made its escape.

The four-toed print photo is unusual. The foot was not unusual in its appearance other than it had only four toes. I know there are groups in the bigfoot community that believe any number of toes other than five on a foot is a hoax; I disagree. There have been many legitimate sightings where four-toed prints were found. I don't have an explanation as to why some feet have different numbers of toes.

GREG ZOGHBY

In my years of bigfoot research I have always canvassed areas for witnesses, and I make it a point to always check with lodges, hotels, and motels in rural areas. The people in these places like these locations for a reason, and it's usually the outdoors and associated beauty. Many times they spend considerable time in the woods exploring and understanding their general area so they can pass on this information to their guests.

I had made my way through Smith River, Hiouchi, and Gasquet and met many great witnesses. I decided to make my way further north and follow Highway 199 as it meanders through the forests on its way to Oregon. One of the first locations I came to is the Patrick Creek Lodge. It is located on Patrick Creek at its junction with the Smith River. To the south of the lodge is the Siskiyou Wilderness Area where you can walk for 20 miles and not see a road, a vehicle, and probably not a person. To the north is national forest for almost 15 miles until you enter Oregon, where there is even less humanity. It is a gorgeous spot with the equally beautiful Patrick Creek Lodge in its midst. The buildings on the property look as though they may be 60 to 70 years old, but they have been refurbished to almost-new condition. Patrick Creek flows out of the mountains to the rear of the lodge and comes from a pristine environment.

Patrick Creek Lodge.

I parked my car in front of the lodge and made my way into the front lobby, where a hostess greeted me. I explained who I was and the reason for my visit, and she escorted me into the bar area where I met the owner, Greg Zoghby. What immediately struck me was the absolute beauty of the bar interior and the setting outside the bar windows. Greg is a contractor by trade and he and his wife have lived there since 2003. They moved to Patrick Creek after selling another lodge they remodelled in Kauai. They had wanted to move to the Northern California area and again restore a lodge, so they found Patrick Creek and spent the last several years fixing it up. I complimented Greg on his outstanding craftsmanship and great taste.

I explained to Greg that I was in Del Norte County investigating bigfoot incidents and wanted to know if he knew someone who

227

had an encounter. Greg's eyes immediately got large, and he said, "Yeah, as a matter of fact."

Greg said that one day in either February or March 2005 in the very early morning hours a trucker stopped by the lodge to report a very strange occurrence. The trucker approached him and asked if he had seen any strange creatures around his lodge, to which Greg said no. He said he wasn't sure where the conversation was going, but he was becoming more interested. The trucker stated that he was traveling eastbound on Highway 199 towards Oregon and when he was approximately a quarter mile west of the lodge he saw a huge, hairy creature walking upright on two feet eastbound on the south side of the road in front of him. The trucker passed the creature and saw that it walked like a human, but had dark hair over its entire body and it was much larger than any man; he guessed it was over eight feet tall and very broad in stature. The trucker pulled over to the side of the road and looked in his rear-view mirror to get a better look. To his astonishment the creature was now almost running towards his vehicle. The trucker said that he was getting scared, so he put the truck in gear and left the area and drove to the lodge (This trucker was later identified as Travis Cover; see his story below).

Greg said he felt that the trucker was describing bigfoot, and this was the first time he had heard of a sighting near the lodge. The trucker calmed down and left the lodge without leaving his name.

The second incident occurred a few months after the trucker incident, in May 2005. Greg stated that he knows the outdoors well, has spent many hours hiking and working in the Sierra environment, and is familiar with most of the animals that make these mountains home.

Greg remembered that it was after midnight and he and his wife were in bed at the lodge almost asleep when they heard the loudest scream they had heard in their lives. He said there was no way it was a human because it was much too loud for any person to make. He knows there are animals in the nearby woods that make loud screams and calls, such as elk, owls, mountain lions, and others, but that there was no way any normal animal could have made the screams they heard. The calls came from the area outside their bedroom window, which is on the fringe of the woods. He said he was too afraid to leave the room and check to see what was making the screams, but they eventually stopped and he hasn't heard them since.

Greg took my card and ensured me that if he heard anything about bigfoot in the area he would give me a call.

I should state here that I had an interesting conversation regarding this area of Highway 199 while staying at the Red Lion Hotel in Eureka. Often when I am checking into a hotel and there aren't others around, I'll engage the clerks in a conversation about the area, especially if they were raised in the region. On one trip up the coast I was talking with a clerk, a young man around 23, about bigfoot in the region. He seemed optimistic that they existed and said he had a friend who claimed to have seen bigfoot on Highway 199 outside Gasquet on his way to Oregon one night. He said his friend was "freaked out" and didn't want to report the sighting to anyone, but appeared to want to talk to a friend about it. This was the only time the clerk had first-hand knowledge of a bigfoot encounter with anyone. He stated that his friend moved from the area and he lost contact with him.

I've taken many drives in the wilderness behind the Patrick Creek Lodge and I can guarantee that the region is remote, lonely, and ideal bigfoot country. Patrick Creek has its start many miles back in the mountains and the effort to get to its banks in the area is huge. The creek runs through a steep, very deep canyon that would take climbing equipment to get into. It would probably be a fun weekend trip to work your way back into that canyon and see what creatures might live there.

JAN WYATT

When I first get into a new town I like to look for some type of government official (other than USFS) and try to make friends. Gasquet (pronounced Gas-kee) is a very small town with one market, a small post office, and an equally small airport. I first went into the post office and introduced myself and told the lady that I was interested in knowing anyone who had a bigfoot sighting in the community. She said that there was a woman who knew others and had some experiences, and I should ask the storeowner next door about "Jan." I went to the store and the owner directed me to a local neighborhood where I caught up with Jan Wyatt.

Jan Wyatt is a 74-year-young (I mean young) woman who has done a lot in life. My daughter was with me on this trip and she wanted to see how I interviewed people and how I found witnesses, and this was a good learning experience. Jan saw us pull into the driveway and greeted us with a huge smile. I explained who I was and what I wanted to talk about, and she became even more excited and invited us into her home.

Jan Wyatt and her dog.

Jan explained that she was born in San Diego and was a mother of five, and had been a homemaker when she was younger, but when her kids left home she started to work for the Forest Service. She held a variety of jobs, with the last being a fire lookout, and is now retired and enjoying her free time exploring the mountains with her dog. She said she has always been fascinated by bigfoot and the associated stories, and has always tried to be aware of sounds and impressions when out picking berries or exploring.

Jan's first incident occurred on Tuesday November 19, 2002. She was picking huckleberries with her sister up French Hill Road. The road heads south into the hills just west of Gasquet. It is a very good gravel road that is one of the major routes into the backcountry south of the town. Jan explained that she saw a bare spot on the ground where there was no ground cover, and a very large footprint in the dirt. The print was approximately 13 inches by 7 inches. The track was made next to a newly cut road and there was a slight incline in the area that allowed the water to drain. Jan came back to the area two days later and made a cast of the print. She did the best she could making the cast and you can distinguish five toes, but there were a lot of bubbles in the cast and it didn't photograph very well. She searched the entire area for additional tracks, but didn't find anything. The print did look similar to a large human footprint and similar to bigfoot prints.

Jan's second incident occurred in November 2005. She was at Shelley Creek Ridge behind the Patrick Creek Lodge at Patrick Creek and Highway 199, again picking huckleberries. She was trying to concentrate on the picking and to get home, as it was getting cold. There were two large huckleberry bushes behind the ones she was working and she saw those bushes to the rear start to shake violently. Then she heard loud, bipedal footsteps. Jan said the footsteps were so loud she could almost feel the rumble on the ground. Her dog was with her and he immediately took off running back to the car. She tried to see what was shaking the bush, but thought it would be more prudent if she also left the area. Just as she was about to enter her car, she heard a very loud wood-hitting-wood sound. She said it sounded like someone hitting a tree with a baseball bat, but she knew that she was the only one in the area for miles. She estimated that the sounds were coming from a few hundred yards away.

Jan returned to Shelley Creek Ridge last year and visited the same spot, and said she found some things about the area very odd. She found a very large pinecone that had been twisted open and the insides taken out, and she found several large branches from nearby trees that were twisted and pulled off five to six feet from the ground.

LOCATIONS

Shelley Creek Ridge is in an area that has had significant bigfoot activity. The owners of the Patrick Creek Lodge have reported sounds related to bigfoot, and a trucker who saw a bigfoot on the highway adjacent to their lodge contacted them. Just south of this location and across the mountain is the Tyson mine and French Hill Road, all with significant related bigfoot activity. This region of Del Norte County has a lot of water, a lot of isolation, and is prime bigfoot country. I have actually spent several days exploring this area. There are several creeks in this region that have gouged deep gorges that make them extremely difficult to explore. The area three miles behind the lodge suffered a severe forest fire several years ago and still suffers from lack of canopy, but there are enough trees near the lodge and the surrounding area to make this region very compatible to bigfoot. As most woodsmen know, areas that have been forested or burned have significant new growth of young herbs and plants, and this brings in animals to feed.

French Hill Road is a USFS road that was eventually supposed to connect to the Go Road coming out of the Bluff Creek and Orleans area. This road does eventually end at the fringe of the Siskiyou Wilderness Area, also a region with a lot of bigfoot activity. The Siskiyou Wilderness Area is probably the least visited wilderness area in California. It is located at the extreme northern part of the state and there isn't easy access to the area. If bigfoot wanted to be left alone, this would be the place. If you refer to the NABS Northern California Bigfoot Map (available at our website) you can see how the incidents and sightings intertwine. Bigfoot has a relatively short trip from the fringe of the wilderness area to the French Hill area and eventually Gasquet.

DEL GERMAIN

I got in touch with Del Germain through Travis Cover (story below) from southern Oregon. Del grew up on the north coast of California and completed high school in Del Norte County. After high school he joined the Marines, completed a four-year stint, and utilized the GI bill to enroll at Cal Poly San Luis Obispo. Del said he was initially interested in chemistry and physics, and completed three and a half years, but then dropped out of college. His interest in and passion for the outdoors drove him back home. He liked the forests and animals on the north coast, and he ended up in the low-pay arena of logging, working as a surveyor for the lumber company, a job he relished.

While in high school, Del played baseball and traveled throughout the tri county area playing games against various local schools. He distinctly remembers one trip he made with friends to play Hoopa High School in the late 1950s. Before the game they went to visit a friend, Kenny Matts, at his uncle's house in Hoopa to relax and socialize. This was just about the time bigfoot tracks first

232

appeared at logging operations around Hoopa, and the kids started to talk about it. Kenny's uncle was the oldest and most knowledgeable person in the room, and the kids asked him about a huge creature that might be lurking in the woods. Del remembers the uncle looking at the ground, hesitating, and then telling the story. He said that when he was younger he and a friend had packed into a lake on horseback to go hunting. He was lying on the ground to watch for deer and other animals, and was trying to be very quiet. He was at the lake for many hours when he heard a rustling in the bushes and saw a huge, hair-covered creature come out of the vegetation toward him. He was scared to death because he had never seen anything so huge, walking on two legs. It looked menacing. The boys asked if he had shot at it, and the uncle said his rifle wasn't big enough to bring down a creature of that size. Apparently it never saw him, and it left a short time later. He got his friend and they left the area that night, never to return. He said that he had never been so frightened in his entire life.

South Fork Smith River.

Del's family owned a small cabin off the South Fork Smith River, just opposite Kirk Stewart (story below) off French Hill Road. The best route to the cabin was to travel out of Hiouchi, take the South Fork Road along the Smith River, cross the river, and take the French Hill Road up into the mountains. The area was very desolate while Del was growing up, but he always loved the isolation and the beauty of the area.

In the late 1970s Del's mother, Dorothy Germain, was living on the cabin property on a full time basis in a 35-foot trailer the family had put on the grounds. Part of the trailer stretched over a small embankment so that its windows stood high off the ground, while the remainder of the trailer was at ground level. One night shortly after she moved into the trailer, just as it was just starting to get dark, Dorothy heard a sound at the back of the trailer. She looked toward the rear window that was over eight feet off the ground and saw a human-looking face staring in at her. The face almost had a copper tinted complexion. She instantly knew how big this thing must be; it had to be over nine feet tall. Dorothy was very frightened and locked herself inside, waiting for morning so she could leave. She called Del, explained that she was leaving and would never return, and she never did. In later conversations with his mom, Del recalled she said that the face wasn't menacing and it didn't exude anger, but the thought of how large the person or thing was scared her.

One item about the property that always caused Del anxiety was a constant humming noise. He said the noise sounded like a propeller plane flying over at 1,000 feet, but at a lower tone. It was a hum that would go on for hours and then suddenly stop. He walked throughout the mountains and could never determine where it was coming from. He even went to neighbors miles away and could still hear it faintly. He stated it was a real sound because there were times that everyone would hear it, but it sometimes suddenly stopped, like a motor being turned off. I asked Del if it was possible the sound was coming from beneath the ground, and he stated it was, as there had been placer mines in the area years earlier, but there was nothing operational around the times he heard the sounds.

The property the Germains own is quite unusual. It is 90 acres that is almost like an island. The United States Forest Service owns the land completely surrounding the property for many miles in every direction. The Germains were able to obtain the property years ago because they had worked land in the area as a claim and got the land from the claim procedure.

I asked Del if there were any other unusual events that occurred on the property, and after a brief hesitation he said there were other things that happened on the land that he could never explain. There were times he would walk outside at night and would see bright lights, similar to orbs that were floating on the hillsides across the valley. He would watch them work an area of the mountain, and then suddenly the light would shut off and apparently vanish. The first time he saw this he couldn't believe what he was seeing. After many instances of seeing various odd encounters in the sky in Del Norte County, Del wasn't surprised by much.

Another odd aerial encounter occurred when Del was on a back-packing trip in his younger days. He and a friend had hiked for two days into Island Lake at the far eastern end of Del Norte County where they wanted to stay and fish. The first night they were at their location they were lying in their sleeping bags looking at the stars when they saw jet fighters chasing a white light in the sky. He said it was obvious that the light was much more mobile and faster than the jets, but they were obviously chasing it for a reason. It was almost like something out of a movie and was an incident that he will never forget.

Just as the conversation was coming to an end, Del offered one more story. He stated there really wasn't much to it, but it was something he felt he needed to say. He remembered one time he was alone at the cabin in the late 1980s. He had gone outside near dusk and found it to be very quiet. Del has spent a lifetime in the woods and nothing really ever scared him, even the observation of a mountain lion, but on this particular night he walked out onto the front area of the cabin and for some reason the hair on the back of his neck went up. He immediately became very frightened; it seemed something had overcome him. There were no sounds, nothing distinctly done to frighten him, but something was telling him to get back inside as something bad was outside. He went back into the cabin and the feeling immediately subsided. He said that was the only occasion where he felt very uncomfortable at the property.

Conclusion

The location of the Germain cabin is intriguing. On the opposite side of the mountain is Gasquet and the Kirk Stewart property (where there have been multiple bigfoot incidents). French Hill Road on the Gasquet side has a long history of bigfoot sightings from individuals who do not know each other. The area of the South Fork Smith River is very desolate, yet teaming with wildlife. One day I was driving the South Fork Road and I saw one of the biggest and healthiest looking black bears I had ever seen, easily over 600 pounds and a glistening jet black in color. The Smith River is loaded with steelhead, salmon, and smaller trout all year long.

This location is on the northern side of the Siskiyou Wilderness Area, which is the region that separates Bluff Creek from Gasquet. The wilderness area has had many sightings throughout the years, including the famous Patterson–Gimlin film of 1967. On the opposite side of the mountain, as you go towards Gasquet, you cross into Oregon and the Kalmiopsis Wilderness Area that also has a long history of bigfoot sightings.

The sighting of a human-like face in the window of Dorothy Germain's trailer has significance. Many of the bigfoot or wild man sightings made over the years have described the face as being peaceful and human in appearance (refer to chapter 12, Oklahoma, Ina Faulkenberry). The copper coloring could be attributed to the low glow of the sun or the dark skin coloration that is often attributed to the facial area. If she saw just the upper portion of the face and not the lower cheekbone area, it would be possible to not see any facial hair, and it would almost have the appearance of a normal human.

Del's narration of the incident at the cabin when he was frightened is a classic bigfoot encounter. People become very frightened, they don't know why, and they want to leave the area. This is usually associated with a sound vacuum, and people are usually alone. His description of the event mimics many others I have heard over the years, some of which are followed by a bigfoot sighting and some not. The sighting merely depends on whether the person stays in the immediate area or has the ability to retreat. In Del's incident he had the chance to escape into the cabin and leave the feelings behind.

The entire French Hill Road area is an ideal location for long-term study of bigfoot. Its proximity to two wilderness areas and the significant amount of food source in the region make it ideal habitat for the biped.

AARON CARROLL

Gasquet is a very small community along the banks of the Smith River in far Northern California, the last town eastbound along Highway 199 as it meanders its way to Oregon. Gasquet has a population of less than 500, a small store, a post office, and two coffee shops. All of the kids from Gasquet, Hiouchi, and Smith River bus together to Crescent City (25 miles away) for high school. Del Norte High School in Crescent City has kids from a 30-mile radius all attending the same school, and thus almost all the high-school-age kids from throughout the county know each other. This is how I met Aaron Carroll.

Once the word goes throughout a community that a person is visiting who wants to hear bigfoot stories, names start to flow and phones ring. Through a very widespread web, I was given Aaron's name and was told that he now lived in Oregon. It took me three days to track Aaron to Grants Pass and another two days to actually catch him home. We finally ended up meeting in Cave Junction, Oregon and were able to discuss his sighting in detail.

Aaron is a 22-year-old line cook at a local Grants Pass restaurant. He is married and they live with his wife's sister. He grew up in a family where his mom and dad divorced, so he split his time between Crescent City and Gasquet. He graduated in 2003 from Del Norte High School and continued to live in Gasquet the following year. Aaron said that some

Aaron Carroll

of his best friends lived close to him and they spent a lot of time hiking, hunting, and fishing in the hills surrounding the community.

Aaron said that he had heard rumors his entire life about bigfoot, and people reportedly seeing it in and around Del Norte County. There were very few kids that spent as much time in the outdoors as he did, so he felt that if he hadn't seen it, bigfoot probably didn't exist. He thought people were fabricating their stories, and maybe some people who didn't live in the mountains and didn't know what a bear looked like perhaps mistook a bear for bigfoot.

It was in June or July of 2004 when Aaron was 19 that he and two of his friends decided they would head up into the woods and shoot something. Aaron explained that it really didn't matter if they shot anything; it was just important for all of the friends to spend some quality time together. He stated that his gun was a 25-year-old 22-caliber rifle that jammed after every shot, so it wasn't reliable, not very accurate, and, in essence, was just above junk status. The guys piled into one car and headed up French Hill Road.

Aaron's friend knew the area of French Hill Road very well and knew an area they could go to that had diverse terrain and would be perfect for plinking birds, squirrels, or even a small deer. It was about 2:00 p.m. when they arrived in a place known as Tyson Mine, an abandoned claim from 50 years ago. It did have diverse topography and even a meadow near the mine. The guys decided to split up and meet back at their vehicle after about an hour. If someone found a good spot they would all go back to that location.

Aaron remembers that he went over a small hill and found a small open area with about 20 small pine trees in the center of the meadow. The trees were not mature but they were large. He decided to crawl up a steep hill on one side of the meadow and take a vantage point to watch the area. It was pretty warm and he wanted to just lie back in the dirt with his eyes open.

He reached the point on the hill where he wanted to lie down and when he looked down at the pine trees he thought he saw something moving in the center small grove. He moved a little and saw a huge figure sitting on its knees moving it arms frantically. Aaron said it looked like it was bent over digging through the dirt. His mind started to go through a deduction process to decide what he was looking at. The creature was big and hairy like a bear, but bigger, much bigger, than a bear. The hair coloring wasn't anything like he'd seen on a bear; it was a salt and pepper color that he had never seen in the wild. The creature's arms were moving like a human's, but no human has hair covering all parts of its body. Aaron was confused and wanted an even better view of what he was looking at, so he decided to move a few feet. He took his eyes off the creature as he maneuvered on the steep hillside, and when he looked back to the pine grove, the creature was gone.

Aaron was now on the hillside, hearing leaves and large branches breaking, and loud and very heavy footsteps amongst the trees below him. He decided to move in the direction the noise was moving. Because of the slope of the hillside, he lost sight of the trees and continued to move in the same direction as the sound. He moved parallel along the hillside when all of a sudden he saw the same creature sitting on the hillside approximately 45 to 50 feet below him. He immediately froze and watched as it rolled boulders down the hillside with both hands. Aaron explained that it was acting almost like a small child would act when it wanted to have fun and throw things down a hill. The creature reached behind while sitting, and with each arm and hand grabbed a rock and then threw its arms forward, launching the rocks down the hillside. Aaron couldn't believe what he was seeing. He said that there didn't appear to be any rationale for why the rocks were being rolled down other than having fun.

Aaron described the creature as huge: large shoulders, long legs and arms, and a massive upper body. Its head was equally large because it was so wide that it came within one to one-and-a-half inches from each shoulder. It had a flat face and nose, and it had hair flowing from the top of its head to back of its shoulders. It also appeared to have hair flaring from the area where a human would have sideburns and whiskers growing from its chin area. It was sitting with one leg tucked under its body while it was throwing the rocks, and its hands were black in color. The color of its hair—salt and pepper—was uniform over its entire body.

During the entire time that Aaron was making his observations, he had his rifle at his shoulder and aimed at the creature's head. He contemplated shooting it, but decided that he probably shouldn't. He said that he was in a position that wasn't perfect and decided to move again, but just as he started to move he saw it turn its upper body and head in his direction. It appeared to Aaron that the creature saw him through peripheral vision, and Aaron thought that he actually saw the corner of its eye as he looked at him. At this point Aaron moved slightly on the hillside and took his eyes off the creature for a few seconds. When he looked again, it was gone. Aaron felt that the only way it could have escaped so quickly was by jumping straight down the hillside and out of sight. He never saw it again.

Aaron went back to the truck, met up with his friends, and took them to the location where the creature had been digging in the ground. They discovered that it wasn't digging, but was moving leaves around on top of the ground in an apparent attempt to locate mushrooms or grubs. (Note: In my first book, *The Hoopa Project,* I described how Native Americans believed that bigfoot hunted for mushrooms in areas that were heavily covered with leaves.)

I asked Aaron if he could describe the creature's appearance or age in any distinctive manner. He said that it looked extremely healthy, as though it was in the prime of its life. He said that best description he could give me for the creature was that it looked a lot like the creature on the early nineties television show, *Harry and the Hendersons.*

Aaron reiterated that he is positive he saw bigfoot, and that is exactly what he told his friends. He said that he's seen many bears in his life and this looked nothing like a bear.

Aaron signed an affidavit covering all facets of his sighting.

CONCLUSION

I definitely believe that Aaron saw a bigfoot. His description of the creature fits well with other witness descriptions. The salt-and-pepper color Aaron described is unique, but not unheard of in bigfoot sightings. The notation of hair flowing off the top of the head and back towards the shoulders is one specific aspect that witnesses in this area seem to always explain.

I have been to the Tyson Mine and have observed bigfoot tracks along a ridgeline just above the mine.

KIRK STEWART

When I go into a new county to conduct research, I always try to make alliances with someone in the community who is outgoing and knows the region. In far northern Del Norte County my contact is Jan Wyatt (story above). Jan is a retired Forest Service lookout who knows the mountains, rivers, and people very well. She was very eager to help me get the contacts I needed in the area.

One of the first people Jan put me in contact with was Kirk Stewart. Kirk runs an organic ranch above Gasquet, California in a beautiful spot just on the north side of a wilderness area. Kirk's property is remote, beautiful, and perfect for bigfoot. Jan gave me the directions to his property, made a phone call to say I was enroute, and told me to enjoy the beautiful ride.

Kirk lives in an area off French Hill Road above Gasquet. French Hill has a very long history of bigfoot sightings and the road runs the ridgeline at the top of the mountain. It is a dirt road with many residents and an ideal location to see bigfoot. Kirk has a huge meadow next to his residence and lives in a gorgeous setting. They raise small farm animals, organic produce, and have two dogs and two great kids. I would truly enjoy living in the setting they own.

I arrived and was met by Kirk's wife, Susan. Kirk wasn't home yet, so we had a few short conversations about bigfoot and things they had seen in the area. I told her I had heard they found bigfoot tracks in a nearby ridgeline and asked how far away it was. Susan offered to take me there.

Kirk and Susan Stewart.

Susan drove with her son as I followed in my vehicle for the 15-minute trip to a very isolated area near their residence. It was obvious from the landscape that there used to be a mine in the area; there was a lot of old rusty mining material sitting on the ground and some of the trees and brush had been cleared away. Susan's son led me down several paths and trails while Susan kept asking him, "Do you know where you're going"? He may have seemed lost to her, but it appeared to me that the 12-year-old knew his way around the mountains. After going over and under some obstacles we arrived at the edge of a mountaintop. The boy pointed to a ridgeline that ran at an approximate 45-degree angle down to what appeared to be an older mining site (later determined to be the Tyson Mine). It was immediately clear what he was pointing to.

The ridgeline contained imbedded footprints of a huge creature.

241

There was green grass topping the ridge and the ground was moist from recent rains. I would guess that the prints were made before the last rain, but within the prior two weeks. The prints were huge, 17 to 19 inches, but I couldn't see detail. I pushed my size 12 boots into the ground next to the print and could barely break soil. The footprints I observed indented three to four inches into the soil and appeared to be similar to human footprints, except they were huge and heavy. It appeared from the prints as though the creature was heading uphill and had a stride of 45+ inches.

I spent considerable time taking photos and examining the tracks for additional types of evidence. The ridge was steep and tough to walk because of its moist condition. After almost an hour on the hillside we decided to go back to our vehicles and make our way back to the Stewarts' house.

We found Kirk waiting for us. His son explained where we had been and Kirk got a big smile on his face. He said the tracks emanated from the Tyson Mine, which is located at the bottom of the ridge. Three years earlier he had been checking out the old mine and found huge footprints at the bottom of a small spring located adjacent to the mine. He wasn't quite sure what he was looking at because the feet were so big and they appeared to almost be human. It didn't make sense to him at the time that someone would be foolish enough to walk around an old mine site in bare feet. Kirk realizes now that the prints were bigfoot.

Kirk's second story occurred in 1995 or 1996, he wasn't positive. He asked me to meet him at a location in Crescent City the following day. He explained that his parents owned an old home that backed up to the city swamp and he would take me there.

Crescent City is located on the Pacific Ocean in far Northern California. Much of the town is located below sea level, and as such has a region where it's almost a bog. Visitors and tourists will not ever see this portion of the town, as it is located approximately three miles inland and not noticeable from any of the major roadways. The area around Crescent City is some of the most rural of Northern California, and a region known for huge redwoods and bigfoot sightings. I was getting a good education from Kirk on the town and what to expect. He said the boggy area was almost something out of Louisiana and really isn't talked about by locals. It contains bears,

Stewart footprint.

Stewart ridgeline.

deer, and almost any kind of wildlife you can imagine that is indigenous to the area.

I have always thought there is some connection between bigfoot and Native Americans, and following Kirk to the house, I saw several signs indicating that the local Indian casino was just down the road from the residence. Coincidence? We left one of the main roads in Crescent City and drove down a long dirt road until we reached a gate. Kirk exited his car, unlocked the gate, and it was like we were entering a different world.

The old residence was not in great shape. Large bushes had grown around the house, berry bushes were everywhere, and there was barely enough road to park our vehicles. Kirk explained that there is nothing behind the house for miles other than a swampy bog. He said it was almost impossible to make it across that area because of thickets, water, marsh, mosquitoes, etc. He pointed to many hucklberry bushes, which ripen in October, around his house.

Kirk stated that in 1995 or 1996 he and his wife were living at the house, and in October of those years, in the very early morning hours, he and his wife were awakened by loud screams coming from the area of the swamps. They could tell the creature was moving, as the location of the screams changed. Kirk talked to neighbors about it and everyone felt that it sounded like a woman getting attacked, but the sound was too guttural to actually be human. He also stated that he personally killed a lot of deer in this immediate area as he was growing up, so he knows there is abundant wildlife in the region.

Kirk and I walked a significant distance into the swamp. It was hard to believe that a mere two miles away was downtown Crescent City. There was no possible way that anyone could walk through this area in the middle of the night. We made our way out of the swamp, back to our cars, and drove into town.

We went to a local spot and sat to talk about his first bigfoot incident. It had occurred in 1990 when he was hunting with his friend Blaine Jolly (name changed by request; see his report below). Kirk explained that he and Blaine had been friends since they were much younger and that Blaine was now a correctional officer at Pelican Bay State Prison.

Kirk and Blaine had decided that they would hike into the

244

Siskiyou Wilderness Area for deer hunting. In was in October and the weather fluctuated from fairly warm to cold at night. He remembered hiking and being very warm and sweaty. They were approximately five miles from the trailhead and in an area where they might spend the night. He stated that the hike in was unusual compared to prior years when they made the trip, as they did not see or hear any wildlife, and that was very unusual for this region. The two friends walked another half a mile and made it to the top of a ridgeline that surrounded El Capitan Lake. They both stood on the ledge surrounding the lake and their attention was drawn to the lake below.

They heard water splashing at El Capitan Lake and saw two odd figures standing 200 yards below in the water. The figures appeared to be in two feet of water, they were facing each other while standing on two feet. The creatures were dark in color, covered in hair, and splashing water on each other, like two children would play in a lake. Kirk and Blaine stared in disbelief at what they were witnessing. Both men knew what bigfoot was and what it represented, but they couldn't believe they were seeing two at the same time. The play and splashing carried on for five minutes until for some unknown reason one of the creatures turned toward the men's location and obviously saw them. Both creatures immediately took off running across the water at an unbelievable pace. It almost appeared as though the creatures were running on water, they were moving so fast.

After the creatures ran out of the lake, the men could hear them running through brush in an area just below them. They could tell that the creatures were trying to quickly escape further detection.

Kirk and Blaine picked up their packs and rifles and made their way to where they'd set up camp for the night, just downhill from the lake. Both men say that they fell asleep quickly. It was a cool night and they were lying in their sleeping bags. Sometime between 11:00 p.m. and 2:00 a.m., something huge ran through the middle of the campsite. It was so heavy that the ground was moving, and they could hear its bipedal footsteps as it was leaving their area. They awoke and stared at each other in disbelief. Just then the loudest ear-piercing scream they had ever heard came from the forest. It wasn't far away, and it was obvious that something didn't want them in the area. Kirk and Blaine didn't get much sleep that night, and they both felt that it was the bigfoot from the lake keeping them awake.

Kirk looked up bigfoot websites and one had a recording of a bigfoot scream. He listened to the sound and stated that he was 100 percent positive this was the same scream he and Blaine heard at their campsite, and the same scream he and his wife heard at their residence in Crescent City.

Kirk had originally told me that there were only three bigfoot related stories that he would describe for me. At the end of the last story I thought we were done when Kirk asked if he could tell me one more story that had always bothered him.

It was in the late 1970s when Kirk was on a trip deep into the Siskiyou wilderness. He explained that when he was younger he was fearless, and he enjoyed getting away from everyone and would always attempt to get into areas that seemed to be impossible to access. It was one of these trips when Kirk was on a game trail in the middle of nowhere. He had walked up a small canyon and found himself under a huge tree in the shade. It was quite hot, so he decided to take his backpack off and rest.

Kirk was relaxing in the shade when he casually started to look around at his surroundings. One item that immediately seemed odd was that the height clearing of the tree and surrounding bushes was very high, almost eight feet. Normal game trails have a four- to five-foot clearing. He continued to scan the area and saw something unusual on the back of a huge Douglas fir tree in front of him. He got up and started to walk around the tree and was immediately stopped. At the base of the tree and leading up to a height of four feet was the largest pile of scat he had ever seen in his life. He explained that the pile was consistent with that of a horse, but there were piles upon piles upon piles. It became quickly obvious that something was living in the tree and defecating down to the ground. The scat had grass and other leafy material along with appeared to be bones. Kirk looked up into the tree but it was huge, over 300 feet tall. He also looked at the base and the surrounding area for footprints, but found none. He said that as someone who has lived in the forests his entire life, he knew that in no way were these piles of scat made by a bear. The circumference of some of the piles was much too large for any bear, and the consistency and look was very different.

At this point Kirk started to put the pieces of the puzzle together: a game trail with a very high clearance, a remote location away from any other trail, and a pile of scat bigger than he had ever seen in his life. He knew he was in a bigfoot area and he had to get out. He was so far from anything that he spent the majority of the next two hours looking over his shoulder at what might have been following him. He was able to make it back to a dedicated campsite and spent the night without incident.

Kirk signed an affidavit.

LOCATION

The Siskiyou Wilderness Area sits in probably the best location in the world if you want to study bigfoot. It is located between Bluff Creek and the end of the Go Road (the location of the Patterson–Gimlin movie) and the region in Del Norte County of Gasquet and Crescent City. This region is remote. There are no vehicles allowed and I have personally never seen anyone take horses into the region. It is a seven-hour drive from San Francisco where the closest mass of population lives in Northern California. Not only is this area remote, it is treacherous. The weather conditions can change in an hour. One minute it can be 80 degrees in September, and the next it is 40 degrees and raining like you're in a rainforest. The next hour it could be snowing at 2,000 feet. Many of the mountains have sides that are nearly vertical, making roaming the hillsides almost impossible. Without a good trail map and some knowledge of the area, it is easy to become disoriented and lost. If all of this isn't enough, there is the real chance you could accidentally walk into marijuana grow area and be shot at by the illegal farmer. There have also been people jumped and almost killed by mountain lions in Del Norte and Humboldt County. A recent attack involved an older man saved from the attack by his wife who was hiking with him. The threats are real and ever present.

I have always wanted to be dropped by helicopter into the middle of the Siskiyou wilderness with three other researchers and a month's supply of food. I truly believe that the right group with the right equipment, patience, and persistence could walk out of that wilderness area with enough evidence to startle the world. I know you are probably thinking that there must be a lot of people that hike that area every summer. Wrong. There are too many other options for hikers and people who want sheer beauty to make the trip this far north. California has Yosemite, Sequoia National Park, Lake Tahoe, and all have more beauty than Siskiyou.

The area around the Tyson mine is also an important location to this book. Independent of Kirk's footprints found in that area, I also found another witness, Aaron Carroll, who while he was hunting, actually saw a bigfoot in that immediate area. If you count two incidents where prints were found several years apart and the sighting by Aaron, that is three bigfoot incidents at the same mine in a matter of five years. The location of the prints on the ridgeline fits the theory we have heard from the Yurok: bigfoot is a ridgewalker. The prints above Tyson Mine go directly up the ridge; they don't deviate.

The bigfoot sighting at El Capitan Lake is one of the most unusual I have ever heard. I met with Kirk's companion, Blaine, and he related the story to me in the same manner as Kirk. Blaine is a correctional officer for the State of California and as such is very concerned about the safety of his family. He requested that his name be changed for this document and we complied. He signed an affidavit about his experience with Kirk. He agreed that it was the most unusual incident he has had in his entire life; he couldn't believe what he was seeing. The behavior that bigfoot exhibited would be comparable with two four-year-olds playing in the water. This specific behavior has been reported before, which shows that bigfoot has a playful side to its daily life.

UPDATE

One evening in 2008 I received a call from Kirk asking for some assistance in handling an issue at his property. Kirk stated that earlier in the year, near the end of May, he had an unusual incident in his melon patch. Kirk explained that he had planted honeydew and cantaloupe in 977 Skip Loader Tires that he had raised off the ground. He placed them in three rows of six and placed them at the far end of their property approximately a quarter mile from his residence.

This area was then surrounded with seven-foot-tall deer netting that was supported by seven-foot-tall six-inch posts buried 18 inches into the ground. The idea behind the netting and the melons being placed up off the ground was to keep bears and deer out of the garden.

His son had a birthday party on May 29 and they had friends over to the house. It was approximately 3:00 or 4:00 p.m. when the loudest scream he had ever heard came from the area of the melons. Whatever made the scream did it three more times. Kirk said it was not any animal he knew that made the screams. Many people attending the party felt a bigfoot made the noise, and Kirk said the screams were very similar to those he and Blaine heard the night near El Capitan Lake.

In September Kirk made one of his many trips out to the melons to check on the growth and ensure they were being watered. What he found was truly baffling. There were approximately 50 melons lying on the ground in a perfectly straight line. All of the melons had their centers eaten out, same as a human would consume. It was extremely odd to see that whatever had eaten the melon lined them up perfectly after they were cored. He now had only 60 melons left of the original number he had grown. Kirk stated that the ground around the tires was extremely hard and compacted, and the possibility of finding tracks was nearly impossible. He examined each of the melons and found something truly fascinating. One of the melons that had been picked but not consumed had a finger hole in it where you could actually see a fingernail indentation. Obviously not a bear! Kirk also found one tire that had a huge indentation in the middle where it appeared the creature sat and ate the melons.

The process of elimination shows that few creatures could have raided the melon patch. The entry point to the melons had a seven-foot tall post pulled to the ground and the netting pulled to the side. It appeared that whatever got inside the patch had hands to manipulate the post. To gather the melons and place them in a straight line is a very unique trait. Bears eat the entire melon, not just the inside. The fingerprint, the melons placed in a straight line, and only the insides eaten—all point to bigfoot. Remember, the screams that were heard came from this exact area, but there weren't any melons to eat at that point.

Following this, other unusual things have happened. On the opposite side of Kirk's property from the melons is a pen that held 15 peacocks. The pen was a very sturdy facility with a six-foot-high fence held up by the same type of posts used around the melon patch that would keep out bears, mountain lions, and even smaller game like raccoons and skunks. To enter the peacock pen something had to step over the six-foot-high netting surrounding the birds. Nevertheless, the pen was raided and several of the birds were taken. The netting showed no signs it had been damaged in any way. When Kirk got inside he found that 10 of the birds were missing and it appeared from the amount of feathers inside the pen, the birds had been plucked. There were no bodies in the pen, just a lot of feathers. If a bear, mountain lion, or other mammal had entered the pen, it would have consumed the peacocks on the spot and not plucked and taken them for later consumption.

(Note: I have seen this same type of behavior in Hoopa. There was an incident where chickens were taken from a residence. I found several small piles of feathers 200 yards from the residence, as though something had sat down and plucked the feathers on the spot. This is not the behavior of any mammal that is recognized by science in these mountains.)

Kirk said that the theft of the peacocks was something that disturbed him greatly. He knew a bigfoot had taken the peacocks based on the huge step it had to take to get over the netting. He searched the area and found a clump of hair that was attached to the bottom of the netting where the creature had placed its foot. The hairs were three to four inches long, very fine, and had a reddish tint. Kirk said he could guarantee they were not bear hairs because of their fine texture, and also the hair looked very odd because they were also very curly, almost like pubic hair.

(Note: The description of the hair is consistent with bigfoot hair. I took a multiple-sighting report of a bigfoot approximately 10 miles from Kirk's property 12 months earlier; the description given to me of that bigfoot was that it had a reddish tint to its hair.)

There are interesting factors relating to Kirk's melon and peacock losses. Whatever took and ate Kirk's melons knew that he had almost twice as many melons left that were not consumed, yet the creature never returned to eat more. Very unusual behavior for any creature, considering it was an easy meal. The peacocks were not consumed on site, yet the feathers were removed in the pen, very unusual behavior for any creature.

FOLLOW-UP

Kirk gave me the hair he had recovered from his peacock pen and it was sent to a laboratory that NABS utilizes. In every other previous purported bigfoot hair sample that NABS has submitted, the hair has always been returned as bear, deer, cow, etc. The first report that we received from the lab indicated that the hair was synthetic or it was hair that was treated with a dye. The lab said that there were inhibitors in the hair that didn't allow DNA extraction and this most likely was the result of the hair being treated with an artificial dye.

Kirk sitting in a tire in the melon patch.

By pure chance, I was reading a lab report from the 1960s regarding a researcher named Jim McClarin. The lab had written that they felt that hair that was submitted by McClarin might have been artificial or treated with a dye. Knowing that McClarin's sample had been taken only 30 miles south of our location, I realized that there was something odd about the scenario. I trusted that Kirk, his wife, and kids would not be walking around the woods of their ranch wearing artificial fur coats. There had to be something in common between the two samples. I forwarded a copy of the earlier lab report to our lab and asked that the hair fiber be re-examined by a fiber expert.

The lab that NABS work with is very interested in understanding the bigfoot paradox. They forwarded the hair to a law enforcement hair expert for his opinion on whether this was a synthetic fiber or a real hair. The expert stated that this was a real hair and classified it as "unknown primate." He stated that the hair doesn't have a core or medulla and it had a worn tip, as though the hair was old and had never been cut. If we believe the law enforcement expert, the Jim McClarin sample was probably a real hair as well.

Our lab now ordered additional mitochondrial testing kits to continue their efforts to extract DNA. Their efforts at extraction have not been successful to this point because of a lack of core or medulla. The lab has been in contact with world experts on mitochondrial DNA testing and they are diligently working to overcome the inhibitors.

FOLLOW-UP NOTE

Just as this book was going to press, NABS continued to receive updates from our lab regarding the hair fibers sent to them from the Kirk Stewart incident of 2008. The director's report from the lab that specializes in DNA identification is shown on the opposite page. We are excluding the name of the lab and the director since we've had people contact them in the past requesting information and stating they were NABS personnel. The report is on file and we'll gladly show it at any conference.

December 19, 2008

Mr. Dave Paulides

Dear Mr. Paulides,

Our laboratory has been working on the samples you submitted for species identification over the past few months. The first sample was extracted and when quantified, it also appeared to have viable DNA in the sample. This extraction was our standard mitochondrial hair extraction that normally amplifies fine with all animal hair except feline. When we amplified the sample and tested the amplicon on an agarose gel, there were no results for the sample. We therefore were under the impression that the hair must contain an inhibitor of some type like in felines. With that in mind, we performed a forensic extraction designed to purify any DNA and concentrate it, thus removing any inhibitors. This extraction also failed during testing.

We then sent the hair to the hair expert at the ███████████████████████████ ████████████ to determine if the hair was 1) real hair from an animal 2) if the hair had been tanned or tampered with and 3) if any type of prospective species ID could be made on the hair. We submitted horse mane hair along as a control as the unknown hair more closely resembled the size of horse mane hair. The analyst did this as a favor for me personally and in no way did this in the capacity of an analyst for the ████████ Using the horse hair and a large set of non-human standards, he was able to determine that the hair was 1) real hair from an animal 2) that the hair was not chemically altered and 3) the hair was primate in origin and that no other species of animal resembled the microscopic appearance of the unknown hair other than primate.

With this in mind, we are now of the conclusion that the unknown hair submitted to our laboratory has an inhibitor level much higher than previously seen in hair in other species. We also believe that further testing is warranted in light of the hair analyst's findings. As a result of these findings, we are in the process of using novel extraction protocols and kits to hopefully achieve an absolutely pure DNA sample from the hair. Once we achieve amplification of the mitochondrial DNA from the hair, we will use several sets of DNA primers to attempt species identification. We also will use human primers to check for any cross amplification as occurs sometimes in primates. Assuming we find an unknown species and are successful with both the amplification and sequencing, then we will attempt to place the mitochondrial sequence into the primate phylogenetic tree. A phylogenetic tree or evolutionary tree is a tree showing the evolutionary relationships among various biological species and their relatedness and time of emergence.

If you have any questions or comments concerning our strategy for completing this case, please feel free to contact us.

Sincerely,

The lab report stating the hair sample was from
an unknown primate.

On March 13, 2009 NABS received additional information from the lab regarding our hair sample. They had been having difficulty with the sample because of inhibitors within the hair that were not allowing DNA extraction (this has occurred to many researchers who've attempted extraction). In 2009, new DNA extraction kits became available that were much more advanced than earlier renditions. The lab went to work and started to pull DNA from the hair fiber. We received a call late in the day stating that they were getting preliminary results that showed that the owner of the DNA was possibly a female and that they were getting results along the human line, but one extra peak near the non-human area. They were going to continue to work with the sample and attempt further analysis.

BLAINE JOLLY

When I had asked Jan Wyatt, one of my main contacts in Del Norte County, to forward me the names of all credible people who claimed to have seen a bigfoot, she hit the nail square on with Blaine Jolly.

Blaine is a very humble, quiet, and polite family man who lives in Northern California. I am purposely leaving specifics about Blaine a little murky because of the job he holds as a correctional officer for the State of California at the toughest prison in the state, Pelican Bay. It sits in a serene location just north of Crescent City very near the coast just off Highway One. This is the prison where the state sends its toughest, most abusive, and aggressive inmates, and where inmates are sent who have assaulted other correctional officers at other institutions. This is not the place you want to be a correctional officer unless you want to live on the edge. Blaine does not fit the mold of a Pelican Bay officer.

Jan gave me Blaine's name and address and said that she had called him and he was looking forward to meeting me. I drove to his house and knocked on his door, but nobody answered. I found a garage door opened and walked over to see if anyone was there. It wasn't until I was there several minutes that a family member came by to see who I was and what I wanted. This family is very cautious and sensitive to the type of people that Blaine works around. After a few formalities and introductions, the family member said that Blaine was at the store and I could return in an hour. An hour later I came back to the residence and met Blaine. Again, Blaine does not fit one's idea of that tough, physical, no-nonsense correctional officer; he's just too darn polite!

I immediately liked him. From the start of our conversation I could tell that he wasn't going to embellish, and he wanted to ensure that he got everything absolutely accurate in his story. Blaine started his story by explaining that he was raised in Del Norte County, graduated from Crescent City High School, and enjoyed fishing and hunting. He liked salmon and steelhead fishing in local rivers and always enjoyed going with friends into the wilderness area and hunting for big game deer, elk, and bear. He never went to college and made a living in his younger years as a construction worker. Six years ago Blaine was hired by the Department of Corrections, went through their academy, and requested a position at Pelican Bay. That's right, "asked for a position at Pelican Bay." He said that the prison offered a good wage, great benefits, and steady employment. He likes the work, and the people he works with are "really great people."

Aerial view of Pelican Bay State Prison.

It was in October 1990 when Blaine was working construction that he decided he needed a break from everyday labor. He contacted a local friend, Kirk Stewart, and they decided to go deer hunting. They had made trips into the Siskiyou Wilderness Area several times in the past and always enjoyed the isolation and the challenges that this presented. They drove from the Crescent City area north on Highway 1 and turned northeast on Highway 199 to Hiouchi and on to Gasquet. After passing Gasquet the road gets pretty lonely, as there isn't a lot of traffic. This part of California is very far from population centers, and it takes a determined traveler with a lot of time to get to it. It is a solid seven- to eight-hour drive from San Francisco to reach Gasquet.

Once past Gasquet, the wilderness area is just over the southern hills where you have 25 miles of nothing until you reach the Bluff Creek area. You did read that right: Bluff Creek. The Siskiyou Wilderness Area is the region just north of Bluff Creek, and the next major roadway you hit is Highway 199. Blaine and Kirk took the highway until they reached an area they knew well, the Young's Valley Trailhead. This is one of the most remote locations you can drive into from the northern reaches of Highway 199, an area you have to drive, park, and hike into, where you'd likely never see another person, even on a holiday weekend.

Blaine and Kirk parked their cars, loaded their packs and rifles, and started their six-mile hike into the middle of nowhere. The area they were hiking is rarely utilized by hikers and almost never hunted because of the huge effort to get there. By early in the afternoon they arrived at one of their stops, Sanger Peak on the upper ridge of El Capitan Lake. They were on a ridge that overlooked a bowl that contained El Capitan Lake, 5,600-foot elevation. They were going to rest in this location where there was a good breeze and a great view of the lake, and they put their packs down and walked over to the ridgeline to look down. Blaine explained that the lake is 10-foot deep at its deepest location and is surrounded by dense forest at the base of the ridge.

They looked over the edge of the ridge simultaneously and saw something they couldn't believe—two bigfoot standing fairly far out from the bank, standing on two feet and splashing each other with water. Each creature looked to be black in color, covered with hair or fur, and each was over six feet tall. They were about 250 yards from the creature at their closest point and they decided to get a better look at the creatures by viewing them from the riflescopes. Blaine believes they were able to watch the creatures stand on two feet and use their hands to splash each other with water, like two kids playing in a lake. This went on for five or six minutes.

Blaine said that he and Kirk just watched these creatures in complete amazement, knowing they were probably the only humans for many, many miles. They were both very quiet and hadn't made any noise, but the wind must have shifted. The creatures each stopped splashing at the same time, turned and looked at the two men on the ridge and immediately started to run towards the shore, very similar to the way a human would run. The creatures were running in water that was probably two or three feet deep, but they ran as though the water offered no resistance; it was almost as though they were running on the top of the water. They both made it to the side of the lake and Blaine said that he could hear the creatures thrashing through the brush below, but he never saw them again. Blaine and Kirk agreed that they had just witnessed two bigfoot, and they also agreed that it was one of the strangest experiences of their lives. They put their backpacks on and started a short hike into Young's Valley where they would spend the next few nights.

They spent three days hunting the surrounding area and never saw a buck and never took a shot. During the first two nights they were camped in Young's Valley they were awakened each night by a very loud whooping sound that appeared to be coming from the area on the ridge above them, maybe 100 yards away and almost as though it was coming from the area where they had seen the bigfoot. They had both been in the woods hundreds of times, camped out many, many times, and neither of them had ever heard a sound like those whoops. One night it sounded like something huge had run through the middle of their camp, although they never saw it. It seemed as though something was very angry and didn't want them camping in that area. After being yelled at by a creature they hadn't seen and not knowing what it was, and having not seen a buck anyway, they decided to get out of the area and hike back to their cars.

Blaine said it was a few years after this incident that he was browsing the Internet and came across a site that purportedly had a bigfoot audiotape. He listened to the tape and was surprised to hear that it was almost an exact replica of the sound of the whooping that he and Kirk had heard in Young's Valley. He thought that since they had seen the bigfoot in the lake, maybe they had taken over its territory. Blaine also stated that he has seen many bigfoot videos and photographs since he witnessed the creatures in the lake, and confirmed that the creatures he and Kirk witnessed matched the general description of the bigfoot creatures on many of the video clips that are publicly availably.

Blaine has spent hundreds of hours in the forests and woods of Del Norte County since his bigfoot experience, but has never seen or heard anything like he experienced with Kirk in October 1990.

He has only told a few people of his experience and he doubts most people would believe what he saw and heard.

Blaine can still be found working at Pelican Bay, and he is still hunting and fishing the most remote locations of Del Norte County.

MIKE CUTHBERTSON

There is an old saying, "bad publicity is better than no publicity." While I don't subscribe to that logic, publicity has been a good thing for North America Bigfoot Search. The release of *The Hoopa Project* gave me a venue to describe what the organization does and to ask for further assistance in locating witnesses to bigfoot sightings and encounters. In August 2008 I gave an interview to *The Triplicate* newspaper in Crescent City, California (Del Norte County). Subsequent to the article being published, I received a series of e-mails and phone calls describing various incidents in the county. The most interesting of those contacts was an e-mail from Mike Cuthbertson.

Mike Cuthbertson.

Mike is a retired California Department of Corrections officer who spent a majority of his career at the state's toughest prison, Pelican Bay. Mike has spent a vast majority of his life living in Del Norte County and graduated from Crescent City High School. He had a variety of jobs after high school, working at Safeway in Redding and Los Angeles, and working in post-production video processing in the LA area.

My conversations with Mike were very revealing and productive. I could quickly tell that he was a very intelligent individual who enjoyed the outdoors and recreational activity that his county offered. He told me he had spent many hours roaming the hills of the county and being in some very desolate regions. He explained that he was specifically contacting me about an incident that occurred, to the best of his recollection, in the summer of 2002.

Like police officers working the street, correctional officers

Raspberry Lake.

working a prison have a bond that is rarely found in other professions; they rely on each other for their very lives. Trust is a huge issue. That trust leads to strong relationships that go far beyond the prison walls and usually envelops their personal lives. Many officers feel that it's only their work partners who truly understand what their lives are about and what they face on a daily basis.

With a basic understanding of how officers think, it's very clear why Mike and three other correctional officers decided to go on a backpacking adventure into the northern Siskiyou Wilderness Area. Mike's three friends wanted to climb the highest peak in the region, Preston Peak, 7,309 feet. Mike would go along, hang behind at the camp, and ensure all was accounted for when they returned.

Mike and his friends took the 45-minute ride from Gasquet to the Young's Valley Trailhead and then started the three-hour hike into Raspberry Lake. One of Mike's friends, Rick, had purchased two goats that they were using for backpacking and carrying gear into the wilderness (vehicles, bicycles, and motor-driven cycles are prohibited in a United States wilderness area). Arriving at the lake, the guys found a pristine environment with the lake sitting at the

259

bottom of a giant bowl. Huge cliffs surrounded the lake and kept the region quiet and gorgeous.

The first night was uneventful. Mike's three friends left early the next morning for their push to the summit, and Mike slept. He woke up to find one of the goats had chewed threw its rope and was staring at him in his tent while he slept. He got up, tied the goat back up, and went back to sleep. Several minutes later the same goat had chewed through the rope again and was back staring in the tent. This time Mike emerged slightly irritated, again tied the goat up, and recovered a few energy bars from his backpack for breakfast. It was a warm summer morning and the lake looked enticing, so he took his fishing pole with a few tied flies and went to the shore to try his luck. He was walking towards the lakeshore when he saw a black bear running up the far side of the lake's steep cliffs; he watched in awe. He continued to the shoreline and it wasn't long before the first trout hit the fly and he quickly landed the first course of lunch. Several minutes after landing the first fish, Mike landed several more. After all that hard work he decided to take a dip in the lake. It was a little cold, but refreshing, and felt good after the long hike the day before. He cleaned the fish, got some fresh clothes, and decided to take a hike.

Raspberry Lake sits in a deep basin, very similar to a bowl at the bottom of a volcano, except this bowl is filled with huge trees. There is one stream that empties the lake; it's small, but appears to run year round. Mike decided to walk out of the basin following the stream.

He checked the goats; they were fine. He grabbed his .38 revolver and strapped it on and started walking slightly up from the creek on its right side as it flowed out of the lake. The first 45 minutes of the hike was fairly steep. He had been walking through a grove of old growth pine and Douglas fir and had a very clear view for 100 to 200 feet on the basin floor. Mike guessed that this area had never been logged, as the trees and floor were in perfect condition. He had walked quite a distance to this point and knew he would be walking uphill for the return portion. He walked down to the creek, crossed, and took a position on the opposite side about equal distance to the creek that he had on the trip down.

Just as Mike was taking the first steps back towards the lake he heard a *bahhh, bahhh*—a goat. He immediately thought that a goat had chewed through the ropes again and was following him. The sound was coming from across the creek near the exact location of where he started to cross the creek. The goat sounds were immediately cut off by one of the loudest and longest screams Mike had ever heard in his life. The screaming reminded him of something killing a goat. Mike's thoughts now went to a cougar attacking the goat that was following him; he wasn't sure. The scream was followed by a very deep and loud guttural sound, similar to deep growling, and that was immediately followed by something that almost sounded like a language similar to gibberish. By now Mike was equally confused and scared. He knew he was far enough from camp that there was no way he was hearing the goats from the campsite location, so they must be nearby.

Mike explained to me that he has always loved the outdoors, and knew the sounds, threats, and predators that lurk in the wilderness, yet the sounds he heard that day defied every bit of knowledge he had accumulated in his 45-plus years. He had stopped and was trying to understand what was occurring, trying to rationalize something that had no rational explanation. As he was standing and absorbing the sounds he heard, "Hey, Hey" like a friend calling him; it was bizarre. The "hey" sound was interspersed with more screaming, and then everything went quiet. Every bit of noise normally found in the forest was absent, like a vacuum. There was no sound. Mike was so scared that he raced back towards camp. Still silence.

After our initial conversation and his explanation of the events, Mike sent me a follow-up email clarifying the level of his fear: He had pulled his .38-caliber pistol and immediately thought that this caliber was much too small for whatever was making those sounds; the hair stood up on the back of his neck and on his arms; he said he was as frightened then as he had ever been in any environment. He knew he was alone on the creek with whatever had been yelling at him. He said it was a very confusing scenario because whatever was yelling had mimicked a goat and a human. It was very bizarre. Mike believed that the screams were no more than 100 feet away, but he was never able to see what was doing it. He continued to look around the trees trying to see what might be following him, but he saw nothing. He continued a slow pace back to camp with his pistol in his hand and the thoughts of how useless the .38 would be.

Mike made it back to camp and immediately checked on the goats. Both were still tied up exactly how he had left them. His friends arrived back in camp at 8:00 or 9:00 p.m. and Mike related what happened to him. They checked the vicinity for tracks and found nothing. One of Mike's friends said that he probably ran into an angry bear, but Mike said there was absolutely no way a bear made those sounds. The guys spent an uneventful night and hiked to their cars the next morning.

Once Mike was home he went straight to his gun dealer and bought a .44 magnum. He now carries this whenever he is traveling the woods of Northern California. Mike signed an affidavit.

LOCATION

When Mike was telling me his story, I immediately thought of Blaine Jolly and Kirk Stewart. They were at another lake close to Raspberry Lake in October 1990 when they looked over the crest of the bowl and saw two bigfoot playing in the water. The locations of the events and the circumstances of the sightings and encounters are extremely close. The men knew each other but had never spoken about their incidents. The question that came to my mind was, "How many other people have been backpacking in this region and had encounters?" If someone said that nobody but these guys had an event of this type, I'd say that was near impossible. Neither group had reported their events to the local Forest Service office, but you'd think someone might.

The northern Siskiyou Wilderness Area is the one region of the state that has the least number of visitors. It takes a monumental seven-hour drive to reach Crescent City from the San Francisco Bay Area. It's another 90-minute drive to reach the USFS road that leads to the trailhead, and then another 45 minutes to reach the trailhead. Most people are unwilling to spend one day of their vacation driving to a location.

ALEX ROMMEL

Every time I go into a new county to conduct research, I always start with a periodical and newspaper search at the main library. I did that in Del Norte County, but unfortunately the only microfiche reader in the county was broken, so I drove back to Humboldt County's main library and was able to recover most of the articles related to Del Norte. During that search I came across a newspaper account of a sighting involving Alex Rommel. Through another contact in Hiouchi I was able to track down Alex and take his statement.

The Rommels live in the main trailer park in Hiouchi, which is set in the middle of town. It is probably one of the nicest trailer parks I've ever visited, with the mountains to the north, Highway 199 in front, and the Smith River across the road. The park primarily caters to tourists who are vacationing in the area, but it also has permanent park residents. The mountains to the north have nothing in them other than beauty, some wild animals, and of course bigfoot.

It was a Friday at noon when I was finally able to visit the Rommel residence. Alex opened the door and I explained who I was and asked if he would talk to me about his bigfoot sighting. We stepped into the front yard of the lot on a beautiful spring afternoon and started to discuss what Alex had observed.

Alex stated he actually had two bigfoot sightings, but only one had been highlighted in the newspaper. The first sighting occurred in July 2003, when Alex was 12, near dusk in the trailer park when he was with his friend, Max, from Crescent City. Max was spending the night and the boys were out riding their bikes in the park. They rode to the far northeast corner of the park where there was a small circular drive in a corner with forest on two sides. Alex said he and Max heard leaves breaking up above them in the woods and when they looked up they saw a bigfoot partially hiding behind a small tree. The boys could see its lower body, but its upper body was partially hidden by the branches. The lower body was very stocky, black in color, with hair or fur covering everything they could see. They were approximately 35 feet from the creature when they first saw it, and they initially thought it was a stump because of how

Rommel second sighting in woods.

Rommel trailer park.

large the lower body was. At this point Alex and Max rode quickly back to Alex's house to tell the family.

Alex's mom apparently didn't believe the boys and forced them to eat their dinner. Alex said that they ate in five minutes and rode back to the site with his sister, Sara, who was 15 at the time, and Max's sister, Sonja, also 15. Alex said Sara took binoculars along and had a very good look at the creature. The creature had moved further up into the hill and was by then close to 60 feet away. They all watched the creature for a very short time until it ran parallel with them for approximately 20 feet, and then downhill to their elevation, and then turned and ran off into the forest adjacent to a creek. Alex stated that it had a very large torso, and ran somewhat with a hunched back, but generally ran like a human bent over on two feet. According to Alex, it didn't have much of a neck compared to a human. Alex claimed his sister had the best view of the creature's face, as she had the binoculars for the longest period of time.

At the end of the sighting all of the kids ran back home and told their folks. He said that his sister has had nightmares about the creature since the incident, and sometimes doesn't sleep well.

Alex's second sighting occurred directly behind their residence at the base of the mountains sometime in April or May 2006. He was home alone one day cleaning his room when he found a knife he didn't know he had, so he went into his backyard to see how accurate he could become throwing it into the ground. He thought it was just before his parents got home from work, at about 5:00 or 6:00 p.m.— it was fairly well lit outside—when he heard some branches cracking up the hill behind his trailer. He looked up to see a very large creature stand up from behind a stump and immediately run westbound and out of sight. He said that this was also a bigfoot, as it was covered in hair or fur, was very large, and swung its arms as it ran. He stated that this sighting was much shorter than his first, but he could definitely determine this was also a bigfoot. It was black with a tint of gray, and this time the creature ran much faster than a human. It left the area by going over a barbed wire fence behind their trailer park property line.

Alex said that his father called a bigfoot organization and they sent out an investigator who took a statement and also looked at the fence. He found hair stuck on a barb where the creature went over, but Alex said the investigator threw the hair on the ground and didn't retain or test it. When Alex asked me why the investigator wouldn't have kept the hair, I had no answer. Alex also said that the investigator told them he put gorilla pheromone in the creek area near the first sighting.

I asked if he ever returned and checked the area or recovered the pheromone, and Alex said no. I found Alex to be extremely believable, conscientious as to the facts of his case, and very patient with his explanations.

The Daily Triplicate, a Crescent City newspaper, ran an article on October 20, 2003 about Alex's first sighting. The facts in the newspaper account are identical to what I have described, but the description of the creature is much more detailed here than in the newspaper account.

Alex signed an affidavit outlining the details of each sighting.

LOCATION

The Hiouchi Hamlet Trailer Park is located directly in the middle of the town of Hiouchi. The park faces a southerly direction, towards the Smith River. This entire region has had a history of bigfoot sightings going back many years. The sightings have occurred on each side of the river in a 20-mile radius. The Rommel trailer is the only residence that sits far up into the hillside, as the others all sit at the base of the park on flat land. The location of Alex' first sighting was also at the base of the mountain, but at the opposite end from his family's trailer. This entire region around the trailer park has a very large amount of vehicular and pedestrian traffic. Directly next door to the park is the lone market and gas station for the town.

JENNIFER CROCKETT

Witnesses to bigfoot come to me from a variety of sources. Even after years of interviewing, I am amazed how the witnesses find each other and get to the point of talking about bigfoot. I believe that bigfoot is a bigger part of people's lives on the north coast than many realize.

One of the many questions I ask when interviewing a witness is if they know of anyone else who has seen bigfoot. When I asked Kirk and Susan Stewart this question, they said I should talk with Jennifer Crockett. They knew her as the woman who had owned the coffee trailer in front of the Hiouchi Market. She had sold the trailer, but they gave me the name of the new owner to contact for help in tracking her down. The coffee trailer was closed, so I asked the people at the store for help, and through their assistance I was eventually able to contact her father, Barry Brown. I explained my mission, and he said he would let her know and ask her to get in touch with me. Jennifer had recently given birth to twin girls and she often left her phone off the hook so they could sleep. Shortly after my conversation with Barry, Jennifer called.

Jennifer Crockett.

Jennifer is 36 years old and grew up in Humboldt and Del Norte Counties for the majority of her life. She graduated from Eureka High School and later attended The College of the Redwoods. She operated the coffee trailer in front of the Hiouchi Market for several years before selling it.

Jennifer said that it was either December 2005 or January 2006 at approximately 8:00 p.m. when she was driving westbound from Hiouchi on Highway 199 to see her fiancé. She came to the fork in the road with Highway 197 and turned westbound on 197. It was somewhat light outside and a bit foggy as she was approaching 3700 Highway 197. She saw a fairly tall figure, approximately six to six-and-a-half feet, thin in build, but with fur or hair over its entire body.

It was kneeling in the lane in front of the house. She slowed to a stop and turned on her high beams and saw that the creature was light brown in color, ragged (possibly wet), and thin looking. It stood up after a few seconds and walked on two legs to the south side of the road (opposite lane). The creature continued walking until it was near the ditch in the opposite lane. Jennifer moved ahead at a very slow pace, and as she passed the creature, made eye-to-eye contact, and then looked at it in her rear view mirror. She saw the creature turn and look toward her vehicle and then jump the fence into the field on the south side of the road. Jennifer continued on to her destination.

She told me that when she made eye contact, she felt a sense of calm about what she was looking at. She knew the creature was not going to harm her and that she was in a good presence. She said she knows that the creature wasn't human, but it appeared very close to human. She was as close as 10 feet from the creature.

Jennifer said the creature walked much like a human, but had the appearance of a bigfoot, based on the videos and photos she has seen of bigfoot on television, except that this creature had shorter legs and was thinner than most bigfoot depicted on television. She

Berry bishes along roadway near 3700 Highway 197.

also stated that she had never heard of a bigfoot that was the color that she observed, almost golden.

After taking Jennifer's initial statement I drove to 3700 Highway 197. As I drove up towards the residence, I actually got a smile on my face. I hadn't seen so many berry bushes lining the roadway since I last left Hoopa. There were huge berry bushes surrounding the field opposite. I drove up the long driveway to the house and contacted the residents. They are owners of one of the largest private businesses in the county and were extremely cordial. They said they have lived on the property for 33 years and have never had any type of bigfoot incident. They have seen mountain lions, deer, and elk in their yard. They claimed to have had only one strange occurrence that they could not explain. They had a ground rodent problem and placed a trap in a hole covered by plywood. They returned days later to find the wood moved and the trap gone. They never found the trap, but their rodent problem subsided. Could bigfoot have taken the trap with the rodent captured inside? Maybe bigfoot had lunch?

I wrote up Jennifer's statement and met her the following day at her house. She reviewed the statement and signed an affidavit. We accomplished this without waking the twins.

The Location

3700 Highway 197 is located approximately half way between the intersection with Highway 199 and the intersection with Highway 101 adjacent to the Smith River. The river is located to the side of the vacant field where the bigfoot went after being seen by Jennifer. Behind the residence at 3700 Highway 197 is a vast mountain chain that goes on for miles and continues into Curry County, Oregon, where they also have bigfoot sightings. Jennifer's sighting parallels many others I have taken in the past. In my book, *The Hoopa Project,* there was a similar sighting near berry bushes on a roadway with the creature escaping through a field. When I took Jennifer's statement my first book had not yet been released.

It would appear from the sightings that have been reported

269

along Highway 197 and Highway 199 that bigfoot is living in the mountains surrounding the valley. It has yet to be determined if the creatures are nomadic in the region or live permanently in the mountains. There is not enough vehicular traffic or hunter/tourist traffic in the hills to determine the volume of bigfoot activity over a calendar year. This area gets heavy rainfall in the winter with some slight amount of snow on the valley and slightly greater amounts on the hilltops.

FORENSIC SKETCH

Several months after I initially met Jennifer, Harvey Pratt and his wife, Gina, made a second trip to California to meet with bigfoot witnesses. Jennifer agreed to meet at 1:00 p.m. on a Thursday in April 2008. We met at her residence in Hiouchi on a beautiful clear and warm day, and sat in her backyard.

Jennifer & Harvey.

Jennifer was apologetic about the time span between her sighting and the date of the meeting, and was slightly afraid that some specific aspects of the sighting might have faded from her memory. Harvey told her that he had done sketches when the sightings were over 25 years old, so two years was really nothing. He gave her subtle encouragement and said he would help her memory through a series of questions that stimulated thought.

After Harvey's initial questions, he started to draw a figure that was standing and turning slightly back to look at Jennifer. She told Harvey that the creature had to turn its entire upper body, as it didn't appear to have a large neck, and that it looked thin and underdeveloped compared to larger creatures she had seen on bigfoot specials, She also said she had never heard of a bigfoot of the color she saw, which

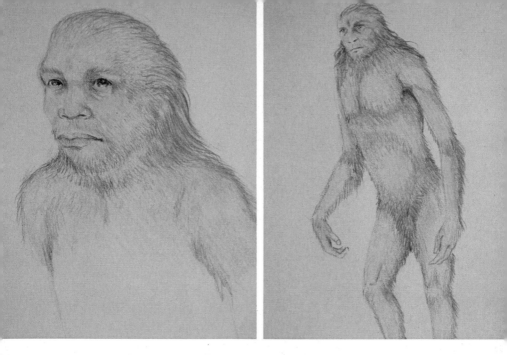

Jennifer Crockett bigfoot sketches—portrait and body.

was close to a darker golden retriever dog, and that it looked much more human than animal.

Note: It's interesting that during Harvey's first trip to California we had never seen or heard of anyone describing the creature in the colors Jennifer used. On this trip, however, Jennifer was the second witness to use the same tone of gold to describe the creature; Marlette Jackson (see chapter 10 for her report) also described the same color. Jennifer and Marlette do not know each other and live 90 miles apart.

After 75 minutes of drawing, changing, and sketching from Jennifer's descriptors, Harvey had put her creature on paper. Jennifer looked at the drawing and then looked at me at the other end of the table and stated, "How did he do that, this is remarkable, what a drawing." She stated that the creature Harvey drew is what she saw that night on Highway 197. She explained that she drives that road almost every day, and each time she thinks of that creature. Jennifer thanked us for the time and reiterated that the sketch was a great likeness of what she had seen on the road.

MELISSA HENRY

Part of my investigation into Jennifer's sighting included knocking on doors in the neighborhood. I always like to get a feeling for what has occurred in any region and the best way is to engage locals in conversation and understand what they've been experiencing. One of the residences in the area was that of Melissa Henry. She and her family have lived in the area for several years and actually have had some strange occurrences. Melissa asked her brother to come outside and they both spoke to me about their event.

Melissa Henry.

Melissa explained that it happened in the winter months of 2006; she can recall that it was cold. The windows were closed and it was very late, maybe 2:00 a.m. She was awakened by the loudest scream she had ever heard in her life, initially one long scream followed by another very loud, long scream. She was so scared she called to her brother in the next room and asked if he had heard it; he had. They stated it was very frightening to hear anything scream that loud for that long. They felt that it was coming from a nearby ridge where there is now a quarry. They both told me they had no idea what could make such a loud scream for such a prolonged period of time. I asked them if they had heard the scream ever again, and they both stated they had not.

It is important to note here that the scream and Jennifer Crockett's sighting were in close proximity. The ridge mentioned is down the road from the Henry residence and extends from inland to very near the Smith River. There is a large quarry at the point where the ridge hits Highway 197.

I checked bigfoot databases on this region to see if there were any sightings or incidents for this area, and found one on Highway 197 one mile west of Highway 199 in a small park called Simpson Grove. It occurred over 25 years ago, in 1979, along the Smith River in a redwood grove. The four witnesses were sitting around a campfire when they heard a very loud rustling in the bushes adjacent to them. They turned to see what was making the noise and they observed "red glowing eyes, actually glowing and not reflecting." It watched the people for approximately one minute and then left. A description of the creature was difficult to obtain because of the darkness, but they estimated it was eight feet tall, based on the height of the eyes. The creature left the same way it arrived, crashing through the brush. (BFRO Report, Del Norte County).

This report indicates that bigfoot has been visiting this area for many years. The difficult part of the bigfoot phenomena is that it appears in one area for a few years and then disappears for 20 years—no reports, no activity, nothing. There was a long period of inactivity in this region, and then there was a small blip of activity around Melissa's scream and Jennifer's sighting.

TRAVIS COVER

Travis is a 37-year-old truck driver who hauls milled lumber from Brookings to Grants Pass in Oregon. He has spent the majority of his life in Brookings, graduating from Brookings High School, but has also lived in Medford and Grants Pass. He is divorced and does not have children, and lives near his parents in a house near the mountains of Brookings. He has always enjoyed the outdoors, and as a youngster always camped and hiked in the mountains of California and Oregon with his family. Since Brookings

Travis Cover.

is the first Oregon city you hit after crossing the California border at the coast, it's natural that Travis and his family became very well acquainted with Northern California.

In my first book I did a search of all bigfoot sightings in every California County. The county with the greatest per capita number

273

of sightings was Del Norte, the California County immediately adjacent to the Oregon border and Brookings. The quickest roadway from Brookings to the interior of Oregon is Highway 199 from Highway 101 just south of Brookings. This is the route most residents of Oregon must take to get to Grants Pass, Medford, and Cave Junction. This is also the route that Travis's parents took to go camping when he was younger.

In the summer of 1982 when Travis was 12 years old, his parents decided to spend the weekend camping. The family had gone to the Siskiyou Forks campground many times and had always enjoyed the serenity, big trees, large creek, and general lack of people. The campground is located approximately five miles north of the Patrick Creek Lodge on Highway 199.

In the morning on the second day camping, Travis was riding his banana-seat bicycle downhill from the campsite when he heard rustling in the bushes to his left. Travis slowly applied the brakes, looked into the foliage, and saw a creature standing on two feet spreading the limbs of the trees with its hands. Through the branches he could almost see a human-type of face, but he could also tell that it wasn't quite human. The creature had hair over its entire body (excluding part of its face), was seven to eight feet tall, and had a massive chest and waist. Travis could also see hair on the backs of its fingers, but did not smell anything. He didn't see any hair on the forehead area, but there was hair on the head that sometimes flowed onto the forehead. He got so close he could actually see wrinkles on the face just under the eyes. Its eyes were almost black, and it generally looked more human than animal. Travis estimated that he got within 15 to 20 feet of the beast.

As the creature pushed its way through the brush, it appeared to a young Travis that it didn't want anything to hit its face, as it kept that area of its body clear of leaves. The creature was almost down to the bottom of the hillside before Travis made a u-turn and went directly back to his parent's campsite. He claims that he told his parents about the sighting, but they laughed at him.

I could tell this sighting had a profound effect on Travis, and that it was something he would truly never forget. He said he has a distinct recollection of that entire day.

My planned meeting with Travis centered on a story I had heard about his days of driving a lumber truck on Highway 199. This incident occurred on September 22, 2005 at 4:30 a.m., 23 years after his first sighting. Travis was driving a big rig truck for South Coast Lumber and was just pulling out of the California Transportation vehicle yard that sits on the north side of Highway 199, heading northbound. It was still dark outside when Travis turned a corner and saw a huge creature standing on the right shoulder of the road adjacent to a road sign. The lights of the big rig illuminated the creature and Travis immediately hit the high beams. The bigfoot instantly reacted by putting its arm up in front of its face as though to shield its eyes from the light of the truck.

Note: I have heard this exact behavior on multiple roadway sightings. It is obvious that the creature does not appreciate headlights illuminating it in darkness.

As the creature lifted its arm Travis saw hair several inches long hanging from the forearm. The truck came within five to 10 feet of the bigfoot as it stood on the shoulder, and Travis got a good view of it and its stature. He didn't get a great view of its face, but could tell it was different than the animal seen in the Patterson–Gimlin movie in that this creature didn't have a lot of hair on its face and it looked more human than animal. He made it clear that this creature was not human, and based on its size—almost eight feet tall—huge shoulders, and hair covering its entire body.

Immediately after passing the creature Travis pulled to the side of the road and dialed 911 to report the sighting. He initially got Oregon Highway Patrol and was transferred to the California Highway Patrol. He was told they received the report and would forward it to their patrol units. Travis put the rig back into gear and continued his trip. On his return to Brookings he stopped at the location of the sighting and found large impressions in the ground, but no distinct tracks worth casting.

LOCATION

If Del Norte County has the largest per capita number of bigfoot sightings in California, then Highway 199 has to be the "bigfoot freeway" of Del Norte County. To put Travis's Highway 199 sighting into perspective, I will place it in relationship to other events in the area:

Cave Junction: Prints found in sand—14 miles away
Patrick Creek Lodge: Yelling heard—3 miles away
Patrick Creek Campground: Tracks seen—3 miles away
French Hill Road: Sighting—10 miles away
Kirk Stewart: Ranch incidents—9 miles away
Kirk Stewart: Lake sighting—7 miles away
Siskiyou Forks: Travis Cover camping sighting—2 miles away
Jennifer Crockett: Roadway sighting—18 miles away
James Renae: Forest sighting—16 miles away

The location of each of Travis's sightings is just outside the northern fringe of the Siskiyou Wilderness Area. This wilderness area gets very little tourist traffic because of its remote location from every population center in Oregon and California. The region north of Highway 199 is nothing but open wilderness for a hundred miles into Oregon. This region is vast and open, with few people and lots of precipitation. The Bluff Creek site is approximately 20 miles south across the wilderness area and on the south side of Del Norte County.

Harvey and Gina Pratt and I drove to the location of Travis's sighting on Highway 199. One thing was immediately evident about this exact spot: two ridges, one from each side of the roadway, converged on this exact spot. As I have stated on multiple occasions, the Yuroks believe that bigfoot is a ridgewalker, and with two ridges converging on Highway 199, this is one reason that this spot was a sighting location. On the north side of the roadway we spotted a very heavily used game trail that disappeared at a very wide opening at the summit of the road cutaway. This was much wider than most game trails in the area.

FORENSIC SKETCH

Several months after my first meeting with Travis, forensic sketch artist Harvey Pratt met with Travis at a restaurant in Smith River where I briefed Harvey on the two sightings.

The first sketch Harvey started was the creature that Travis saw on Highway 199. Since Travis hadn't got a great look at the face, it was decided that Harvey would actually draw the creature's hand up blocking the light from the truck. What was captured was a massive bigfoot with a large barrel chest, six- to seven-inch long hair hanging from the forearm and long hair flowing from the back of the head. The creature was mainly dark in color, yet the forearm and part of the upper arm was very light and almost golden in color. Travis explained that the creature was so tall it almost reached the window height of his big rig, and the lights of the truck illuminated its upper body, so he couldn't specifically remember looking lower than the mid-section.

Travis & Harvey.

The second sketch, very similar to the first, was of the creature moving branches away from its face. It was interesting that Travis specifically remembered the bigfoot had hair on the back of its hand, a fact never confirmed before now. There weren't a lot of details revealed by the sketch, except that it's obvious this wasn't a human—too large and too hairy.

The session ended with Travis thanking us for conducting a professional and thorough investigation of his sighting, and saying that the drawings were exactly like he remembered the creatures on the days he made his observations.

As we were driving away from our meeting, Harvey made the comment that he felt Travis was very credible and that the sighting he had as a child still had a profound impact on his memory. It was generally felt that what was drawn is what Travis had observed.

An interesting fact about each of Travis's sketches was the arm

Travis Cover bigfoot seen on Highway 199.

Travis Cover bigfoot observed at campground.

of the hominid coming up at an angle where the forearm is parallel to the roadway and the hand is in front of the eyes. There are several witnesses I have spoken with who have described this exact reaction when light is shined in the bigfoot eyes. This little known fact about bigfoot behavior would only be known by someone who had shined a light in its eyes and seen the response. There is absolutely no way that any witness would have learned about this specific behavior without having observed it, since neither our first book, *The Hoopa Project,* nor the sketches in it, were released prior to the second round of drawings. It's also important to note that a bigfoot's eyes obviously do not respond quickly to a change in illumination, thus the sensitivity.

JAMES RENAE

One of the nicest people I have ever met is Jan Wyatt of Gasquet (story above). Jan knew a lot about local bigfoot stories, but she also told me that her son, James Renae (aka Wyatt), had some interesting encounters and said I should contact him. She explained that he was on disability, living in Crescent City.

I arranged to meet James at a local McDonald's restaurant where I hoped the Happy Meals would lighten the mood of the gloomy winter day. James, wearing a heavy coat and cap, was sitting in the corner, looking very serious. I walked up and introduced myself and immediately had the feeling that James was judging my credibility and integrity. He started to question me about my expertise with regard to bigfoot, its physical description, habits, sighting locations, etc. I wasn't offended, in fact it was almost a relief as I could tell that this man knew more about bigfoot than almost any witness I had met. One of the first questions James asked me is what I thought the face of bigfoot looked like. I told him that I'd never seen the creature, but I had an excellent idea based on the many witnesses I've interviewed. He immediately blurted, "It doesn't look at all like the Patterson–Gimlin film, and it looks a lot more human." At this point I knew James was going to be a good witness.

He began by telling me that he and his brother had seen a bigfoot cross in front of their vehicle while they were traveling on French Hill Road above Gasquet. It was a quick encounter and he never got a great look at it, but he saw the creature run by on two feet, covered in dark hair, and moving very quickly. He said from that point he was hooked on trying to find out everything he could about bigfoot.

James is one of those interesting people that you only come across occasionally. They seem to have a unique knack for finding things that others would never see. In my first book, Inker McCovey

James Renae.

was another of those people who had the ability to always be looking for, observing, and understanding what isn't normal in the woods.

Between May 19 and June 20, 1984 James was hired by an individual from Southern California to walk into the Bluff and Blue Creek drainage area and conduct a formal bigfoot search, with the goal of eventually capturing one. They thought they were well stocked with food and supplies, and they also felt they understood the landscape. Well, they didn't. They almost starved during their first outing as there wasn't the food source present that they thought was there. They hiked 10 hours into the high country and saw a few bigfoot tracks during their first few days, but not much the following days. They eventually walked out, got restocked, and headed back in near June. James said there was less snow present on the second trip and it was a little warmer. Numerous times during the following 10 days he and his partner became separated and spent the day searching the area alone. It was on one of those days alone that James saw something that can only be described as amazing.

James said he was seated in one spot for a very long time just outside the Blue Creek area. He spent the time scanning the valley, trees, and ridges, attempting to make little or no noise. At one point he turned to the left and saw a large female bigfoot with a two- to three-year-old child sitting on her hip. He was absolutely astonished, and he and the creature just stared at each other for several seconds.

James felt that the hominid was as surprised to see him as he was to see it. They were approximately 25 feet from each other and he could clearly see what he thought was a small bigfoot on her right hip. The larger bigfoot had breasts and was a medium shade of brown, similar in color to the smaller creature. James described the attitude of the larger creature as relaxed, and she appeared to be chewing something calmly as she stared. James thought she stood about seven to eight feet tall and would have weighed close to 700 pounds.

They stared at each other for several seconds until in one quick second the smaller bigfoot was flung onto the female's back and she leaped over a huge snow mound and was gone in an instant. It was absolutely amazing how quickly the incident was over and the creature disappeared. James found some partial tracks in the area, but none that were excellent. He couldn't find his partner until several hours later and the creature was completely gone from the area.

The partnership and the expedition between James and his friend lasted until the end of June 1984, at which point James stated he had enough. He backed out of any future endeavors and decided to stay near home for the remainder of the year.

In approximately 1986 James was hired by a San Francisco Bay Area college professor to be the caretaker of 83 acres he owned just outside the hamlet of Hiouchi. The site was nearly surrounded by state park property, and the only trail into the location was through the state park. The property is accessed through Stout Grove in the Jedediah Smith Park. The walk in starts near the Smith River and meanders up through giant old growth redwoods. Some of these trees are over 2,000 years old and reach over 300 feet. The ground in this area is almost always damp with a large number of ferns growing on the floor. From the parking lot to the cabin takes a brisk 30-minute hike.

During James's 20 years of making the hike to the property, he had several bigfoot sightings. He rarely encountered park visitors while hiking, as the trailhead is on the opposite side of the river from the main highway. The trail has a big parking lot, but it is somewhat isolated from main traffic routes. There are several residences in the general area next to the river.

James explained that the owner allowed him to live at the cabin while he was teaching. James would take friends up to the site and spend the night. In 1998 or 1999 James took his friend, Vince, with him to the property. Vince had never seen huge redwoods, and since he was from Missouri he wasn't used to the dense fog and clouds that are normal along the Pacific Coast. This specific location is approximately 12 miles from the Pacific Ocean. It was in the time frame of September to November, and near dusk, when they started their hike. They were approximately one and a half miles from the cabin when they started to hear rustling in the trees and bushes near their perimeter. James said he had heard this before and thought it might be a bigfoot, except there were several locations around them that were making sounds. James stated it was a little unsettling. They continued on the trail and could still see very clearly. Approximately 150 feet in front of them a bigfoot stood up just off the trail and stared at them. Then they started to hear sounds in a 360-degree circle around them, as though they were surrounded, and then they both saw six bigfoot encircling them. The bigfoot that was standing near the trail moved slightly more into the forests, and this gave James and Vince an avenue to walk off. They slowly and calmly walked straight towards the cabin and past the bigfoot they had just seen. Once they broke the perimeter they made a quick trip to the cabin and locked the doors. Vince was scared out of his mind and couldn't believe what he had just seen. James tried to keep a calm demeanor, but he couldn't believe that he had just seen six of the creatures at one time.

The description James gave of the bigfoot is the intriguing part of this story. He stated they did not appear anything like the Patterson–Gimlin footage, but looked more human than animal, almost a cross between a human and ape except their face wasn't flat and dark. The facial tone looked to be grey or tan, and they did have some type of hair or beard growing below their chin. He also said they had hair over almost their entire bodies and they had massive chests. James did not hear them make any sounds and they did not have a detectable odor.

Vince did not sleep at all during his night at the cabin. Very early the next morning Vince awakened James and demanded to be escorted off the mountain. They made the walk down to the parking lot without seeing anything unusual, and that was the last time James saw Vince. He is believed to be living an urban life somewhere in Missouri.

James maintains he has had continuous bigfoot incidents through-

282

out his 20 years visiting the cabin. He believes that the trail and the cabin cross several main game trails that run from the wilderness area to the coast and thus supply food for the bigfoot. The creature stays near the trails and the food source. James explained that the trail he described is an elk path that runs from Elk Valley, Orick, and along the Smith River. Depending on the season, the elk can be in any one of many locations in that area. It should be noted that anyone who thinks these forests cannot sustain a large animal should look at a Roosevelt elk. These animals can weigh over 400 pounds and they are throughout the county.

Roosevelt elk.

James's most recent bigfoot encounter occurred approximately 18 months ago, in November 2005. He was walking alone to the cabin at approximately 4:00 p.m. when he started to hear something walking in the woods just off to his side. Parts of these forests are extremely thick and you can't see 20 feet. In other parts where there are redwoods, you can readily see 300 feet. The location James was in was very thick, with poor visibility through the foliage. He stopped on the trail and watched as 20 feet in front of him a huge gray-colored bigfoot slowly crossed the path. The bigfoot turned to look at him, but continued at its pace as it made its way back into foliage. James explained that this was a quick sighting, but it was unusual as the color of the hair on the bigfoot wasn't normal for what he has seen. The normal hair color is a reddish hue, similar to the bark on the redwood. This bigfoot was a tone of gray.

At this point James started to feel comfortable and related other stories he had heard and some he had seen. He feels that bigfoot lives in the canopies of the old growth redwoods. He explained that you could follow the trail to the cabin and see several older redwoods where there are wear marks on the bark of the trees where the bigfoot climbs. If you look up into the trees, there is a large canopy 200 to 300 feet up that inhibits you from seeing what's at the top. He said that living in the trees allows them a place to ambush predators and a location to watch for humans.

James pointed out the exact spot of the cabin and property on a U.S. Geological Survey map. The property is nearly surrounded by California State Park and the region around it is very well maintained. The park property immediately across the river from this location has some historical bigfoot value. There was a film crew in that location several years back filming a commercial when they allegedly saw a bigfoot walking through a park. There was also a sighting by two kids in a nearby trailer park of a bigfoot watching them from the forests (see Alex Rommel report above). The area of French Hill Road where James and his brother spotted a bigfoot is located approximately 12 miles upstream adjacent to the Smith River and is another location where there have been several sightings.

In February 2007, I made the trip on the Redwood Trail up to the cabin where James was the caretaker. James warned me that his friend had recently died and the cabin and property were in probate. The court process put the property in a state of legal flux. James said that he no longer had the legal right to maintain the property, so he moved out and squatters had moved in. James felt that the new residents were probably doing something illegal. This area of California is known for illegal marijuana cultivation, and hikers are routinely warned to be extremely careful about walking away from main trails.

The trail location starts and stays inside a California State Park, which doesn't allow firearms and that made me a little nervous. You cannot carry or possess a firearm in any California State Park. I normally carry a firearm any time I am in the woods so I am always protected against whatever may want to eat me.

I was approximately 20 minutes into the hike when I noticed several huge redwood trees that looked quite different than their surrounding "friends." These trees appeared to have bark that was much smoother than the others. It almost appeared as though the bark had been rubbed smooth. I made the 50-yard walk off the main path to the base of the tree that was smooth. At the base of the tree I found soil that appeared to have been stomped firm, almost as though something had jumped on the ground from a higher plateau. The ground outside the immediate base was much softer, filled with leaves and needles, and unable to hold a print. I spent considerable time in this area and was unable to locate any prints. I found this spot to be highly unusual.

I continued up the path and started to climb in elevation. The trail narrowed considerably and I got to the spot where James had described seeing the gray bigfoot. I stopped for several minutes and, unbelievably, heard something walking down the path toward me. Remember, I hadn't seen anyone on the trail or in the parking lot for two hours. I held my ground and watched as a derelict looking hippie walked up to me and asked what I was doing there. I was a little miffed, but I couldn't tell what kind of weapon he might have had under his tie-died sweater and coat. He wasn't a large man, but he smelled and looked like he'd been in the woods too long. I politely told him that I was taking a nature walk and was interested in the fauna. He firmly told me to stay on the trails as there were dangerous animals in these woods. I told him that I would, and thanked him for the warning. It was a little nerve wracking having that encounter unarmed. I have always questioned how he knew I was on the trail, or was it pure luck that we ran into each other.

I continued to walk towards James's property and reached the point where it would be located. There were handwritten signs along the trail to stay on the path, no trespassing, violators will be shot, etc, etc. I took the warnings and stayed on the trail and was a little nervous of the hippie encounter. I walked approximately half a mile past the property and near the top of an adjacent ridge. This all appeared to me to be optimum bigfoot country. One part of me wanted to go back to the hippie and ask him about bigfoot.

On my way back down the hill I found two other redwoods that had the characteristics of something climbing up its sides. The bark was unusually smooth on both of these trees and the base ground was very, very compact. I spent considerable time looking for prints in the area and I even got out my magnifying glass and spent over an hour at each tree looking for hair fibers, but no luck. When I say that these trees are huge, they are really gigantic. They have been called the largest living things in the world, and sometimes the oldest living things in the world.

Several witnesses have told me over the years that bigfoot lives in the trees. This incident probably adds more credence to this allegation than any other I've investigated. If bigfoot did weigh 800 to 1000 pounds, then it would be these trees that could support that weight. If bigfoot has been in North America for hundreds or thousands of years, it would be these types of forests that could maintain the population in secrecy. While bigfoot hunters and researchers are looking in caves and dens, bigfoot may be living at the top of the trees laughing at their efforts below. If we think back to the famous Patterson–Gimlin incident and what prompted them to go to Bluff Creek in the first place, it was the tree cutters and road builders. Could it be possible that loggers who were clearing a path for the Go Road out of Orleans accidentally cut down a tree that was harboring a family of bigfoot in their canopy? Could that have been what got bigfoot angry when it threw oil drums over cliffs at the construction site? That theory may have some relevance.

FORENSIC SKETCH

It was an absolutely beautiful day in May 2008 when Harvey, Gina, and I made our way to James's residence in Crescent City. He lived near the Pacific Ocean, so we sneaked out and caught a glimpse of a calm and enduring sea. It was actually one of the warmest days I had ever seen in the city and when we pulled up to his house it was probably near 80 degrees. He invited us inside and I introduced Harvey and Gina. James seemed excited to get started.

I explained to Harvey the different scenarios where James had observed a bigfoot and then let Harvey question James about what he would feel comfortable sketching. After a few minutes of discussing each incident, they came to the conclusion that Harvey would draw the incident at Blue Creek where James had seen a mother and baby bigfoot. This was something that nobody on our team had documented, and we had never met anyone who had seen something even similar.

As someone who had been to the Bluff and Blue Creek area dozens of times, I understood the geography as James walked Harvey into the incident. This is tough country and parts of the regions are very remote and not friendly to visitors. Of the times I have been to those areas, I have only seen deer on one trip, not a normal count considering how remote the area is and the amount of food source available. I have seen dozens of bears, some as big as 600 pounds. The area where James made his observation is also considered the sacred high country to the Native Americans from the area, and as such is considered to have religious significance.

James told Harvey how the creature just appeared to him as he was sitting in an area for a very long time. The creature didn't move for several seconds and he got a great look at both creatures from 25 feet away. James specifically stated that the larger creature looked much more human than animal, never really showed any emotion of fear, and stood watching him without moving, just chewing something. When the creature moved suddenly and threw the younger hominid on its back (similar to a piggy back carry), they were both gone in an instant.

Harvey sketched a little, asked questions about color and size, and then paused. During each pause James would add a little significance to the story, the size of the smaller creature relative to the mother. He thought that the smaller bigfoot was near two years old, based on his knowledge of a human baby's growth pattern. After more than an hour of constant conversation and drawing, Harvey's sketch started to come to life. The first glimpse that James had of Harvey's work showed he was excited. You could hear the tone in his voice change, as though he was mildly shocked how Harvey had interpreted his words and placed them in a sketch. In the end Harvey had completed a sketch of a female adult bigfoot holding a baby bigfoot on its right hip. The face of the female was just as James described, more human than animal but not exactly beautiful, at least in human terms. In fact, the face looked more male than female, but the breasts are a giveaway that this was a female.

James Renae sketch of female bigfoot holding baby on her right hip.

287

During our 90 minutes with James and working the sketch, his story stayed consistent, and he was calm and collected until he saw the nearly completed sketch. James stated that the sketch was excellent and a very good interpretation of what he saw that spring day in 1984.

SIGNIFICANCE

The importance of this sketch is huge. It does show that a female bigfoot does care for its offspring at least to a point where they are as large as described by James. The sighting sheds light on the manner a female cares for its young and also the way it carries its young when leaving an area in an urgent manner, swinging it onto its back. This sketch also confirms that there was breeding occurring in the Blue/Bluff Creek area during the early 1980s, which is important in that the creatures still felt that it was a comfortable and safe place to live. Bluff Creek was the first location where an actual bigfoot was filmed in 1967, and Blue Creek is just one valley north, very close in proximity. It's also important to note that James specifically did not assist in the sketch of a creature that had hair over its entire face, as did the creature filmed in 1967. Again, an important point because anyone who wanted to fabricate a believable story would obviously describe a hominid that matched the creature in the 1967 film. James's facial description is consistent with 95 percent of the sketches we have completed, and is very different than the creature in the 1967 footage. *This number continues to rise as NABS completes more forensic sketches.* James was never privy to our first-round sketches and nobody ever showed them to him prior to his meeting with Harvey, adding even more credibility to his story.

RICHARD GEHR

During one of my trips to Gasquet, I stopped to see a friend who has always supplied significant detail about the community. On this trip she told me she had heard of an individual in her housing area who

Richard Gehr standing next to blind.

had a bigfoot sighting several years prior. After a few attempts, I met the individual and he told me he would be glad to relate his story, but the person I really wanted to talk with was Richard Gehr, and he would attempt to get him to Gasquet. Two days later I met Richard.

Richard Gehr is 56 years old and employed by the California Department of Corrections as a correctional officer at the toughest prison in the state, Pelican Bay, the prison is where the corrections department sends some of their most at risk inmates. He was raised in the foothills of the Sierra Nevada Mountains, Roseville, graduating from Oakmont High School, and attended Sierra and American River College where he studied civil engineering.

In the early 1990s Richard had a part-time contractor working for him doing odd jobs. The contractor lived off Hunter Creek Road near where the Klamath River empties into the Pacific Ocean. One day he came to Richard and told him that his kids had seen some huge, human-looking tracks in the area of their residence. Richard went to the residence and walked out to a small field neat the base of a hill where he saw several large prints that were too big for human footprints. They walked up a hill just to the south of the residence and into a redwood grove. Richard saw a bigfoot 50 to 60 feet from them, somewhat concealed behind a log. The creature had dark inset eyes and honey-oak color hair covering its entire body. There was a small amount of hair under its eyes that got thicker as it got near the chin. There was thick hair from the chest up and the creature's head almost appeared to have a cone shape, which Richard feels could have been caused by the shape of the head or by the hair on the back of the head. The contractor was 15 to 20 feet behind, so Richard turned towards him to alert him to what was ahead. When Richard turned back, the creature was gone. The creature had been very near the forest line and it could have easily disappeared into the woods.

Richard and his friend went further into the woods that afternoon and found bigfoot tracks in a redwood grove further up the hillside. Richard also found hair in the area of the tracks, which they confiscated and later sent to the Academy of Sciences (Dr. David Bernal). They received a report back that it was classified as "an unknown hominid." The two guys made many trips into this region over the following months and years, and each time they would find footprints varying in size, shape, and depth.

On one of the trips to the hillside, Richard found a small shelter made out of twigs and branches and shaped into a small dome. It was not big enough for even a human to lie down inside, but when he entered it he found two sets of prints, a large set near the perimeter wall and a smaller set between the larger set, as though a larger bigfoot was straddling a smaller one between its legs. It appeared that the structure had been made as a blind or ambush position to jump prey as it came by. It was very close to several game trails that all came together near the opening of the blind.

From 1990–95 Richard owned a "Rent to Own" business in Crescent City. Many of the clients of the business were Native Americans from local tribes and reservations, and some were just good friends. He explained there were many times where he would just sit and talk with the Native Americans about bigfoot and the stories that the Indians had about their existence. Richard said that most of the Native Americans in Northern California believe that bigfoot can cross dimensions and disappear at will. He explained that the Yurok believe bigfoot lives in the "Valley of the Big

People," but they would never say exactly where that was located. Another Native stated that in 1993–94 he and his son were driving French Hill Road above Gasquet when they saw a bigfoot cross the road in front of their truck.

In late 1993 Richard received a phone call from John Fay who was then the under sheriff of Del Norte County. John was a friend of Richard's and knew he was interested in bigfoot. On this call Fay told him that the sheriff's office had received a call of an "ape-like" creature seen near DeMartin's Beach at Wilson Creek near the ocean. Two hours after receiving the call Richard was at the location. He found a grassy area near the creek that was heavily matted down and an embankment area that had very clear 16-inch tracks. Richard distinctly remembered a photo being taken of him holding a track on March 27, 1993. During the course of the early 1990s Richard specifically remembered getting casts of bigfoot prints from Damnation, Hunter, and Wilson creeks. Several times Richard received calls from the sheriff about bigfoot-related incidents because they never wanted to respond to those types of calls.

Richard Gehr with footprint cast.

One of the last items Richard wanted to communicate to me was that on each occasion he either observed a bigfoot, or saw their blind, or believed he was in close proximity to a bigfoot, there was always a stench in the air. It reminded Richard of rotten chickens; it was putrid.

LOCATION

Hunter Creek Road forks off Highway 101 north of the Klamath River. If you drive to the end of the road you come to a gated drive that you cannot drive down. The water company owns it and people can walk the road. It is beautiful. I walked for several miles down this stretch several years before I met Richard, and thought to myself what wonderful bigfoot habitat this would be.

Proceeding down Hunter Creek Road and looking directly south, the large hill where Richard had his bigfoot sighting is visible. The elevation of the sighting is probably only 200 feet, but that hillside has no visitors. All the land is private and visitors are not allowed access. This region of the Klamath is known for being Yurok territory and the region east of this location is known for being very isolated. You can only drive a short distance eastbound on the Klamath from Highway 101 until the road stops. Then there is a stretch of the Klamath where there are no roads for almost 15 miles until you reach Pecwan. That stretch of the Klamath has always been rumored to be bigfoot country.

The hillside where Richard has his sighting has significant food source throughout the area. The Klamath and its tributaries all carry salmon and steelhead; the hillside has berries, pine, Douglas fir, and old-growth redwoods. Mushrooms grow in this climate in large quantity, partly because of the 60 inches of rain they get annually in the area.

Do not attempt to enter the area of the hillside, as you will be arrested for trespassing. If you want an adventure, go to Prairie Creek Redwoods State Park just south of this location or Del Norte Coastal Redwoods State Park just north. Each of these locations has a history of bigfoot sightings and you can search in the comfort of knowing you are on state or federal property. If you are going to actively hunt for bigfoot prints, stay near creeks and rivers and keep your eyes on the sandbars and banks that can retain a print. Much of the forest floor in this area of Northern California is not conducive to absorbing a track for visibility.

Just to the north side of Hunter Creek Road is the Yurok Redwood Environmental Forest that is owned by the Yuroks for the Yuroks. Again, you cannot go onto Native American property without a permit, which you can apply for at their local offices. If you do not have a permit and are seen off a paved roadway, the Yurok police will arrest you, and there are a lot of them in this region of California.

10 HUMBOLDT COUNTY

Located on the Pacific Coast, 200 miles north of San Francisco, 410 miles south of Portland, Oregon, and 110 miles south of the Oregon border sits the bigfoot capital of California, Humboldt County. That title may upset some residents of the region, but its alternative name is "The Marijuana Capital" of California. Marijuana has been considered the number one cash crop in the county for over 20 years. The social and environmental factors of the county make growing the green leafy matter almost ideal. The strongest marijuana (based on THC content) grown in the continental United States comes from Humboldt.

The county's 127,000 residents are scattered, with the largest number living in Eureka, with 42,000 residents. Eureka is a harbor city that sits in a beautiful spot next to a bay and is home to many salmon and crab fishermen. The town had notoriety as home to a thriving lumber industry in the early 1900s. As environmental concerns and cheaper locations to log trees were found, the mills closed, the jobs evaporated. The Pacific Lumber Company was a huge employer in the county and essentially was the town of Scotia until cheaper lumber was found in Japan and Canada; now it's almost a ghost town.

The county is home to one national park, Redwood, located just north of Eureka along Highway 101 and extending east into the Bald Hills. It's a beautiful setting along Redwood Creek, and has few visitors compared to Yosemite in central California.

Humboldt State University and College of the Redwoods are the two institutions of higher education located in the county. The university is large, specializing in forestry management and related degrees, and calls Arcata home. The Arcata/Eureka Airport supplies jet transportation for the county residents and is one of the few growing entities in the region.

The employment outlook for the county hasn't been good for

293

many years. Twenty-five percent of the employed have education, health care, and social services as their profession, while 18 percent have their primary occupation with the government.

Here is a list of the annual events in Humboldt County:

Redwood Coast Jazz Festival—March
Redwood Acres Fair and Rodeo—June
Fourth of July Celebration—July
College of the Redwoods Fair—July
Blues by the Bay—August
Bigfoot Days–Willow Creek—Labor Day Weekend
Truckers Parade—December

One of the biggest events in Humboldt County is Bigfoot Days in Willow Creek over Labor Day weekend. Willow Creek calls itself the "Gateway to Bigfoot Country" and the bigfoot museum in the middle of town is a must-see on any bigfooter's agenda. The weekend celebration usually brings the biggest names in the bigfoot world, along with sidewalk vendors, acts, and the usual street-fair atmosphere. The beautiful Trinity River flows just on the outskirts of town and makes for a great late-afternoon spot to cool off during the normal hot summer days. If you like fishing, bring your pole. Fishing and rafting are huge tourist attractions in this portion of the county.

WILLOW CREEK

Facts 2007:

Location: Six Rivers National Forest
Elevation: 610 Feet
Population: 1,743
Median Income: $27,246
Nickname: Bigfoot Capital of the World

In the 1930s to 1940s, Willow Creek was a small village (population 167) without phones, or electricity, or jobs. Loggers, truckers, and foresters used to roam the nearby mountains, harvesting giant redwoods, Douglas fir, and pine, but much of the millwork has since dried up. Many of the residents would spend their weekends away from the woods fishing the Trinity River for steelhead and salmon. The Trinity flows through the heart of Willow Creek and makes a beautiful backdrop for the town. The other recreational release for the workers was hunting the nearby hills for bears, deer, grouse, quail, and other small game. Sportsmen in Willow Creek in the 1930s and 40s didn't have the Internet, television, or even a daily newspaper. It was a somewhat isolated community, even though it was 35 miles from Arcata and Eureka.

There was plentiful game in the mountains surrounding Willow Creek in the 1940s, but many of the mill workers were too tired on the weekend to hunt and look for game to shoot. Others fed their families with what they shot around the city. Small game wasn't the only thing being seen by hunters and loggers. Stories of a giant ape that walked on two feet started to surface in Willow Creek in the late 1940s. Many people didn't take the stories seriously, but others remembered similar stories emanating from the rail-line build by the Northwestern Pacific Railroad (later bought by Southern Pacific) along the Eel River in 1914. The stories seemed too eerily similar to completely discount.

HISTORICAL BIGFOOT

Most people associate the 1967 filming of a bigfoot at Bluff Creek as happening in Humboldt County, but that is incorrect. That event occurred in Del Norte County, but the only way to get to the location is through Humboldt County because of the Siskiyou Wilderness Area separating the regions.

In 1958 a road-building crew was constructing a road north from Orleans, California, and the crew began noticing huge, human-type footprints in the ground when they returned to the work site in the mornings. One of the crewmembers named the creature that made the tracks "bigfoot." The name stuck and the creature in the western United States has been called bigfoot since.

In *The Hoopa Project,* I chronicled a variety of bigfoot sightings occurring in and around the reservation just south of Bluff Creek. The Hoopa people have claimed they have had bigfoot encounters for the last 200 years, and they consider the creature a caretaker of the woods and call it, "Oh-mah." There are stories I've chronicled from the early 1900s where young, disabled Hoopa people were left in the woods to be cared for by bigfoot. There were other accounts where Hoopa women were abducted by the creature and raped.

During my research on the Humboldt County region I found a great resource, *To The American Indian, Reminiscences of a Yurok Woman,* (Heyday Books) originally printed in 1916 (and reprinted in 1991) and authored by a Yurok woman named Lucy Thompson. Lucy lived in Pecwan just downstream on the Klamath from Weitchpec and approximately 35 miles from Hoopa. This area is part of the Yurok and Hoopa reservations, and the Natives share and utilize the land in harmony. I chronicled a bigfoot sighting in Pecwan in *The Hoopa Project.* Pecwan sits just downhill from the Blue Creek/Bluff Creek area and Blue Creek empties into the Klamath very near Pecwan. Just across the river from Pecwan are the Bald Hills. I chronicled another bigfoot sighting in those hills in *The Hoopa Project* where one of the women in the vehicle at the time of the sighting claimed the bigfoot was an "Indian Devil." Inker McCovey and Ed Masten are both Hoopa Natives, and explained to me that many of the Yurok and Hoopa people still call bigfoot (Oh-mah) by the name Indian Devil, even though the Indian Devil is known as a shape shifter (has the ability to change from a bigfoot to another animal). It's a name from the early 1900s and still used by elders and some tribal people. The name still refers to the same creature we all know as bigfoot.

Chapter 9 of Lucy's book is titled, "The Indian Devil." I nearly dropped the book when I first read this. She went on to state, "say that some were made to be good and honorable, some bad: and some were real bad and mean, which they termed devils, or Oh-Ma-Ha." (pg. 129). She explained that these devils are living human beings that are alive and have left the tribe. The tribe has a conception that Satan is invisible and is a real devil that walks the Earth. She made these statements all in the same paragraph, almost as though she was aligning the devil with the Indian devils.

On page 129, paragraph 2 Lucy stated, "These Indian Devils would sometimes watch the camps of the Indians very closely and follow them about as they moved from place to place, watching for an opportunity to seize one of the young women and carry her off to make her his wife. The women of the tribe had great fear of them, as they had great horrors of becoming the wife of a wild man." The "wild man" phrase was one that was used in the 1896 article "The Hermit of Siskiyou" when describing the creature the hunter saw eating berries in the Marble Mountains (refer to chapter 8, Siskiyou County).

Lucy also explained that there were times when women would escape from their captors and make it back to the tribe. In many of those instances, the child would not come back. On page 130, "the children would inherit the wild habits of their father, as they would be whistling, making strange noises, romping wildly about and always on the go." Her descriptions mimic bigfoot activity, as we know it today. They are known for their loud whistles, they are known to roam vast distances, and they are definitely known for their strange sounds.

On page 133, paragraph 2, "We are always afraid of the visible devil (Oh-ma-ha)- that is, the living devil here on Earth, as we are compelled to guard continually against these monsters in keeping ourselves from being harmed." Lucy continually went back and forth in her writing, calling the creature a devil, monster, and human being, very close to what witnesses today describe as bigfoot. If her 1916 book is true—and there is no reason to doubt its validity—then there is a possibility that Oh-mah (bigfoot) has bred with humans.

I had conversations with Inker McCovey and Ed Masten on the topic of Native American Hoopa and Yurok women being kidnapped by a bigfoot, and asked them if this was a possibility. Both stated there had been many stories throughout their tribe and the Yurok that women along the Klamath had been abducted and raped by bigfoot, and many were forced to have their babies.

The consequences of bigfoot breeding with a human, and the subsequent offspring born to a human mother, would be significant to the evolutionary development and changes to the hominid. This could be the major factor as to why many of North America bigfoot forensic sketches have a significant human quality. The idea that there are offspring to the Hoopa or Yurok roaming the hills of Northwest California is fascinating. In *The Hoopa Project* there are accounts where Hoopa members were told to talk in their native language to the bigfoot if they should ever encounter the creature (refer to Damon Colegrove sighting, chapter 5, *The Hoopa Project*). When I questioned these individuals about why they thought speaking their language to the creature would have any effect, they had no answer other than their elders had told them it would understand. The answers are now quite clear and speak volumes about the relationship between bigfoot and the Native Americans. This relationship may also account for the many bigfoot sightings around the Hoopa/Yurok reservations. Bigfoot feels comfortable in this environment, as it may be their home.

In this chapter I have included an interview with Al Hodgson, who was friends with a freelance reporter, Betty Allen. Betty had become good friends with the Native American women in Pecwan, and alluded to Al that bigfoot had kidnapped women in Pecwan; and it seemed during Al's travels with Betty up to Bluff Creek that she was also afraid of the creature. She told Al that many of these women were embarrassed and that she would tell him the details when they passed on. Al told me he always felt that Betty would have confirmed that these women had been raped and/or abducted by a bigfoot.

If you are starting to doubt this, consider these points: Lucy Thompson first wrote about this in her book in 1916; Inker McCovey and Ed Masten claimed that elders said that tribal women had been kidnapped and raped by bigfoot; Al Hodgson (whom Inker and Ed have never met or spoken to) had formed the same opinion, based on information alluded to by his friend, Betty Allen, when she interviewed Native American women about the historical significance of the creature. These are three independent sources all claiming the same activity in a confined region around Pecwan. This information cannot be ignored or discounted.

298

AL HODGSON

Al is "Mr. Bigfoot" to the Willow Creek community. He was one of the first locals to offer support to John Green on his visits to Bluff Creek. He was also one of the first researchers ever to go into the Bluff Creek region and view tracks observed by road crews. Al has been a consistent presence in the bigfoot arena by assisting in the establishment of the Willow Creek Bigfoot Museum, being an organizer of Bigfoot Days in Willow Creek, and most essentially, being that careful observer of facts and a great conduit to the big-foot community.

During the great depression, Al Hodgson's family was farming in Albion, Illinois and having a very difficult time. By 1932 his parents were nearly broke, so they decided to leave Illinois and head for California with their eight children. They arrived in Eureka in 1933 with their kids and their pet German Shepherd dog riding on the running board of their 1929 Chevrolet Truck.

Al Hodgson.

Al wasn't much interested in school and dropped out of high school at 17 and joined the Navy. He spent five and a half years as an aviations machinist mate, the first three years at Naval Air Station San Diego where he suffered through bouts of depression because many of his friends were shipped out to fight during the Second World War, and several died, while he was kept stateside. The last 10 months of his tour of duty were spent working on a seaplane tender (KV 14 USS Kenneth Whiting) crew on a small atoll off Okinawa. He said it was beautiful but there wasn't much of any-thing to do other than work and sweat. He earned a 60-day leave, went home, and helped his dad. He returned to the Navy and report-ed to the CASU 5 at Alameda Naval Air Station across the bay from San Francisco.

In 1948 Al moved to Willow Creek, and his first job was brushing rights of way (clearing brush) for Pacific Gas and Electric. One day Al and the crew were returning from work when they came upon a Navy TBF torpedo bomber that had just crash-landed on the highway in front of one of the local schools. The plane was a total loss, with engine pieces and plane fuselage scattered over a wide area, but the pilot and co-pilot were unharmed. Years later Al met up with the pilot who told him that he had been lost and was running out of fuel and had felt the highway was the lone place he could try to put the plane down. He said he was "picking apples" as he made his approach, and was doing well until he clipped a wing on a madrone tree and crashed.

In 1950 Al landed a job at Hanson Mill where his brother-in-law talked him into driving a lumber truck covering a route from Willow Creek to Arcata. In those days the road was a composite of dirt, gravel, and sand with a lot of potholes, washouts, and landslides making this road very dangerous. A trip from Eureka to Willow Creek could take as long as three hours; now it's a 60-minute trip. After a short time with the lumber company, he left to drive a dump truck for the Burnt Ranch convict camp. It paid better money, but it was just a seasonal job and he lost it after a few months.

During those months that Al was driving a truck he heard stories about a giant, wild, hairy man or giant ape, and other descriptions of creatures that people had seen in the mountains around Willow Creek. The creature was described consistently, but it was always called by different names. One locally famous story involved a Eureka surgeon, Dr. Burre, who had a summer home in Willow Creek. One time when he was driving back to Eureka he saw a huge creature walking on two feet and covered with hair cross the road in front of him. It was also during this same time period that various other stories filtered to Al about people seeing the creature and hearing it scream.

Since Al was out of work, he decided to return to Albion, Illinois where he moved in with his sister and worked as a welder at a small airstrip adjacent to the Wabash River. At the first local dance Al attended, he met his future wife, Francis Elfritz. They dated for almost a year when they were approached with an unusual offer—the Edwards County Fair was looking for a couple to get married at the fair. They decided the time was right, and tied the knot in front of county fair attendees.

They moved back to Willow Creek to enjoy the California weather, and Al worked for his brother at his welding shop where they fixed loggers' trucks and tools. His brother asked him if he wanted to join him in looking for the "ape man" reported in the mountains surrounding Willow Creek; Al declined. This was years before the Patterson–Gimlin footage and media blitz. Al said there were scattered reports from credible people about seeing creatures, seeing tracks, and hearing screams. Some people were scared of what they saw and heard; some people were embarrassed to talk about their incident in a public setting, but still the stories persisted in small circles. Most of the news about the creature was confined to the Willow Creek region.

On one occasion when Al and Francis were in Weaverville, they walked into a local variety store that had dry goods, hardware, some clothes, but no food. The Hodgsons had built a new home and were looking for a new opportunity. Al and Francis felt that they could open a similar store in Willow Creek, so they first went to the North Town 5 & 10 store in Eureka where they were put in touch with a salesman for Skaggs Store, and the rest is history. In 1956 their store was opened and it thrived. They built an 11,000-square-foot building and supplied the area with much of its clothes, sundries, sewing supplies, cards, etc. for over 40 years.

Hodgsons Department Store

The time period 1948 to 1958 was a decade-long gap absent of any reports of the "ape man" or similar stories about strange creatures. It was as though the creature retreated deeper into the wilderness and didn't want humans to see it. This was also a time where the local economy had suffered, people moved away, and there was little activity in the region. (A possible explanation for the lack of sightings during this decade could have been the economy: a lack of visitors, and locals having less free time to vacation in the woods.)

In the late 1950s Al met Betty Allen, a freelance writer who had authored pieces for the *Times-Standard* newspaper in Eureka. She soon became the newspaper's representative for Willow Creek news and was a regular contributor. Betty covered a huge area for the newspaper, from Willow Creek to Weaverville to Orleans and onto Pecwan. Native Americans—Yuroks, Kuroks, and Hoopa—inhabited much of Betty's coverage area.

She also became a good friend of the Hodgsons and visited their store on a regular basis, talking about the creature in the woods and routinely trying to get Al interested in joining her in journeys into the forest. Al always declined; he thought the creature was a hoax and never really existed.

In 1962 Betty told Al that huge, human-like tracks had been found north of Hoopa near the confluence of Notice and Bluff Creek, and she invited him to drive up and view the impressions. Al brought along Francis and his kids, and they made the long drive to the site with Betty. Al remembers arriving at the location and seeing giant, human-looking footprints in a freshly cut road. Ray Wallace's sister-in-law was there and explained that the tracks led down to a nearby creek where they were lost in the foliage. Al followed the tracks for quite a distance to the creek, and the entire time he was thinking they could be fakes, but that whoever had done it had made a huge effort because of the distance and foliage they had to work around. Francis said she thought the tracks were fakes, because they looked too perfect. Al brought along a large supply of plaster of Paris and cast several of the good prints.

One unusual aspect to their trip was that Betty Allen wouldn't walk down to the creek with them. She said she was too afraid and would wait in the car with the doors locked until they came back. Al said that he thought this was a little unusual for Betty because she always was willing to do anything for a story. After checking out the creek and the tracks, Al and his family went back to the car and found Betty locked inside.

Several days after the trip to view the tracks, Al saw Betty in town and asked her why she had been too scared to go to the creek and view the tracks. Betty explained that, out of privacy for the victims, she couldn't give Al all the information, but there were Native American women who had told her there was reason to be afraid of some bigfoot. Betty was very protective about the specific information, but she inferred that a bigfoot had been raping women in Pecwan and they were afraid for their safety, thus Betty's reluctance to leave the protection of Al's vehicle. This wasn't specifically stated to Al, but that was the implication. Al asked her if she could ever tell him what she was afraid of, and she said she would tell only after all the victims had died.

After hearing this story, I approached several tribal members in Hoopa about it. They confirmed they had heard the same stories from grandparents about bigfoot and his affinity for women. There were stories from the Pecwan area of women who were kidnapped and held against their will, and later gave birth to deformed or odd-looking babies. Many times these women were sent away to live elsewhere.

Shortly after making the trip to the creek, Al had a tourist come into his store and ask questions about a big hairy ape. The person was camping in the Horse Linto area (northeast of Willow Creek) and had found a footprint. The Horse Linto area is on the far side of the Trinity River and is very desolate. It is now adjacent to the wilderness area and in close proximity to Tish Tang Creek.

In 1966 John Green contacted Al and asked if he would meet him, Bob Titmus, and their tracker in the Blue Creek Mountains just above Bluff Creek. John and Bob had chartered a private plane and were flying into the small airport in Orleans. (The airport no longer exists. There is a very nice airport in Hoopa, approximately 25 miles west of Orleans.) John told Al that he had the toughest, meanest, and best tracking dog in the world. They had heard that tracks had been found on the mountain, and they would pick up the scent there. Al met the guys at the site and saw that they had a German Shepherd in the car with them. John and Bob had only brought Canadian dollars and asked Al if he could loan them $100 US, which he did ($100 was a lot of money then, but that's the kind of person Al Hodgson is). Al had brought his son, Mike, with him. It wasn't long before Mike had made friends with John's dog and was sitting in the back of the station wagon with his arm around him. Al said that the dog handler became "unglued" when he saw Mike with the dog. The handler said, "That dog is a killer."

They made an early start next morning. The dog was on the track and appeared to have a good scent. After going very deep into the mountains, the tracker suddenly stated that the dog had lost the track. Al was very surprised, as it seemed that the dog was doing a great job and hadn't lost any interest. It appeared to Al that the tracker felt they were getting too close; the tracker wasn't armed and Al thought he got scared, so he pulled the dog back stating it was getting too warm. After John Green left, Al called Roger Patterson and told him of the incident involving the dog.

Roger Patterson was a cowboy and outdoorsman in Yakima, Washington who was also fascinated by bigfoot. Reports were being made in Canada and Washington State similar to what was occurring at Bluff Creek. Roger discussed with Al the logistics of setting up a trip to the Bluff Creek basin with a view to finding and filming a bigfoot.

Directly at the bottom of one of the forestry roads into Bluff Creek stands the Bluff Creek Resort, located adjacent to the Klamath River. At the time, the resort had a fully furnished store and small dry-supplies area. Lucien Sanders owned the resort and made quick friends with Patterson and Gimlin. Sanders agreed to work with them and supply their team for the trip into the basin. The roads into the area weren't paved and weren't gravel, they were unimproved dirt. Depending on the road, washouts, and snow, the roads were sometimes good and sometimes horrible. Rain in this region at any time of the year was always a possibility.

Roger called Al and told him that he and his friend, Bob Gimlin, were going to be in the basin from October 18 to October 20, 1967. He had rented a professional model 16mm camera, and they were going to be on horses so they could cover a lot of ground. They were also going to be carrying rifles, but had agreed they would not shoot the creature if they saw it.

Bigoot family pictograph at Painted Rock.   Beaver pictograph.

Large bigfoot pictograph.

(Above left) Sketch of female
bigfoot seen by Shirley Fork.

(Above right) Jennifer Crockett golden
bigfoot.

Jennifer Crockett bigfoot observed on Highway 197 west of Hiouchi.

Travis Cover sketch of biped in campground.

Travis Cover sketch of biped observed on Highway 199 east of Gasquet.

Mother and infant bigfoot observed near Blue Creek by James Renae.

HP 4-16-08

James Re

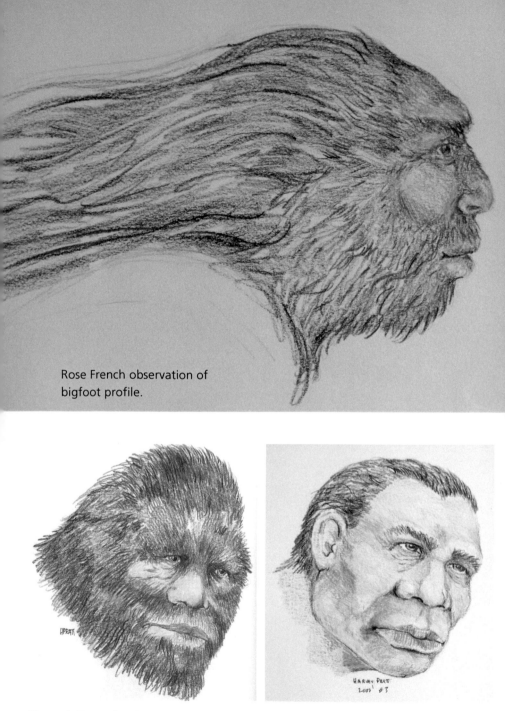

Rose French observation of
bigfoot profile.

Harvey's Patty sketch.

Patty sketch without hair.

310

Marlette Jackson sketch of biped observed on Highway 96 north of Willow Creek.

HP 4·15·08 #7
Marlette

Julie Bradley sketch.

Eaton Wood sketch.

Keith Lumpmouth sketch.

Debbie Carpenter
sketch.

Charles McCovey sketch.

Patrick Holliman bigfoot observed on Bear Creek near the
South Canadian River, Weatherford, Oklahoma.

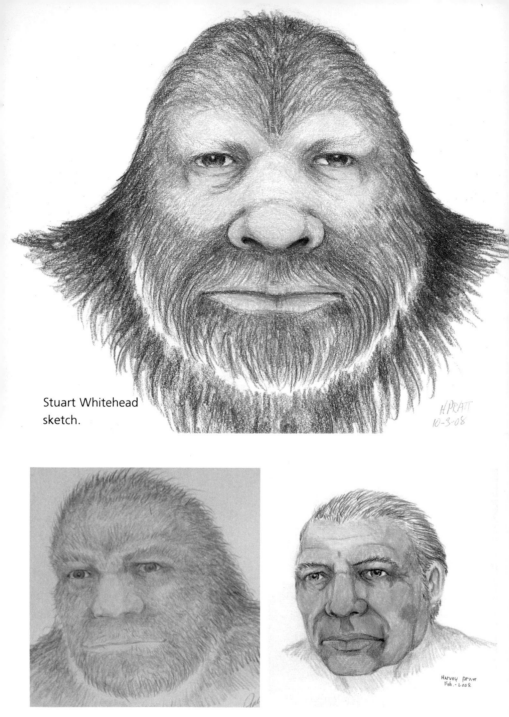

Stuart Whitehead
sketch.

Masten drawing.

Masten drawing without hair.

316

Redbird bigfoot facial sketch.

Mary Lonebear sketch.

Roy Russell Long photo and composite sketch comparison.   Ina Faulkenberry bigfoot.

Mary Lonebear facial sketch.

Ina Faulkenberry black face.

Rose French bigfoot facial sketch.

Rose French biped showing age progression, middle age (left) to elderly (right).

PATTERSON–GIMLIN FILM FOOTAGE

October 20, 1967 was the last day that Patterson and Gimlin planned to be in the mountains. This was also going to become the most famous day in bigfoot/sasquatch history. The two started early, on horseback, and decided to work an area approximately two miles northeast of Louse Camp. Bluff Creek in this area was very remote and there were no roads into that specific region, supposedly. They rode out of the timber at about 1:30 p.m. and down into the creek area and immediately saw a large, hairy creature bent down near the creek with what appeared to be something in its hand. Their horses started to rear (Patterson partially fell), but both men being cowboys, they settled them down, dismounted, and started to film. The creature walked on two legs briskly towards the east and made its way into the adjacent mountains. The creature appeared to be a female because of its large breasts. They eventually lost sight of the creature after 35 seconds of filming.

After tracking the creature they went to the spot where the creature had been and found multiple tracks. They followed the tracks into the woods and then returned to their horses and rode back to their camp where they had plaster of Paris that they had brought with them. They rode back to the creek, made plaster casts of several of the tracks, recovered the casts, and made their way back to the horses. They rode back to their truck, fed and watered the horses, and then started their drive to Eureka. Patterson and Gimlin have stated that they filmed the creature at 1:30 p.m. If after seeing it, they tracked it, went back to their horses, investigated the tracks, rode the horses back to their base camp, got the plaster of Paris, rode back to the site, cast the prints, waited for them to set, recovered the casts, and rode back to base camp again, that is a very significant amount of time.

An interesting side note to this story is that Bob Titmus later told Al Hodgson he went back to the area of the filming and found a spot where he believes the creature sat down and watched the men work the area for evidence after the filming. He said it appeared to be well matted down and in a location where you could watch the Bluff Creek spot.

The road to Eureka from Bluff Creek, even in 2007, is long,

Bald Hills Road above Redwood National Park.

dusty, unpredictable, and grueling. There are two ways to get to Eureka from this area: through Willow Creek and over Lord Ellis and Highway 299 (the best road); or over Bald Hills, through Weitchpec, and through Redwoods National Park. This route is long and winding with many forks in the roads, and it is very, very easy to get lost. Patterson and Gimlin took the Bald Hills route. They drove dirt roads out of Bluff Creek, over Bald Hills, down through the park, and then met Highway 1 at Orick. At Orick the road turned to pavement, and they drove straight south for 40 miles to reach Eureka. Once in Eureka, they went directly to the post office and mailed their film back to Roger's relative, Al Deatley, in Yakima for developing.

Patterson and Gimlin got back into their truck and started their drive up to Willow Creek by driving Highway 299. In 1967 the road was a two-lane highway without a lot of passing room. In 2007 there are parts with passing lanes, but you will get stuck behind trucks; it's inevitable. In recent times the fastest I have ever made it from the post office into Willow Creek driving a 2007 Chevrolet Tahoe at 60 to 70 mph is one hour, and I was flying. I don't think the vehicle that Patterson and Gimlin were driving could have made it faster than 90 minutes, under ideal circumstances.

Al's store closed at 6:00 p.m. every night. At 6:15 p.m. Al

received a call from Roger in downtown Willow Creek. Roger stated that they were in town, he had gotten footage of the creature, and they wanted to meet. Al said he met the guys in front of his store and they talked for quite awhile. They explained the trip they had just made and the events as I have recounted it.

Al felt that he should contact a local official of the USFS for the area, Syl McCoy, who said he would meet the guys at his office. The meeting at the Forest Service office went on for many hours. Al said that Syl appeared to believe the creature existed and he asked a lot of pertinent and applicable questions (refer to statement by Mary McCoy, below, for specifics on Syl). It was late at night when the meeting ended and Patterson and Gimlin headed back for the camp above Bluff Creek to look after their horses.

That night after the meeting it rained very hard at Bluff Creek. The truck that Patterson and Gimlin had driven got stuck on the roads and they needed assistance getting dug out. They found a backhoe used for road building that they could use. The backhoe belonged to Charlie Wittson, but he wasn't around because of the bad weather, so Patterson and Gimlin couldn't ask his permission to use it. They needed it and borrowed it, but Roger did call Al after the event and asked how much Charlie wanted for the use of the backhoe because he wanted to make things right.

Roger Patterson died in January 1972. He went to the grave holding true to his story, claiming that he and Bob Gimlin did actually catch a bigfoot on film.

When I interviewed Al about the Patterson–Gimlin film, I was interested in his perspective of the footage and his recollection of the subsequent meetings. I didn't know that I'd be questioning the authenticity and timeline of the events that Patterson and Gimlin gave. Remember, this is Al Hodgson's version of the incident, and he has remained consistent in his recollection.

I have personally spent many weeks in the Bluff Creek region studying the topography, foliage, wildlife, and general terrain. I've driven every road in the area and have a good working knowledge of places, distances, and roadway conditions. As Al told me the story, there were a few questions that I kept coming back to. Al said Roger Patterson told him that after they had shot the footage, the men got into their truck and drove to the Martins Ferry Bridge. This is located west of Weitchpec and east of Pecwan and is the route you would take over Bald Hills Road. Al was then told that they drove over Bald Hills, through the Redwood Park down to Orick and down Highway 1 to Eureka. That is one long trip, trust me, I have done it many times. If you haven't done this trip five or six times, it is easy to get lost on the many dirt roads that criss-cross the mountains. Under the best conditions it would have taken 30 minutes to drive from Orick to Eureka in the late 1960s. The road from Weitchpec to Orick would take another 60 minutes if you were flying. The area of Bluff Creek, where the footage was taken, to Weitchpec would take another 60 minutes if you were driving very fast. It would take another 30 minutes to find a parking place in Eureka, exit their truck, complete the necessary paperwork and insure the package to Yakima. The total driving time to this point from Bluff Creek to Eureka is three hours and 15 minutes. It would take another hour to drive from Eureka to Willow Creek, again, under ideal conditions, no traffic, ideal weather etc. Total time now is four hours and 15 minutes.

Consider that all reports from Patterson and Gimlin state that they saw the creature at 1:30 p.m. They shot the footage, tied up their horses, looked at the tracks the creature made, and discussed what they had just seen, 30 minutes minimum for all of that to transpire. But all accounts state that they then had to ride their horses back to their base camp to get the plaster of Paris, then mix it up, and then make casts of the tracks they had found. To return to base camp, ride back, mix the ingredients, pour it in the tracks and let it set would take one hour, minimum, probably two hours. They then retrieved the casts and started their ride back to their truck. It is a minimum 30-minute ride for the two miles back to the truck where they then had to dismount their horses, feed and water them. If they saw the creature at 1:30 p.m., then add the following:

FILM CREATURE, FIND TRACKS, DISCUSSION:	30 MINUTES
ROUNDTRIP RIDE TO CAMP, MIX MATERIAL, CAST AND RETRIEVE PRINTS:	90 MINUTES
RETURN TO CAMP, DISMOUNT, FEED AND WATER HORSES:	30 MINUTES
Sub total:	2 hours 30 min
DRIVE FROM BLUFF CREEK TO MARTINS FERRY:	60 MINUTES

DRIVE FROM MARTINS FERRY TO ORICK:	60 MINUTES
DRIVE FROM ORICK TO EUREKA:	30 MINUTES
PARK, COMPLETE PAPERWORK, MAIL PACKAGE:	30 MINUTES
Sub total:	3 hours
Total time to Eureka:	5 Hours 30 min

If it took the absolute minimum five hours, 30 minutes to complete all of the tasks from the time the creature was first filmed until the mailing was completed, there leaves no time to make the one hour drive from Eureka to Willow Creek. Patterson and Gimlin met Al at 6:15 p.m. at Al's store in Willow Creek on the night of the footage. My estimates of the time and distance were very liberal, giving every opportunity that the trip was possible in the times and distances related. It would appear that this trip could not have happened on the day it was indicated.

I should state here that I believe that Patterson–Gimlin caught an unknown creature on film. I don't think the film is a hoax, as some have claimed. I do believe that there is something more to the story, as the timelines for the trip don't make sense.

I brought the timeline issue up to Al and asked him his thoughts. He stated that nobody has ever discussed with him the issue of the casts and the time involved to lay them out, cast, and later retrieve them. Al also reminded me that the casts would have had to have been retrieved that day because the following day there was a horrendous rainstorm that caused Patterson and Gimlin's truck to get stuck.

Betty Allen had started to spread the word to the world that something huge, hairy, and walking on two feet was in the hills of Humboldt and Del Norte County. Many of Al's casts were displayed and people asked to see them. Al made two or three more trips into the Bluff Creek and Go Road area to see if new tracks ever appeared, without telling others he was going.

In the mid 1990s the lumber industry started to dry up in the region and many workers left the area. The Willow Creek economy took a huge hit. Al closed his store in 1998 and now rents the building to a healthcare group. He lives on the outskirts of Willow Creek with his wife and their dog. After his store closed, Al started to volunteer his time at the local museum.

Community members started to say that they should add onto the museum and dedicate some of it to bigfoot. Then, one day John Green called Al and told him that Bob Titmus had passed away and he wanted to donate all of his old casts to the museum in Willow Creek. Locals heard of the bigfoot donation and a lumber mill donated the wood for the additional structure, and through community involvement the bigfoot addition was completed. The museum rents the land for $1 a year from the local water company. Al continues to volunteer his spare time to the bigfoot portion of the museum. He is one of those people in the community that everyone goes to with their bigfoot stories, but he has always wanted to stay outside of the bigfoot spotlight. He isn't on the board for the museum, and has no official capacity or any official association with the museum. He does respond to requests from the museum when asked, and he always will answer questions posed by community members.

Al is an individual of extreme integrity, and although he has an interest in and connection to bigfoot research for decades, he never truly believed the creature existed until fairly recently. He said he had seen all of the tracks in the wild, met many people who claimed to have seen bigfoot, but it wasn't until a close friend talked to him about her incident that he turned the corner on his beliefs. A church friend approached him after a bible service at Al's house and stated that she had never approached him out of respect, and being somewhat intimidated, but she wanted to tell him that bigfoot is a true creature living in the woods because she had seen it. Al said that when Rose French said this to him, he knew for a fact that the creature existed. He said that he has known Rose for many years and she is probably one of the most honest people he ever met. Since she said she saw a bigfoot, Al is now 100 percent a believer (see the Rose French story below.)

It is Al's integrity and honesty that makes him an outstanding person to be involved in bigfoot research. It has truthfully been my pleasure to work around this fine man. Thanks Al for all of your assistance in bigfoot research.

HUMBOLDT COUNTY WITNESSES

MARY MCCOY

During my hours of discussion with Al Hodgson there was one name that continued to come up as both a responsible member of the United States Forest Service and an avid bigfoot researcher during the 1960s—Syl McCoy. Al gave me the contact information for Syl's wife, Mary, as Syl had passed on years earlier. I was able to meet with Mary in Willow Creek and had a great conversation about her and Syl's life in Hyampom and Willow Creek.

Syl and Mary started their family while they were living in Hyampom, California, a very rural town in Trinity County, from 1960 to 1966. In the early 1960s there wasn't much in this small town other than a large lumber mill. The roads in and out of Hyampom were treacherous, and there was only one road out in the winter, which went through Hayfork and eventually ended near Weaverville. Syl made a good living working as a logger, and had a great interest in the outdoors. He became a master tracker for all types of game during the time they lived in Hyampom. He was a very social person and made a lot of great friends in the community.

In the early 1960s there were a series of bigfoot sightings, along with the discovery of huge, human-type footprints, in the Hyampom area. Syl was very interested in learning more about bigfoot, and every time a report of tracks came out, he would go to the scene.

In 1960 someone in the Hyampom community spotted giant, human-type tracks on the banks of the South Fork Trinity River, on the outskirts of town in a rural area. Mary said that when they arrived at the tracks, they could immediately see they were not normal. It was obvious they were huge, much bigger than any human track, and the distance between the tracks (stride) was large. They followed the tracks as they led near bushes and eventually into the

327

river. Syl found some unusual hair that he submitted it to the USFS lab for analysis. The lab could not identify the hair.

Mary went with Syl on two occasions to view tracks, but Syl went out many more times at the request of locals, who knew he was a tracker. Syl had told Mary that the majority of the tracks looked to be from a real creature and some appeared to be forged. Syl made several casts of the tracks that he felt were real, but Mary has no idea what happened to them.

As time ticked by in Hyampom, Syl's notoriety as the expert on bigfoot tracks started to build. He went out on more and more track sightings and the newspapers slowly started to pick up the story. It wasn't long before another bigfoot enthusiast contacted Syl about

his findings. Bob Titmus asked if he could visit and discuss the casts. Bob had been a long-time bigfoot researcher. Bob had spent time in the Bluff Creek area casting prints in the late 1950s. (He also went back to Bluff Creek immediately after the Patterson–Gimlin filming in 1967 and spent nine days tracking the creature and casting prints.) He was considered an expert tracker.

In 1963–64 Titmus visited Mary and Syl, and they immediately became friends. He recognized that the casts Syl had made were authentic and appeared to be quite similar to the ones he made in Bluff Creek. Syl and Bob started to develop synergy on their search, and at

Mary McCoy

one point decided to do something nobody had ever done—and most researchers still are not doing—which is off-ground surveillance.

They both felt that bigfoot was nocturnal, so they would need to spend a cold night outdoors if they were ever to catch a glimpse of it. They went to an area Syl had identified as having significant and recent bigfoot tracks. They brought food, warm clothes, and climbing gear, and Syl brought his logging equipment, and went towards the South Fork Trinity. They climbed one of the largest trees in the area and stayed in it overnight waiting for the creature to appear. Their theory was if they were situated away from the ground then their human sent would be removed enough that bigfoot might walk by. Mary remembers that the only thing that Bob and Syl caught that night was maybe a cold.

Titmus had left after taking photos with Syl and his casts. He was convinced the creature was real, and that the casts that Syl obtained were of a real biped.

In 1965 Syl received another call from locals about tracks on a nearby mountain where another company was doing logging and building roads. They drove to the site with their kids and found a bunch of large, human-looking tracks around the loggers' trucks and bulldozers. This location was slightly unusual, since the majority of the tracks that had been reported by the community had been near the Trinity River, however, the location of these tracks did parallel sightings of tracks along the Go Road near Bluff Creek when road builders reported finding tracks next to their equipment. Maybe the bigfoot were interested in the unusual equipment, or perhaps they were upset that their environment was being altered.

Logging started to slow in Hyampom and in late 1966 Syl moved his family to Willow Creek, joined the USFS, and slowly started to climb the management ladder. While in Willow Creek, Syl still went out every time he heard about a bigfoot sighting, and one day he crossed paths with Al Hodgson. Al and Syl became good friends, as they both had a common interest in bigfoot and both were somewhat skeptical of some people's claims.

In 1967 Syl was given the lead management role in the Forest Service in Willow Creek, in charge of all road building. This was a huge job with significant responsibility, but it also meant that Syl could be outside working in the forests that he loved. On October 20, 1967 Roger Patterson and Bob Gimlin caught a bigfoot on film in Bluff Creek. They supposedly made their way to Eureka to drop the film off for mailing and then went to Willow Creek to call Al Hodgson. When they arrived to meet Al, he called Syl, and the group met that night to discuss the sighting. Mary remembered Al calling the house and getting Syl on the line, and Syl later going to the meeting. She distinctly remembered Syl saying he believed that the creature did exist and it was probable that Patterson and Gimlin caught it on film.

Mary said that one night she and Syl talked at length about the creature and its possible migratory path. Syl believed the creatures did migrate large distances, and probably followed many of the rural forests that encompassed the regions throughout Willow Creek, Hyampom, and all areas in between. He felt that the creature probably made its home in an area like the Yolla Bolly Wilderness Area and then left at certain times of the year, searching for food or escaping certain climate conditions.

Mary signed an affidavit.

CONCLUSION

Syl was a major figure in Willow Creek during the 1970s, 1980s, and 1990s. He spent 25 years working for the United States Forest Service, and he spent significant time attempting to understand bigfoot. His theory about bigfoot possibly living in the Yolla Bolly wilderness is something we had discussed even before I heard it from Mary. The Yolla Bolly is one of the most rugged and wild wilderness areas in the state. It's quite a drive from every major metropolitan area, and it doesn't get many visitors. The Yollas have several 7,000-foot-tall peaks with steep canyons, and two major rivers—the Eel and the South Fork Trinity—have their headwaters in the Yollas. It's interesting to think that something could enter the South Fork Trinity in the Yolla Bolly Wilderness Area and float to Hyampom, Willow Creek, Hoopa, Weitchpec, Pecwan, and the Klamath, all locations with many bigfoot sightings. I think Syl was onto something. Remember, one of the track sightings that Mary and Syl discovered led into the Trinity River.

ROSE FRENCH

In one of my many meetings with Al Hodgson he told me of a long-time friend who claimed to have seen bigfoot. He thought she was extremely credible and a very honest person, and encouraged me to contact her.

It was August 2007 when I called Rose by phone and requested an interview. She is a very upbeat person who had a pleasant response to my request and said she would be glad to meet. I made the drive to Rose and her husband's home, which sits at the base of a huge mountain in Willow Creek. It is the type of setting where you'd expect bigfoot to be walking down the driveway.

When I arrived, Rose was waiting at her front door and she greeted me with a smile. She explained that her husband was ill and asleep in the next room, but he didn't mind the interview at all. I first questioned her about the remote location of her summer house, the supposed location of her sighting. Rose said she had seen a mountain lion and bears on the property, but she liked the solitude it offered.

Rose French.

She explained that she grew up in Eureka and graduated from high school in 1956. She and her husband, a retired letter carrier in Eureka, have two girls and one boy. Early in their marriage they lived in Eureka and later moved to Willow Creek. Rose said that, "Grandpa wanted to purchase a 40-acre summer house on Bee Mountain off Capell Road." This was in the early 1970s, and located just down from Fish and Onion Lake and not far from Bluff Creek. Grandpa had wanted to use the location for hunting and as a retreat, but couldn't afford it on his own, so Rose and her husband decided to assist in the purchase. Note: This residence is in very close proximity to the location of Eaton Wood's sighting (see report below).

The summer home was burglarized three different times in the first three years they owned it. Sometimes the burglars broke a loft window for entry, and other times they took windows out. The house was in a very remote location, but there were a lot of "hippy" types that lived off the nearby lands and grew marijuana to support their lifestyle. After the third burglary, they decided to install a gate at the entry to their road to stop all vehicular traffic. It was also during this time that her husband routinely walked the 40 acres as an exercise. On one of these walks he discovered that someone was growing marijuana on the corner of their lot, which made them very angry, so they immediately cut the plants down and burned them.

Rose explained that the house soon became a place they went regularly to relax and enjoy the view from their front deck. Grandpa regularly hunted deer from the property, but her husband and son never hunted because the brush was too thick.

In August or September 1973, when Rose and the family were vacationing at their cabin, she decided to visit a neighbor a mile away. She was approximately halfway to her friend's house when she started to get a very nervous feeling, like she was being watched. She couldn't explain the rationale for the feeling, but she knew something or someone was watching her. She arrived at her friend's house and explained her concerns, and said she would not go home alone. When she was ready to leave, she asked her friend to walk back with her, and she complied. Once they reached Rose's cabin, her husband drove the friend back home. Rose said she never had the feeling of being watched before, and hasn't had that feeling since.

In the fall of 1974, when they arrived at their property they found the front gate damaged. Rose explained that her husband and father-in-law decided to fix the gate with the help of her son, Rick. Rick wasn't really helping a lot with the mending of the gate; he was more interested in riding his motorcycle up and down their dirt road. Rose said that work on the fence continued most of one afternoon.

It was about 4:00 p.m. when she heard her son racing the motorcycle up the hill. She was mildly upset and was going to talk to him because it sounded to her like he was operating the cycle too fast. She walked out onto their deck that overlooked their road and saw her son make the turn back towards the house and up into the driveway. Rose said she saw a huge creature running on two feet up the dirt road behind her son, quite a distance behind him as he was now parked in front of the house. It was approximately 125 feet from Rose when it stopped and turned slightly to face her. She has no idea why the creature stopped chasing the motorcycle, but thinks it might have smelled her or heard the cycle engine stop. The creature looked young, not fully mature, and stared directly into her eyes and showed emotion in its eyes as though it was sad, as though it was caught having fun. After what felt like several minutes, but was probably several seconds, it turned back and ran over the side of the road and disappeared, never to be seen again. Rose then yelled at her son and asked if he knew he was being chased; he said he didn't. Initially Rose thought that maybe someone was playing a practical joke on her, but soon realized after talking with her son that she had witnessed a bigfoot.

Rose described the creature as running so fast its hair was flowing from the back of its head and shoulders. Its hair was the color of a golden retriever. Hair was covering most of its body, except under its eyes and on its cheeks. The creature's skin appeared to be dark. She could only see it from the waist up because of the brush. It had a flat nose, large brown eyes, and high forehead. Her son and husband retraced the creature's steps and guessed it was close to seven feet tall, based on the height that Rose had seen from the deck. There was no odor associated with the sighting, and the creature made no sound that she had heard. She stated that she only saw the creature walk and run on two feet just as a human would walk or run, except it ran very fast.

During this same time period, there were two other odd events that happened near their cabin. Rose said that once you turned off Capell Road onto their private driveway you could see an abandoned rock quarry. A road construction crew had used the quarry to build nearby roads and had left it when the building ceased. On one of the many trips they made past the quarry, Rose saw huge boulders stacked on top of each other throughout the space. There were four or five huge boulders stacked one on top of another at different locations throughout the open area. There was no equipment on the property to accomplish this task, and based on the over growth on the road, none had been there. The boulders were so huge that no two men could ever lift one of them, let alone stack them four high.

Sometime after the bigfoot sighting, Rose and her husband hiked to a local creek to explore and her husband found several prints resembling human footprints in the snow. The tracks were much larger than a normal human footprint and no humans lived anywhere near where they were. It was also extremely cold in this area, and had recently snowed, making the idea of a human walking around in bare feet ludicrous. Rose's husband wanted to follow the tracks and see where they went, but she said, "No way."

Rose signed an affidavit.

LOCATION

The cabin was in an area notorious for illicit marijuana cultivation. Even today that area is known for illegal harvesting. The property is ideally situated between the Klamath River and the basins of Bluff Creek, Blue Creek, and Fish Lake. The area is also known for a huge number of bears and deer, yet most hunters won't go into the area for fear of being shot at by pot growers.

Less than a week after Al told me about Rose, I met Eaton Wood and took his report about a sighting near Miners Creek and Capell Road, less than two miles from the French residence. The two families don't know each other and weren't told about each other's sightings until well after reports were taken.

Since the sightings on Capell Road, the area has become hazardous for driving. Capell Road was washed out many years ago and the county has never repaired it. I know this has made many residents extremely angry, and they have taken repairs into their own hands and attempted to fix the wash out. It is still very hazardous driving with very steep cliffs. The driving, coupled with the cliffs, pot growers, bears, mountain lions, and some angry locals, makes traveling in this area a very dangerous proposition. I would encourage any possible researchers to steer clear and find another location to visit.

This sighting is very close to Fish Lake (a location of many bigfoot incidents) and Onion Lake, another location of many bigfoot incidents. The French's residence is even close to the Bluff Creek region where bigfoot fever started in 1967. It is also within 15 miles of Pecwan, a location with historical significance for bigfoot and Indian Devil sightings going back to the late 1800s.

Forensic Sketch

When North America Bigfoot Search first started to work with Al Hodgson, he had stated that our idea of bringing in a forensic artist to meet with witnesses was something that should have been done 30 years ago. Al loved the idea. I decided early on in the second round of drawings that I would invite Al to one of the meetings where Harvey met with a witness. Al was very curious about how the sketch process worked. I knew Al and Rose were friends, so I asked Rose if she would mind if Al observed while Harvey worked with her on her sightings. She said that she would enjoy the company. Harvey, Gina, myself, Al, and his wife Francis, all arrived at the French residence, which sits in a very rural setting against the hillside, just on the Hoopa side of Willow Creek. Rose invited us into her kitchen where we sat around her table and Harvey took a position next to Rose.

Rose French bigfoot profile sketch.

Harvey started to review his forms with Rose and we were immediately impressed with her interest level and attentiveness. From the beginning I knew that Rose had a unique vantage point (a deck above ground level looking down on the biped). She reiterated that she observed the creature chasing her son's motorcycle up the hill towards the cabin. She said it stopped short of running all the way to the house, and then it looked at her on the upper deck. They stared at each other for a few seconds and then it turned at almost a 90-degree angle and ran from the driveway and into the woods. Rose's perspective was unique, as she had two distinctive views: one straight on and one at a side angle. She told Harvey she can clearly remember that as the creature was running from the driveway, it had hair streaming from the back of its head and shoulders that was nearly 12 inches long. She felt that it was close to seven feet tall, but not fully developed as it was slight in its build. She remembers the hair as a shade of gold. They were close when she looked directly into its eyes and face. When the creature looked up at the deck and saw her, she saw emotion displayed on its face, and she says it appeared to be sad that the game of chasing was over. She believes the creature had a "calm" soul and would never hurt anyone. She last saw the bigfoot running at blazing speed down the hill away from her driveway.

Harvey drew the sketch while having an ongoing conversation with Rose. She constantly offered small hints about the creature's appearance and her feeling that it was relatively young—of adolescent age, and not fully developed. She added descriptors that showed the creature did not have hair on the forehead or directly under the eyes. It did have a small amount of hair on the lower cheeks, under the nose, and on the chin. The nose was quite visible and was not large or flattened. The face looked much more human than animal, and did not look like the Patterson–Gimlin creature filmed in 1967 even though it did have hair over its entire body.

Rose French bigfoot face.

Harvey eventually completed the first sketch with the creature. When he showed the sketch to Rose she seemed to glow and had a huge smile. She said that it was exactly what the bigfoot looked like when she saw it on the driveway. I asked her if she had to rate the sketch on accuracy, with 10 being absolutely accurate and one being horrible, what would she rate it? She said she had no doubt that the drawing was a 10!

Harvey asked if we could stay longer so he could develop a sketch with a side view of the creature as it was running. He said it would be an interesting drawing because we had never had an optimum view from the side of a bigfoot running to get an accurate reading on the length of the hair. The second drawing didn't take as long as the first, and Rose's response to this was equally as impressed.

I asked Al Hodgson what his feelings were about the sketch process and Harvey's abilities, and Al said he was shocked at how accurate Harvey's interpretive skills were, and how amazing his artistic skills translated into a bigfoot drawing.

As Harvey, Gina, and I were leaving Rose's residence, I asked Harvey how he felt about the drawing and the witness. Harvey and Gina both stated that she was a very, very credible woman who had an amazing memory for detail. Harvey explained that he offered to produce a second drawing because he knew Rose had the ability to complete another amazing sketch.

MARLETTE GRANT JACKSON

One of my several meetings with Al Hodgson dealt with the number of bigfoot sightings in and around Hoopa. I explained that it was an unusual phenomenon because when you compare the number of people in Hoopa to those in Willow Creek, Salyer, Denny, and other small towns along the Trinity River, Hoopa seems to have far more sightings. Al told me there was a woman who worked in Willow Creek and lived in Hoopa who had a sighting many years ago. She was now an employee of Humboldt State University. Al felt she was extremely credible and encouraged me to contact her. He stated it was an unusual mix in that she was working in Willow Creek, yet was a Hoopa Native.

I first contacted Marlette by phone at her office at Humboldt State University and explained that Al had given me her information and who I was, and asked if she was willing to grant me an interview. She agreed.

Marlette was raised in Hoopa and was born to a Native American father and mother who had moved down from Washington State. She explained that she is part Karuk, Abenaki, Shasta, and Yurok. Her dad had several businesses when she was growing up. He once owned an auto parts store in Hoopa and was also a logger. Part of logging history in this region deals with bigfoot, and Marlette explained that she heard bigfoot stories as she grew up, some from her family and some from friends. She never really put much interest or credibility into the bigfoot stories because there was a significant amount of alcohol consumption by some of her friends at those times, and she associated the stories with alcohol.

Marlette Jackson.

In 1986 Marlette graduated from Hoopa High School and wanted to attend college. She was accepted at Humboldt State University (HSU), approximately 60 miles from Hoopa and a good hour-long drive through the mountains. Marlette graduated from Humboldt State University with a bachelor's degree and landed a job in the Indian Teacher and Education/Personnel Department of HSU, where she still works today.

In June 1987, Marlette was home for the summer and living with her parents, and had a job at the biggest restaurant in the area, Cinnabar Sam's in Willow Creek. It is one of the busiest restaurants in the town and has a lot of historical photos on the walls of various miners, loggers, and even some bigfoot events that have occurred in the area. The restaurant is on the main road through Willow Creek and gets a lot of tourist traffic during the summer months. They even get a moderate number of tourists in the fall and spring because of the fisherman coming up for salmon and steelhead.

Marlette was a waitress at Cinnabar Sam's restaurant and often would be asked to stay until late in the evening waiting tables and helping the staff to clean the kitchen. She said it was good money, decent hours, and the people were fun, so she rarely said no to staying late. One night in late June Marlette left the restaurant, entered her vehicle, and made the turn onto Highway 299 westbound. The highway runs through the heart of Willow Creek, a short half-mile ride through the town, passing some huge wooden statues of what people believe bigfoot looks like. At 11:00 p.m. there aren't many businesses open, except the gas stations that are self-service only and maybe one local watering hole. She made a right turn onto Highway 96 and started to make the 14-mile drive to Hoopa. She had made this drive hundreds of times, passing the Bigfoot Motel, state highway yard, Highway Patrol office, and then the road starts to meander its way through the forests as it parallels the Trinity River. Marlette remembers it was a warm night and she wasn't in a hurry to get home. Approximately six miles from Willow Creek, adjacent to the highway, is the local dump that holds huge garbage containers where locals take their trash. The dumpsters are sealed at night and carted away when they are full. Approximately half a mile south of the dump is a very large turnout and small passing area for northbound traffic. This is an area where the road makes a sweeping left turn with a huge dirt turnout where locals sometimes park their cars to access the river.

Marlette was going into this turn fairly slowly and starting to exit it when she saw a huge creature standing in the middle of the road. It was tan or golden in color, almost seven feet tall, very physically fit but not bulky, and she could see hair flowing from it as it started to move. As her headlights hit the creature, it immediately started to move towards the shoulder of the road (river side) in two huge steps. One distinguishing feature she remembers is that as the creature was walking off the road she could actually see the muscles in its legs and upper thigh flexing. Marlette said she has seen many, many bears while living on the reservation, and she is positive this was not a bear. This creature had the body composition of a very large human and not the barrel body or snout of a bear. The creature walked like a man, didn't stoop over, and was very big and hairy.

Marlette Jackson sighting location.

She did not see the creature walk on all fours, only on its two feet. Marlette slowed her car considerably to get a good look at the creature and watched as it took just two strides to make it to the right shoulder of the road and then stepped behind a huge fir tree. As she passed the creature, she looked in her rear view mirror and saw her red brake lights glistening off the creature's eyes. At this point she continued driving straight home.

Once home, Marlette told her parents what she had seen. Her dad said they should go back to the scene immediately and see if they could find any tracks. Approximately 45 minutes later Marlette and her family were back at the sighting location searching the area with flashlights, trying to find some evidence. She said they found some hair on the fir tree that the creature was behind, but she doesn't think anyone kept it. She explained that the tree sits on the edge of the shoulder of the road with no room for anything to stand behind it; it's all cliffs behind the tree. Marlette thought that the creature must have been hugging the tree and not standing behind it when she saw it in her rear view mirror.

340

Marlette claims that she only told her mom and dad about her sighting, and decided to not tell anyone else. It wasn't until her parents contacted Al Hodgson and explained what their daughter had observed that Marlette felt comfortable enough to talk about her sighting to others. She does feel that seeing bigfoot is a positive experience, as many of her tribal members believe.

Marlette signed an affidavit.

LOCATION

Highway 96 borders the coastal range in this area and has a long history of bigfoot sightings. Approximately four miles from Marlette's sighting is the Tish Tang Campgrounds. There were a few sightings in that area and footprints seen on the opposite bank of the river from the campgrounds. Four miles further down 96 you start to enter Hoopa, and there were a series of sightings I documented in my first book that were all close to Highway 96 or directly on the roadway. The time of the sightings (11:00 p.m.) is probably on the early side of normal for seeing a bigfoot; most sightings in this area are later at night or early in the morning. The most unusual aspect of the sighting is the color of the creature. Marlette told me that the color was almost like a golden retriever. It's rare to get a bigfoot sighting of that color. It was also interesting to question her about the body composition of the bigfoot. She stated that it was quite healthy looking and not overly stout. Many sightings in this area have described bigfoot with huge legs, upper back, etc. Marlette made it clear that the creature was proportional to its size but not abnormally large, possibly an older adolescent.

Marlette said she had seen many bigfoot specials on television and had seen pictures of bigfoot, and the creature she saw on the road generally fits the description of what she had seen on television and in pictures. The one major difference she adhered to was that the color of the creature was different from anything she had seen on television.

I found Marlette to be a very credible and excellent witness.

After a couple of false starts we managed to meet with Marlette at a McDonalds restaurant in Mckinleyville. It was a nice April day in 2008 when Marlette was gracious enough to take time away from work at the university to meet with us. She was a very intelligent and cooperative witness who worked well with Harvey.

Gina and Harvey asked Marlette several questions about that night in 1987. She said she could clearly remember the night and was only concerned because the color of the creature was something that nobody had ever described before; it was golden. (Jennifer Crockett [chapter 9] was another witness who observed a golden colored bigfoot.) That there hasn't been media coverage about a golden colored bigfoot doesn't concern NABS. We now have others explaining that they saw it and are concerned whether or not people will believe them. Gina, Harvey, and I all smiled and we advised Marlette not to worry, we had heard these descriptors before.

Marlette saw the creature at a distance of 25 feet for almost 10 seconds, a significant amount of time to absorb details. Even though the incident happened over 20 years ago, the human mind has the amazing ability to remember specific details from major life events; a bigfoot sighting would be categorized as one of those events. She estimated that the creature was over seven feet tall and slender, not fully developed or not full size. She did not see breasts or other genitalia or ears. She did see long hair streaming from the forearms, another detail of many sightings that witnesses recall, yet isn't talked about in the media.

Marlette Jackson sketch.

We spent approximately an hour at McDonald's that afternoon and there was a constant banter around the color, size, age, and brisk walk that the creature had. Marlette could remember that as she passed the bigfoot when it was standing behind the tree, it peeked around the tree at her, and she can remember making eye contact. When Harvey asked Marlette what feelings that she had from the creature, she described it as friendly with no threat presented.

When he finished the sketch, Harvey showed the drawing to Marlette for her opinion. The one thing she said she might change would be to make the creature more golden than Harvey had drawn. She said it was very golden and a color that will stick with her forever. As we were getting ready to leave, I asked Marlette if she thinks about the bigfoot every time she drives past the sighting location, and she said she does, every time; she'll never forget it.

JULIE AND ALLEN BRADLEY

Willow Creek claims to be the bigfoot capital of the world, but residents of this small town still claim a few secrets. The residents have established a bigfoot museum in the middle of town, and various businesses are named after bigfoot, yet many of the residents have refused to step forward and talk about their sightings—until now.

Even though he is long retired, Al Hodgson still keeps his close ties to the community. He attends church locally and often talks with parishioners about bigfoot. It was during one of those conversations that news came to him about the Bradleys.

Allen Bradley spent many years working for a utility company in the Sierra foothills above Sacramento. He said he has worked around all the mammals that anyone would usually see in a mountain environment, and they never bothered him. When it came time to retire from the utility company, he and Julie decided Willow Creek would fit their lifestyle. Allen had inherited 50 acres several miles into the woods outside Willow Creek. The land had been in his family for three generations, and it is surrounded by national forest. It is the only piece of land that hasn't been clear-cut in over 300 years, and there are huge pine and oak trees on the property that don't exist in the national forest. There is a herd of deer that live in the brush nearby, and they have frequent visitors such as black

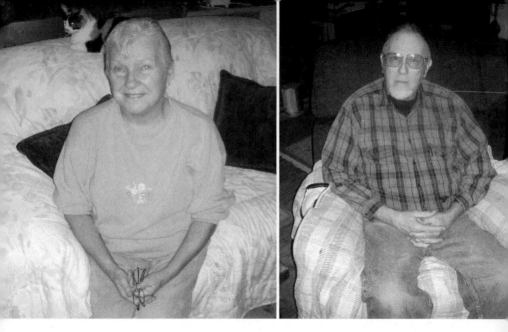

Julie and Allen Bradley.

bears, mountain lions, and raccoons. The land is in Humboldt County, but the Trinity County line is a quarter mile away on the South Trinity River. The elevation of the river in this area is 800 feet, the lower property sits at 1,600 feet and the upper side sits at 2,200 feet. The area just west of the property has three small lakes nearby with an elevation of 2,400 feet. The property is seven or eight air miles from Highway 299.

Their home was 14 driving miles south of town off the South Fork of the Trinity River, and was extremely remote, only accessible by a long, lonely dirt road that Allen built; their closest neighbor was eight miles away. They lived off the grid, made their own power, and had no connections to city or county services. They lived on their property for 12 years. They had fruit trees, raised some vegetables, and generally had a great lifestyle. Allen brought a manufactured home onto the property over the road he built. He said it was quite an excursion to drive the house over and around huge trees and ruts in the road. When visitors finally reached their house they could not believe it had actually been brought up their road.

344

The Bradleys were snowed in at least once a year, so they eventually sold the property and moved into Willow Creek.

Julie and Allen purchased and operated a video rental store in downtown Willow Creek. It was a place where almost everyone in town would eventually stop and rent movies. They commuted into Willow Creek daily to operate their store and that store gave them continual public contact with some very interesting bigfoot stories. Allen remembers one day when an individual he didn't recognize came into town and engaged him in conversation about bigfoot. The store wasn't busy so he decided to listen to what the person had to say.

The individual identified himself as a producer of a famous television series, and said that he had seriously researched bigfoot and he felt it was a time traveler with the ability to walk in different dimensions, and that was why he was never caught or observed for extended periods of time. Allen said he thought the conversation was interesting, but he never told the individual that his wife had seen bigfoot. Nobody in town knew Julie Bradley had a bigfoot sighting. Nobody.

Julie Bradley was born and raised in London, England and came to America to pursue a relationship with an American serviceman. She was a hairdresser by occupation, but retired from that to help with the video store. Julie enjoyed the rural lifestyle and had routinely seen bears, deer, mountain lions, and other smaller mammals. She also cared a lot about what local community members thought of her family and their business. She didn't want to be noticed and didn't want to make any waves in the community. The video rental business was doing well, and she and Allen didn't want anything to go wrong.

In the summer of either 1999 or 2000, Julie and Allen had a group of friends and relatives up to their home for a gathering and barbecue. It was a beautiful day and Allen offered to take the group on an extended hike into the mountainside and show them the beauty that surrounded their house. The Bradleys had two dogs that were always rambunctious and Allen decided to take them along on the walk. The group left at approximately noon and Julie was in the house preparing food for the barbecue and setting the front table.

Nearing 1:00 p.m. Julie was in the front of the residence when for some unknown reason she looked out the large front windows.

She saw a huge creature walking across her front yard. She couldn't believe it! The creature was walking on two feet, was approximately seven feet tall, did not have a snout like a bear, and didn't seem to be in a hurry. It was almost shuffling its feet in the dirt as it made its way across her front yard. The creature had dark brown hair over its entire body, had a head that was slightly thrust forward from shoulders that were slightly slouched, and it didn't appear to have a neck. The creature walked as comfortably on two feet as any human, but she could quickly tell that this wasn't a normal human being. The top half of the body was the same length as the bottom half, but the top was very thick and appeared quite healthy. The arms did not sway as the creature walked, but stayed close to the body.

Up until Julie's sighting of the creature in her front yard, she had never seen the Patterson–Gimlin film, nor seen a picture of bigfoot or sasquatch. She was very confused about what she was looking at. She said initially it looked like a bear walking on two feet, except bears can't walk that smoothly or quickly on two feet, and the creature didn't have a snout or facial structure of a bear. Its arms looked more like human arms than bear legs. She watched as the creature walked across her front yard and disappeared into the forests.

She held back saying anything to the group after they came back from their walk; it was only the next day that she told her niece what she had seen. Allen heard the description and later told Julie she had seen bigfoot. In the years following the sighting, Julie made it a point to become educated about bigfoot, and learned that the creature she saw in her front yard was a match for the standard bigfoot portrayed in the media. Julie and Allen made a pact that they wouldn't tell anyone about her sighting, as they didn't want anyone in Willow Creek to think they were kooks. Years have passed since the sighting, the Bradleys sold their video store, and both agreed it was time to tell the world their story. Julie said she knows what she saw, and she is positive she witnessed a bigfoot.

The bigfoot sighting wasn't the only unusual occurrence on the Bradley property. The property was so remote that they didn't hear noises normally associated with human habitation. There were many times they would eat dinner on their front deck and then sit and enjoy the sights and sounds of their forest. It was during one of those leisurely evenings that Allen and Julie heard a "gong" type sound. The sound lasted three to four seconds and they heard it on two different occasions. They described the sound as almost identical to a gong, and it appeared to be coming from the ridgeline to the south of them. They both agreed that it definitely wasn't a natural sound.

The second unusual sound that they heard in their yard was a "put-put." This wasn't coming from a vehicle, but from some of the most rural areas of their yard. They heard this 3-4 times on their property, but they didn't hear it the last four years they were there. An unusual part of the "put-put" story came from Allen's mother. His mom had spent considerable time on the property when she was very young and distinctly remembered the "put-put" sound coming from the forests; it was never identified. It stuck in Allen's mom's mind to the point that she specifically asked him if they had heard the sound after they had spent a few years living on the property. Allen was shocked that his mom asked him this question, but relieved that he and Julie weren't losing their minds. He confirmed to his mom that they also heard it, and that it was bothering them that they couldn't identify what was making the sound.

Sounds are one part of the unusual nature of the Bradley property, but there is more. Julie told the new owners of their property that she had sometimes felt as though she was being watched when she stopped to open the front gate. That gate was located down near the river and easily 10 miles from any residence. The new owner told Julie that she also had the distinct feeling she was being watched (to the point that it made her feel uncomfortable) when she was in the orchard.

Julie said that, aside from the bigfoot sighting, the most unusual thing she experienced at the property happened in the orchard. In the summer of 1999 she went down to the orchard to check on the pear and apple trees. She walked over the ridge and into the grove and found each tree completely full of ripe fruit and ready for harvesting. The next day Julie grabbed her bags and tools and made the trip to the trees to harvest the fruit. She arrived to find only 20 pears on a tree that had at least 100 the day before. All of the pears up to a height of eight feet were completely picked off. The apple tree was the same; the only fruit left was up very high. Julie went to get

347

Allen. During the walk back, Julie tried to figure out what had happened to all of their fruit. It was very doubtful that anyone would make the hike back into their property to take fruit that would have needed barrels or baskets to move. There was no evidence that a vehicle had been in the area or that there were baskets available for picking the fruit.

Allen and Julie went back to the orchard and tried to determine what had happened to their fruit. They searched the entire property and found only one piece of fruit that was on the ground, an apple. The apple was located uphill from the apple tree in an area where it could not have rolled. Allen picked the apple up and found that it had a giant bite mark on one side. A piece was not bitten out of the apple but something had dented the skin with its teeth, almost as though they wanted to leave its mark in the apple. The unusual nature of the bite was its size—too large for any human to make. The teeth marks were similar to a human's, without any fangs and an almost uniform dent. Allen examined the apple carefully and then placed in directly under the apple tree.

Allen and Julie also examined each fruit tree very carefully. There were no claw marks on the trees and no indication that anything had climbed the trees or caused them any damage. They discussed the possibility that someone growing marijuana illegally nearby came onto their property and stole the fruit. They quickly discounted that idea as there was too much fruit to carry away, and their dogs would attack and bark at anyone coming onto the property. They also talked about bears and deer eating the fruit, but also discounted that idea because they would leave pieces of fruit behind, and they would also cause damage to the branches and leave scratch marks on the trunk. The disappearance of the fruit was very puzzling.

The next morning Julie and Allen went back to the fruit trees to pick the remainder of the fruit that was high up in the trees. As they reached the area of the apple tree they both noticed that the apple Allen had placed under the tree the night before was gone. He walked around the area and found it exactly where he had found it the night before, uphill from the tree. Allen made a mental note as he walked towards the apple that it was lying with the teeth marks face up. Allen said that he got an uneasy feeling as he picked the apple up. He felt that something with intelligence was telling him they were being watched and "we could take more any time we want, but we will leave some for you."

When the new owners of the property told Julie that they felt they were being watched when in the orchard, she immediately remembered the orchard incident. The Bradleys feel that whatever took the fruit from their trees must have been huge in order to carry away all of the fruit from two trees. They don't remember any tracks that were left behind, but they also stated that the ground in that area was very hard.

Both Julie and Allen feel that people should understand that bigfoot is a real creature roaming the mountains of Trinity and Humboldt counties. It's been seven years since Julie saw the bigfoot in her yard, but she knows what she saw and is positive it was a bigfoot.

Julie and Allen signed an affidavit.

LOCATION

The Bradleys' property was extremely remote and just uphill from the Trinity River. It sits directly on the county line between Humboldt and Trinity counties. There have been several sightings in Trinity County just upstream from their residence, near Hyampom. Most of the Hyampom sightings occurred in the 1960s when there were many more people living in the area and more logging and milling taking place. The elevation of the property is very close to the optimum for a sighting in this area, while the isolation adds to the likelihood of a sighting.

One interesting thing Alan Bradley described is a "hum" that he heard occasionally on or near his property. He described it as sometimes sounding as though it was coming from underground. Months after taking this report, 100 miles north from the Bradley property and in Del Norte County, I took a report from Del Germain (chapter 9). Del gave an almost identical description of the hum. It's obvious that something odd is occurring in the hills in Northern California, probably underground.

With this sighting being in such close proximity to the Trinity River, it falls into a category that points to rivers as an associated factor in almost 75 percent of all bigfoot sightings and incidents in the four-county region.

FORENSIC SKETCH

We met the Bradleys at their residence on the outskirts of Willow Creek. They have a very comfortable house that sits against a quiet mountainside just up from the Trinity River. We could tell from the beginning that Julie was anxious to meet Harvey and Gina and work

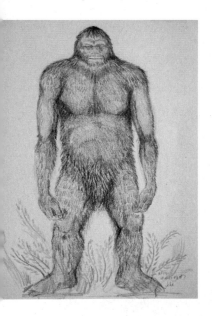

Julie Bradley sketch.

the sketch. Allan stayed in the room and quietly absorbed our discussions prior to beginning the sketch. Harvey explained his role in the drawing and walked Julie through his checklist of questions before beginning.

Julie started by stating that she saw the creature for approximately five seconds (probably the average amount of time a witness is able to view a bigfoot) as it walked through her yard. She thought the creature was nearly seven feet tall and had a golden highlight to the hair. In her mind the bigfoot had a definitive brow with no hair under the eyes but some hair on the lower chin area. She explained that she didn't think the hominid knew it was being watched, as she stood inside her residence and watched through plate glass windows. Julie felt that this creature was a female, probably a young female because there were no sizeable breasts observed. It was large, but not huge, across the body and its arms hung lower than a human's arms.

After approximately 70 minutes Harvey showed the drawing to Julie. She was very happy with the result and offered just minor

350

adjustments. She said it was a little more golden than what Harvey had drawn and was possibly a little thinner in the frame; Harvey made the corrections. The description Julie provided is probably of an adolescent female bigfoot.

This sketch matched others we had completed. Julie's comment that she felt this was a younger female that hadn't fully developed, coupled with her statement about the golden tint to the hair, are in line with our belief that many adolescent bigfoot have golden colored hair. It seems that as a female bigfoot gets older, breasts develop and the hair loses its golden color and becomes a darker tone. We were lead to this conclusion after consistently hearing witnesses describe golden-haired bigfoot that they felt were not mature and were slight in their frame. Everything about Julie's description and explanation about the demeanor of the creature matches the descriptors and information we have received from other witnesses about bigfoot.

JEFF LEWIS

In the realm of personal credibility and integrity, emergency service workers have always scored high in public opinion. Whether they are firemen, police officers, or paramedics, all have advanced training, most are well educated, and each is conditioned to respond at any hour to unusual circumstances. Most of these individuals thrive under extreme conditions, and that's

Jeff Lewis beside his ambulance.

one reason why they choose this line of work. People in these professions regularly testify as expert witnesses in court on a variety of issues, and are trusted by the court for their valuable testimony. In the world of bigfoot witnesses, you just can't find a more qualified individual than an emergency services worker to sign an affidavit on a bigfoot sighting.

During the winter of 2007, I was contacted by Al Hodgson of

the bigfoot museum in Willow Creek. Al and I had become good friends and he had regularly forwarded me information he received about bigfoot sightings and witnesses. Al had been acquainted with Jeff for many years, and knew he was a member of a local paramedic crew that worked out of Willow Creek and Hoopa. Even though, at six feet, four inches, Jeff towered over everyone, he was a very likeable, quiet, and respectful individual. They had never talked at great length about bigfoot, but one particular Sunday Jeff approached Al after church and told him about a strange incident that had occurred when he was working. Al listened to the story and decided it should be investigated. Jeff agreed to talk with me, so Al called me with Jeff's contact information. I phoned Jeff at his residence for the first interview. He was a very friendly and interested person who stayed right with the questions posed, and didn't appear to embellish or exaggerate anything.

There is one paramedic/ambulance group that has bases in Hoopa (at the Kimaw Medical Center) and in downtown Willow Creek. Each rig (Ford F350 four-wheel-drive diesel) carries two medical personnel; one drives and the other stays with the patient in the rear of the ambulance. The ambulance group covers a huge area, from Redwood Creek to Weaverville and onto Happy Camp. This region predominantly contains one-lane, winding mountain roads and drivers sometimes must deal with snowy and wet pavement. It is a very hazardous job, and the professionals must balance the condition of the person they are carrying with speed, road conditions, late hours, animals crossing the road, and regular traffic—a huge task.

In late July 2007, Jeff was working out of the Willow Creek station. It was a clear night, with great weather, and relatively warm for the area. The rig received a request for code three (lights and siren) transport of a patient to the Mad River Regional Hospital in Arcata. This is a relatively long ride up Highway 299 from Willow Creek, over the Lord Ellis Summit and down the valley into the Arcata and Mckinleyville area. It is approximately a 30-mile ride, a long time to concentrate on driving on a winding road in the middle of the night. Jeff was driving as they made the ride westbound up over the summit and started to drive downhill towards the coast. He didn't see any animals cross the road, and there was no vehicular traffic. The ambulance was traveling at 70 miles per hour, and its flashing red and blue lights were activated, but the siren was not on. Jeff explained that having the siren on for the entire trip would drive you, your partner, and the patient crazy, so it's only activated when they come upon traffic.

They had passed the scales just east of Blue Lake and started to reach the flats as they neared the coast. In this spot, the Mad River is just south of the roadway and would be visible during daylight hours. The area of Blue Lake has a history of bigfoot sightings going back 30 years. There is a small contingent of Native Americans in this area and a small municipality adjacent to the river.

A quarter mile east of the Mad River Bridge Jeff saw a very tall, thin figure in the roadway in his lane near the bridge. His first thought was, "What is a seven- to eight-foot-tall black guy doing in the roadway at 3:00 a.m?" This thought quickly vanished when he saw the creature take two huge strides and completely cross both freeway lanes going northbound onto the shoulder. The strides were so long that Jeff could see the lane markers between the legs as it moved. As Jeff got closer he was able to make out more physical features and saw that the creature was thin, very dark in color, and extremely tall, at least seven-and-a-half feet. He was able to judge the height in comparison to the height of his paramedic rig.

Jeff was questioned in detail about the stride, physical description, and movement of the creature. He said he has never seen a human with a stride as long as what he saw cross the road and make it to the shoulder and up the embankment. In some ways it was similar to an extremely tall, thin human wearing a hoodie over his or her head. The initial appearance was of a hoodie and thin man, but as he got closer he saw that it definitely wasn't human. Since the incident he has talked with people, looked at bigfoot photos, drawings, etc., and now believes that he definitely witnessed a bigfoot. The creature was over seven feet, six inches, taller than any human in this area; its movements were not like a human's; its stride was longer than human's, and its ability to clear two lanes of freeway traffic and the shoulder of the road as quickly as it did is just not possible by any known human.

Jeff Lewis signed an affidavit.

LOCATION

Jeff started his trip in the "declared" bigfoot capital of the world, Willow Creek. This is 14 miles south of Hoopa, the center of my last book and a location of many bigfoot sightings for over 100 years. Highway 299 traverses ideal terrain for bigfoot and crosses several rivers as it makes it way to the Pacific Coast. This entire incident occurred in Humboldt County, huge in bigfoot history for sightings during the last 100 years. I have personally documented many sightings on Highway 299, from bigfoot chasing a deer to bigfoot merely crossing the road. It would be amazing to be able to document the number of times a motorist has seen a bigfoot crossing Highway 299 during the last 50 years; I'm sure it would be in the hundreds.

If Jeff didn't have a personal relationship with Al Hodgson, then it is possible that his sighting would have gone unreported. Many people who use Highway 299 live off the grid in the forests, and may not have a venue to file a bigfoot report. Many people may not know that there are organizations (like nabigfootsearch.com) that collect bigfoot reports from throughout the United States. People may not know that there are over 100 reported sightings of bigfoot in Humboldt County. Many people are reticent about reporting a sighting because they fear they may be ridiculed and embarrassed.

When Jeff and I met to take his photo, he offered to meet at the Hoopa station to have his picture taken next to his rig. He thought it important for people to understand that what he saw was just about as tall as the highest point of his ambulance. We talked about his sighting. Jeff never wanted to exaggerate the facts of his observation; he kept me centered on his specific words. He was a little concerned at the creature's thin frame. He said it wasn't the normal physique normally described for bigfoot. This added even more credibility to Jeff's sighting.

The thin frame Jeff described is prevalent in probably 20 percent of the sightings I've personally taken. The famous Patterson–Gimlin footage of a creature crossing a creek bed in 1967 shows a female with a rather large frame. It's easily to see how a story of bigfoot could be told about a big creature with a huge chest and frame after viewing this film. A witness describing a creature that's thin is contrary to the norm and lends itself to a credible sighting. The general consensus among bigfoot researchers is that a thin creature is probably an immature male/female, similar in age and stature as a 15- to 19-year-old human. Humans are tall and thin and generally don't fill out their frames until they mature. There are a few stories in my first book about creatures fitting Jeff's general description. I should note that my first book hadn't been released when Jeff gave me his statement.

The last facet to this sighting that falls into the realm of reality for the bigfoot researcher is the time of day. As Jeff had stated, he hadn't seen another vehicle on the road for his entire transport down the mountain. At 3:00 a.m., bigfoot knows it can wander at ease and isn't likely to be spotted by people traveling along the roadway. While there are many daytime bigfoot sightings, there are significant numbers of sightings at night. Although there are far fewer cars on the road at night, there appears to be more bigfoot sightings then. Conversely, although there are far more cars on the road during the day, there are just a few roadside sightings of bigfoot during daylight hours. The 3:00 a.m. sighting falls well within the normal time for a night sighting. The creature's response to the ambulance coming down the road towards him also fits the profile of every other sighting I've taken of a bigfoot in the road. They will clear the lane rapidly and sometimes stand on the shoulder and watch the witness pass. Other times they will clear the lane and jump off the embankment.

We are not stating that bigfoot is nocturnal. There may be a rational reason why more bigfoot are seen at night on roadways. It may be possible that they understand there is less traffic at night and that may be the safest point in their day to across that region of their territory.

EATON WOOD

Early in my bigfoot quest I met Phil Smith of Bluff Creek Resort. Bluff Creek is where the famous Patterson–Gimlin film was shot, and the Bluff Creek Resort was the business that outfitted the team.

It seemed an obvious place to start the adventure and get my feet wet in the bigfoot arena.

Phil was very helpful, and took his time explaining the history, the players, and the role of the resort. Bluff Creek Resort is a recreational vehicle park that also has small cabins for rent. The resort sits on the bank of the Klamath River in close proximity to where Bluff Creek empties into the Klamath (see *The Hoopa Project,* chapter 5 for Phil's story.)

Phil had actually seen a bigfoot behind his resort, in the woods on the opposite bank. He had routinely taken his rowboat across the river and walked the opposite bank, exploring, and he found odd trails on the bank that didn't fit with the absence of humans in the area. As he was looking up the hillside, Phil saw the bigfoot looking at him around a tree. He also saw trails with very high clearance, not like a normal game trail, and he had also seen huge piles of feces that were much too large to be a bear's, the only other large game in this area. This region of the Klamath is very isolated and is nearly impossible to access unless you are in a boat. The property on this bank is part of the Hoopa Indian Reservation, and technically it requires a special permit to even step foot on it. Phil told me that he would gladly take me into this area during any late summer month when it was easiest to cross the Klamath. I told him I would be back the following summer.

I documented Phil's information and went about the task of accumulating other sightings and related witness information. I was spending a vast amount of my time on the Hoopa reservation, much of that time talking with tribal elders. Many of the elders told me bigfoot was fact, and they frequented all areas of the reservation, but they lived at the far northern end near Shelton Butte (very near Phil's property). They told me it has been known for many years that this is bigfoot's region, and this was an area where the tribe wouldn't travel. After hearing the elders' stories and connecting the dots to Phil's story, it was easy to see that the area Phil had described in his sighting was the exact spot the elders had described as bigfoot's home.

During the early winter months of 2007 the area of Hoopa and Weitchpec (Phil's community of residence) suffered some severe storms. There were heavy winds and a sizeable accumulation of the snow. After one of those storms, several trees around the Bluff Creek Resort toppled and roads in the area were blocked, so Phil's mother asked him to clear some of the roads with his chainsaw. He drove down one of the roads in his pickup and, just as he got out of the cab, a giant tree fell, killing him instantly.

Phil had lived with his girlfriend Vanessa and their daughter. I spoke to Vanessa shortly after his death, and told her I would check in from time to time to ensure all was going well.

In late summer 2007 I stopped by Bluff Creek Resort and met with Vanessa and a new assistant, Eaton Wood. Eaton is a 26-year-old native of Weitchpec who grew up less than two miles from the resort on a hillside location also up from the Klamath River. Phil's death hit him hard. Eaton said he knew a lot about the resort, so it was an easy transition to help Vanessa with tasks that would otherwise be too monumental for her with a small daughter.

Eaton knew about Phil's bigfoot stories, and he said he had also observed a bigfoot when he was a young boy. His family lived west of Weitchpec and north into the mountains, in the direction of Bluff Creek. Their house was on Coppell Road far up the mountain and without many amenities that would be found in a house in town.

Water was one of those utilities not easy to obtain; the family had to lay 6,000 feet of pipe up the hillside until they found the headwaters of Minors

Eaton Wood near Klamath River.

Creek. The water lines were laid up a very steep hill, but not in a straight line; the trail to the top of the pipe meandered in a series of switchbacks several miles up into the hillside. The hike to the top took over two-and-a-half hours and had many serious and dangerous obstacles; Eaton had seen mountain lions and bears, and he had slipped many times in the mud during the winter. It was his job to maintain the line to his family's home. His mom had bad asthma and was in no condition to make the climb to the top of the pipe each time water stopped; his brother was three years older and relegated the task to Eaton. He also had a sister three years younger, but she was too small to make that long hike.

In August 1992 it was very hot. For some unknown reason the water pipe had stopped delivering water, so Eaton's mother told him to hike to the top and figure out why the water wasn't flowing. The previous few times he investigated, he discovered that a bear had bitten on the pipe and chewed it up (apparently liking the sound of the water flowing through the plastic). It was three o'clock in the afternoon and Eaton decided to take his 22-caliber rifle in case he stumbled across a bear or mountain lion. It was so hot he didn't feel like carrying it, but it gave him a sense of security while he was so far from his house.

The hike was very uneventful for the first two hours. It was hot and almost muggy outside with no wind, but he was under huge trees with a good canopy offering excellent shade. He had almost completed all of the switchbacks and was on a relatively level part of the trail. He was getting tired and hot, and was a little bored, as he was walking slightly uphill on the trail with his head down. For some reason he felt as though something or someone was nearby and was looking at him, so he looked up. Approximately 30 feet up the trail was a bigfoot walking directly towards him. Just as he looked up, he momentarily caught the bigfoot looking down, but that was just a second. The bigfoot looked up and reacted first by making a very low but loud "ahh," and then immediately turned and ran straight up the hill off the trail. Eaton also turned abruptly and ran as fast as he could for about 200 yards downhill. When he was out of breath he slowed and cautiously walked the remaining distance back to his house. Once home, he told his mom what he had seen and she didn't believe him. He also told his dad when he got home, but his dad also didn't believe him either.

Eaton said the bigfoot really didn't look like the creature in the Patterson–Gimlin movie, but looked more like the creature in *Harry and the Hendersons*. He said its stature was the same as the Hendersons' and while he only saw the facial area for a few seconds, he does remember that the hair on the creature was as long as the Hendersons' creature. There was a smell in the area just as he saw the creature. The best way he could describe it was a very strong bear odor, very strong. In hindsight, Eaton said it might have been the smell that caused him to look up, but he can't be sure.

Eaton explained that the area where he saw the bigfoot is very remote. It is directly in a line from his house and the region of Bluff Creek where the film clip was shot. A reasonable guess is that this location is six air miles from that filming location, and less than two miles from Bluff Creek Resort.

I asked Eaton if it bothered him that his parents didn't believe him, and he said it really didn't because he knows absolutely positively that he saw a bigfoot and nobody could take that experience from him.

At the conclusion of Eaton's bigfoot story, I told him about Phil volunteering to take me across the Klamath to show me where he had his sighting, and Eaton immediately volunteered to do it. We walked down the path from the back of the resort and made our way to the rowboat, which was sitting in sand next to the river. The spot behind the resort is prefect for a lazy rowboat ride on the huge Klamath River. It is probably the only spot I have seen on this river where it almost forms a manmade lake. The river is slow, deep, and the stretch is wide. There is a reverse eddy that forms and takes you back upstream to make the trip easier. Eaton showed me how to cross and said that he would gladly escort me later in the week. We set a tentative day and I said I would get a permit from the tribal office.

On August 20, 2007 I did further follow-up on the resort's exact location and determined that the property on the opposite side of the Klamath was United States Forest Service Property. I contacted Eaton and we set a date on August 22 to cross and explore.

I met Eaton at 9:00 a.m. on a foggy summer day. It was perfect weather to cross the Klamath, a bit chilly and no wind or precipitation. It's a good 15-minute walk from the resort down a winding trail to the beach where Eaton keeps the small boat. The trip across was uneventful other than a few huge fish jumping around us. There had also been a giant hatching of miniature frogs. There were thousands of them on all the banks, a clear example of the large amount of food source that surrounds the Klamath River.

We docked the boat at a huge rock outcropping slightly upstream from where we started. This side of the Klamath River rarely has humans walking its banks. To access this area of the Klamath from the Forest Service roads is next to impossible. There are no dirt roads

On the left side of this boulder we found huge piles of scat. To the right of the far right tree is a heavily used game trail. On the far side of the boulder is a small den. The boulder has a beautiful view of the Klamath River.

The forked tree where five piles of scat were clearly visible. Some of the piles in this area were piled on top of each other.

leading into the region; the closest one terminates four miles and two valleys away. To make ones way over the mountains and through the valleys would take at least eight hours and nobody would do it. Most of the valleys have almost sheer cliffs protecting their entrances.

When we stepped foot on shore, I explained to Eaton to walk carefully and to keep his eyes open for any prints or other disturbances on the ground. In our first 400 yards of travel we had seen several piles of bear scat. We stopped occasionally and poked through the piles seeing what they were eating and the freshness. One common theme in every pile was the large amount of berry and fruit pits. We made our way up several very steep hillsides and sometimes saw huge indentations 17 to 20 inches long and eight to nine inches wide in the leaves. Many of these appeared to be bigfoot tracks, but amongst the leaves were impossible to photograph and show definition. We also saw many locations where it appeared a huge creature and skidded down the hillside, making the leaves separate and showing only dirt.

The day was separated into two distinctively different hikes, one on the east side of a creek we landed on and the other on the west. The later part of the hike encompassed an area to the west of our docking location and was probably the most interesting.

We made our way to the west by crossing the creek high into the mountains. Down in the creek valley reminded me of something in a rainforest. There were lots of ferns, dense underbrush, and

Bones found adjacent to the boulder.

everything was green. This area was probably 20 degrees cooler than the hillside we were on that was starting to have sunlight. Eaton and I took a break next to the creek and we refreshed by throwing water on our faces.

After sitting and discussing this great location, we made our way into the western corridor and onto the adjacent hillside. This area was predominantly filled with huge, old oak trees with a thick

An example of the size and proximity to each other of some of the scat in the area. The scat had the appearance of horse scat, but there are no horses in this area; five scat piles in an 18-inch radius.

layer of leaves. It reminded me of locations in Hoopa where the Natives harvested tan oak mushrooms. Tan oaks are also one of the main food sources that the Natives believe bigfoot eats. We stayed high on the hillside until we saw a huge boulder (size of a large house) sitting at the edge of the forest line and then headed towards the shores of the Klamath. The location caught our eye because there was a heavily used game trial directly under the boulder. This was also the exact area where Phil Smith had explained that he saw a bigfoot.

We continued walking down the hill toward the boulder and were approximately 200 feet away when we found deer antlers and part of a skull lying on the ground. The skull still had some soft tissue attached and overall it hadn't deteriorated much. We continued walking to the boulder, I on the high side and Eaton on the low. Approximately 15 feet from the boulder I noticed a huge pile of scat lying near my feet, probably two months old. The piles were round and large and filled with fibrous matter, nothing else. I took another step and found more piles of the same type of scat all within several feet of where I was standing. In an area of approximately 150 square feet there were over 20 piles of scat, all the same consistency and all nearly the same age. The oddest scat piles were found at the base of a forked tree. The base of the tree had multiple branches growing at different angles. There were piles of scat at the base of the tree at such an angle that whatever put the piles there had to back its way into the tree to drop the scat. To compound this, there appeared to be multiple piles on top of each other, an extremely odd sight. All of the scat was directly behind the giant boulder. Ninety-five percent of the scat piles had the same fibrous content, but there were a few that had a liquid look to them, as though the creature ate something that was not consistent with its diet and bothered its stomach. Maybe a deer?

After an extensive examination and taking multiple photos of the scat piles, I slowly moved in a circular pattern out from the middle of the scat pile. On the western side of the boulder was an overhanging tree that added significant cover on the trailside, almost shadowing the location from anything on the trail. The area immediately under the tree was clear of branches and twigs and was dirt with a cleared area of eight feet by six feet. Immediately next to this area I found three bones randomly lying nearby. I found one bone that appeared to be a leg bone of a deer, another was a pelvis of a deer and a third was a vertebrae of what Eaton and I thought might be a bear, but definitely not a deer because it was too large. Each of the bones still had soft tissue attached, indicating the bones weren't much more then two or three months old.

I made my way to the top of the boulder and found there was cover while sitting on the top of the boulder and viewing the river and trail. Something could actually sit at the top of the boulder and have a clear view for half a mile down river and up river, a perfect location for an ambush predator. It appeared to have its bed adjacent to the boulder, and it didn't move far to relieve its bowels. One interesting aspect to this site is that it appeared to be heavily used two or three months previously and was then abandoned. This time frame would be the same time when the Klamath recedes to the point where making a crossing by boat possible. Eaton had told me that he had started to make crossings in early June.

I photographed the area, bones, scat, and the boulder, and we headed back to the boat. I left a huge pile of wild apples that I had picked while I was enroute to the resort. It is an old Native tradition to always leave a thank you gift when you enter bigfoot country.

The crossing was uneventful and we made the walk uphill to our vehicles. Once at the resort we sat at the picnic tables and agreed we had never seen scat piles as large or as plentiful as those at the boulder. Eaton said he would keep a watch on the boulder with binoculars over the next several days and contact me abut any oddities he observed. Eaton signed an affidavit.

FORENSIC SKETCH

Harvey, Gina, and I drove to the Bluff Creek resort to meet with Eaton. We caught him doing yard work, but he was very accommodating and made space in his family residence for us. I told Harvey that Eaton's sighting had been just up the road from Rose French and in close proximity to Pecwan, Fish Lake, and the general Bluff Creek area.

Creature seen by Eaton Wood.

Eaton took his time explaining to Harvey the story that led to his sighting, the area, and the time of day. Harvey in turn read Eaton a series of questions off his form to make him comfortable with his process and to allow Harvey to have a good start on what to sketch. From the start I knew that Harvey was going to draw one of the largest bigfoot he had ever sketched.

Eaton explained that he and the creature were walking down a trail directly towards each other and both seemed to look up simultaneously. Eaton was carrying a rifle and I'm sure that added to the creature's fright. Eaton started off by stating that this hominid was tall, eight or nine feet, very large in stature (800 pounds), and appeared to be a mature male. The creature had dark brown hair four to six inches long over its entire body and had an odor that was very strong. The creature appeared very powerful and quite healthy. Eaton described the face as having no hair on the forehead, but hair coming off the top of the head towards the eyes. It did not have hair directly under the eyes or below the nose, and only had a small amount at the bottom of the cheeks and chin, identical to Rose French's sighting just down the road from Eaton's. There were no genitalia, ears, or teeth observed. As with every creature Harvey had drawn on this trip, this one did not have any visible neck and it was described with a head almost sitting on top of huge shoulders.

We spent more than an hour talking with Eaton, and making a variety of corrections, additions, and small changes to the drawing. In the end Eaton was very, very happy with the image Harvey created. The face on this drawing did not look menacing, but it also didn't look friendly and it did appear male. Eaton stated that he felt this was a male just by the size, the way it walked, and its general appearance. He did not see any breasts.

As the creature and Eaton saw each other simultaneously, he said he heard it let out a small scream or yelp, almost out of fright. This was the first time we had ever heard of a creature letting out a sound similar to what Eaton described.

Eaton thanked Harvey and Gina for coming to the resort and working on the sketch. He told me that he wanted to leave an invitation open to trek across the Klamath and check out the opposite bank again this year. He hadn't seen or heard anything odd on that side since we had been there, but then again, because of high water he hadn't spent a lot of time crossing back and forth.

There was a calm reverence to Eaton as we were leaving. He had been keeping this sighting to himself since 1992 and, finally able to get it off his chest and have it put it on paper, he appeared relieved and pleased with the result. As the Pratts and I were driving away from the Bluff Creek Resort, Harvey commented how scary it would have been to be deep in the woods, look up, and see something as large as Eaton described. We were all certain that Eaton had seen a bigfoot, and had no doubt he had been scared.

HOOPA UPDATE

In my last book I concentrated my research on Hoopa and its surrounding region. A majority of my bigfoot witnesses came from the Hoopa area, and sighting reports from that same region are still filtering in. There is something about that valley that attracts bigfoot in a significant way. After spending over two years on the reservation I developed a good network of friends and informants and they continue to call with information about the area and sightings.

Our forensic artist, Harvey Pratt, came to Hoopa for the first book and spent a week interviewing witnesses and drawing sketches that defied the standard description of bigfoot. As with any community, there is going to be a percentage of the population away on vacation, on a business trip, or simply out of town for a break from the norm. When Harvey came to Hoopa last time he missed a few of the witnesses I wrote about in the first book. Their stories were still included, but the sketches were never completed. In April 2008, Harvey returned for a second round of sketches, with the emphasis being areas outside the Hoopa region. In an effort to be thorough and develop a group of sketches that adequately represent the region, I had two witnesses from Hoopa who were unable to meet with Harvey the first time, meet with him on the second trip.

The sketches made from the first trip were tightly held by my organization and never released in any public setting or forum. It was important to us that only the witness assisting on the sketch was allowed to view the finished product. We did not want future witnesses influenced in any way by seeing a sketch. We wanted all witnesses to have a clear mind on what they observed and have no subliminal pressure to draw anything that might resemble something they observed from someone else. The release of my first book was August 2008, and our second round of sketches was concluded in mid April 2008.

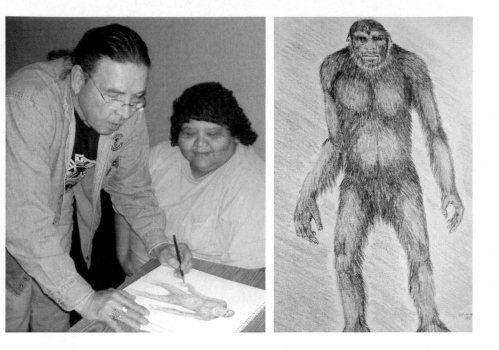

(Left) Debbie Carpenter sitting with Harvey Pratt.
(Right) Debbie Carpenter's bigfoot and full body sketch.

DEBBIE CARPENTER

We were unable to meet with Debbie on our first trip because she and her family had gone to Arizona for a family emergency. Debbie and her husband met us at the hotel in Hoopa and briefed Harvey on her sighting. She explained she was driving home early one morning, turned off the main highway onto Carpenter Lane and was facing the Trinity River. Down the road in front of her was a bigfoot kneeling in the roadway. Once her headlights illuminated the creature, it stood and took a few steps towards Debbie, and then jumped over an eight-foot berry hedge on the south side of the road and escaped through an adjacent field. (For additional details on her sighting, please refer to *The Hoopa Project,* chapter 5.)

Harvey initially asked Debbie details about what she observed, distances, size, colors, mannerisms, and a few other pertinent details.

One of the last of the initial questions that Harvey asks all witnesses is: what type of drawing do they want to draw? Do they want just a facial portrait, full-body shot, or just an upper-body sketch? Once the type of sketch is decided, he lightly outlines what he will concentrate on and shows this to the witness in an effort to get confirmation on the size of what he will draw. It's a consistent dialogue between Harvey and the witness on various aspects of the sketch as he slowly starts to put the pieces of the biped together. Debbie's husband sat in the room in amazement at what Harvey was putting together, but never offered his own opinion on Debbie's description and allowed her to continue her ongoing banter of identifiers.

Debbie chose to work with Harvey on a full-body sketch of the biped. One distinguishing feature that Debbie immediately told Harvey was that the creature she observed was much thinner than the bigfoot she has seen depicted on television. She said she wanted to be as accurate as possible, and it was important to her to tell Harvey this from the beginning. She explained that the entire body was fairly thin, legs, arms, chest, all much thinner than photographs and videos she saw on TV that supposedly depict a bigfoot. Another aspect of the creature that stood out was the length of the arms; they were very long. She said they almost reached the knees, much longer than a human's arm.

Harvey slowly penciled in the outline, then body, face, and hair until the creature started to take life. Debbie reiterated her initial statement that she wasn't positive what the face looked like because she wasn't close enough. She did say that it appeared dark.

After approximately 75 minutes Harvey had finished a majority of the sketch. Debbie looked at it in stunned amazement and said, "That's unreal how you were able to do that. It's great."

CHARLES MCCOVEY

Through a series of unfortunate events, Charles, Harvey, and I were unable to confirm a date to meet on our first trip, even though Charles was anxious to do the sketch. We finally met Charles at his residence and began the sketch at his dining room table. He lives in

(Above) Charles McCovey sitting with Harvey Pratt.

(Right) Charles McCovey's bigfoot.

a region of Hoopa that I coined "Bigfoot Alley" because of the number of sightings in the region. As luck would fall for Charles, he and a group of Hoopa forestry workers were clearing brush in a region of Bigfoot Alley above his residence. They had a fairly easy morning of clearing small trees, as it was cool and a little foggy. The crew decided they would take their lunch break at 11:30 a.m. and everyone except Charles settled into their truck to escape the fog and cool weather.

Charles wasn't hungry when the crew decided to take their lunch break, so he decided to walk further up the hill and look for mushrooms. He was probably 200 yards and a small knoll from his crew when he heard something walking towards him. He looked up and saw a bigfoot walking up a slight hill near him in his direction. Charles stopped walking and started to watch as the creature kept walking towards him.

McCovey explained to Harvey that the bigfoot he observed was huge, with big muscular arms, legs, and a massive torso. He stated it was dark in color and didn't appear to have a neck; it looked like the head sat on top or just down from the shoulders. He specifically explained that the muscles in the arms were so defined that as the creature moved he could see the muscles flex.

Charles said the creature looked up, saw him, and immediately turned and started to walk away. He could see from its profile that it had hair hanging from under its chin and flowing hair coming from the top and back of the head. It could also be seen from the facial profile that the creature had a strong large brow and flattened nose and wide lips. Charles specifically stated that the creature he observed had hair on its face, but none directly under its eyes or high on the cheeks. The hair on the head hung down near the top of the brow but didn't extend over it. It did not have a distinct waist, but was more of a large, barrel-chested creature where the hair and bodylines extended straight down to its massive legs. Charles stated several times that he felt the creature was a male, was massive, and was something that you'd never want to challenge or get upset.

Harvey worked the entire time that Charles explained different aspects of his sighting. After more than 60 minutes Harvey showed Charles the side profile and full body drawing of the creature. Charles stated he never thought anyone could capture exactly what he saw that day on the hillside, but Harvey had somehow done it.

CONCLUSION

The credibility of Debbie and Charles in the eyes of Harvey was very high. He said each witness told him at various times during the sketch development that certain things were incorrect and asked him to change them. Harvey felt that the passion, interest, and precise nature of each of them in their decisions regarding the sketch all pointed to someone telling the truth about what they observed, and someone who was interested in getting the sketch completed accurately.

It's obvious from Debbie's description that her creature is much different than any bigfoot ever depicted on a television special. Most bigfoot are depicted as huge, strong, with large muscles, quite different than the sketch made for Debbie. It was the opinion of Harvey and I that he was probably drawing a teenage or young adult bigfoot, based on an undeveloped upper and lower body. Debbie took considerable risk of ridicule and embarrassment describing a creature that didn't fit the norm for television; but it did fit a standard that we have drawn in the past.

Charles McCovey's drawing also went against the standard description of an American bigfoot. Most bigfoot depicted on television specials show hair completely covering the face, upper cheeks, under the eyes, everywhere. Charles described a face as dark, but with light hair that stopped at the lower cheek and didn't cover the area under the eyes or adjacent to the nose. This description goes against what the general public has been shown in the past, yet follows a similar physical description that we presented in *The Hoopa Project*.

It is the belief of the team that was on scene for these drawings that each witness had high credibility.

RAVEN ULLIBARRI SIGHTING UPDATE

In my first book I wrote about a sighting on the Hoopa reservation at a woman's house on the fringe of the woods, the last house before entering open country. Raven Ullibarri lived in the house and believed she had multiple bigfoot visits, the last being an event where she caught the biped leaning over a metal shed going through her garbage. The sketch of that creature appears in *The Hoopa Project*. Raven had made a comment to me that she felt the visits coincided with her menstrual cycle, as the biped appeared when her sanitary napkins were in the garbage. She knew they had visited on multiple occasions because of the manner that the garbage was moved and not torn, the footprints, and the odor.

In the last incident when Raven observed the bigfoot, Al Hodgson of Willow Creek got involved. He found hair on the top of the metal door where Raven said the biped had leaned over to grab garbage. Al placed those hairs in an envelope and showed them to a taxidermist in Redding. The opinion of the taxidermist was that the hair did not belong to anything living in the woods in Northern California; he had never seen it before. Al held onto the hair and placed it in his closet.

One afternoon I was visiting Al and we started talking about

371

Hoopa, my book, and associated evidence. Al went into his closet, recovered the envelope, and gave it to me, telling me to have it analyzed.

I sent several of the hair samples to our lab that we utilize for species identification and DNA analysis. This was the second sample we had sent to this lab, the first being the Kirk Stewart sample from Del Norte County (chapter 9). The lab found that none of the Stewart samples had a root at the end of the hair, which makes extracting DNA much more complex, but a few of the Ullibarri samples did have roots. The letter below is one we received from our lab concerning the Ullibarri sample. This is a very exciting development for bigfoot research and we anticipate evidence to further evolve quickly as science becomes more finite in its ability to extract DNA.

March 21, 2009

Mr. Dave Paulides
North America Bigfoot Search
15732 Los Gatos Blvd, #438
Los Gatos, CA 95032

Dear Mr. Paulides,

Our laboratory has been working on the samples you submitted for species identification over the past few months. On the newest set of hair samples submitted, it was noted that there was follicular material on the root end of the hairs. We took one hair and extracted the root end and obtained nuclear DNA from the follicular material present on the hair. Since it was nuclear DNA, we amplified the DNA with PowerPlex™ 16, which is a group of STR markers that is used for human forensics and paternity testing. This group of markers is also known to give results for primates as well as humans. Usually, the primates will generate results that are not concordant with human results though. In this way, information can be gleaned about certain primates and in some cases, identity of the type of primate can be obtained.

We did obtain results from PowerPlex™ 16 on this sample submitted. There was certainly an aberrant result for the Amelogenin locus. Normally, Amelogenin is a sex determining marker that has peaks at 106 for female, denoting the X chromosome, and 106/112, denoting the X/Y chromosomes, for male humans. If there is non-human mammalian DNA present, there is normally a peak at 103 but not 106 or 112 since this is a human specific marker. When we tested the sample submitted, the results for this locus were 102/106. The 102 peak was not expected and is unique in that our laboratory has not observed this in the past. There are a number of scenarios that could account for this unusual finding. As a result, at this time it cannot be ruled out that this could be DNA from an undocumented species. Research of this type takes time, but testing of these samples is continuing and now, there are results (data) obtained from testing that are being analyzed at this time and analysis is continuing.

If you have any questions or comments concerning our strategy for completing this case, please feel free to contact us.

Note: The name of the lab has been removed because NABS has had problems with others calling the lab, stating they are with NABS and asking for results on specimens. If anyone has specific concerns or issues with this or the other letter in the book, contact a NABS member at any conference and we will gladly show you the results.

11 MINNESOTA

BOB OLSON & DONALD SHERMAN

In the winter of 2007 I had read about bigfoot sightings at Bena in northern Minnesota. An interesting factor to the sightings is that they were either in or near the Leech Lake Indian Reservation. I knew I'd be in the region because my kids were attending a hockey camp. I contacted one of the people identified in the bigfoot article, Bob Olson in Deer River, Minnesota. He had taken an interest in the sightings and decided to take casts of footprints and interview witnesses. Bob was very gracious on the phone and invited me out to his business in Deer River, so I made the drive north and met Bob, Donald Sherman, and a small group of researchers that worked with Bob. They were all very hospitable and explained there had been many sightings in this area over the past few years. I asked the group what might have spiked the sightings in recent times, and one individual immediately stated that he felt the sightings were related to a cluster of UFO sightings in the area. UFOs had been hovering over various areas with water, and it was there that they were having sightings. He said the connections between the two incidents couldn't be overlooked; he was very passionate about his beliefs.

Bob had a collection of plaster casts that were made at locations of various sightings. There was a good collection; many of the casts had little detail, and a few had excellent definition and appeared authentic.

Don Sherman is a facilities manager at Cass Lake hospital and is a member of the Leech Lake Band of Ojibwe. He explained that some Natives do not like to talk to outside people about bigfoot, but he knew several people in the tribe who would talk to me, and

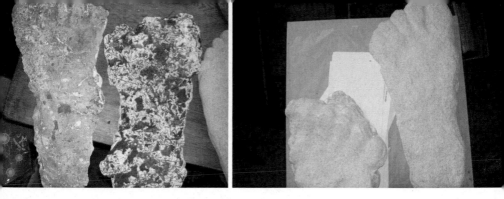

Some of Bob Olson's footprint casts.
Photos permission of Bob Olson.

referred me to tribal member Bill Bobbilink. Sherman told me the word for bigfoot in Ojibwe is bugwayjinini, which means wild man, and that stories about the *bugwayjinini* go back hundreds of years. The most common description is a brown, hairy, large creature that walks on two legs and has a human appearance.

Bob Olson explained that one of the best bigfoot incidents on the reservation occurred in June 2006 on Six Mile Lake Road just off Highway 2. A Cass County road grader had leveled the dirt roadway; when he returned to the site the following morning he observed several huge, human-type footprints near the shoulder of the roadway where it had been graded the day before. Bob Olson and a Department of Natural Resources (DNR) representative responded to the report and took casts of the prints. The DNR representative supposedly told the group that the new casts matched prints found in the 1960s in nearly the same area. The group searched the area thoroughly and found 30 prints leading into the adjacent swamp area. It had rained the day before and these prints appeared to have been made since the rain. The grader said he felt it was unreasonable to believe the tracks were faked because they were in an area where nobody would be walking and it was pure luck he saw them.

A search of the BFRO website indicates that two weeks after the discovery of footprints on the grading site, an Ojibwe woman was driving on Highway 2 towards the Indian casino at 11:00 p.m. When she was just outside Ball Club she observed a large, brown, hairy creature standing on two feet on the right side of Highway 2, and it

375

Another Bob Olson cast.

appeared about ready to cross (BFRO Report #17545).

During the summer months in the 1950s in adjacent Itasca County, a young girl was picking berries (another berry association) near the town of Bovey, Minnesota. She was in the yard of her home, which was surrounded by the forest, when she felt something looking at her. The girl looked up and saw a creature staring at her. She described the following event to BFRO, "something really, really tall standing by a tree. It looked like a huge man covered with hair. When I saw it, it ducked behind a tree." (BFRO #2677). The girl's brothers went back to the area with guns and found the area was matted down and smelled horrible.

In a literature search I found two small books specifically about Minnesota bigfoot that were published in the early 1990s. Mike Quast authored, *Creatures of the North: The New Minnesota Sasquatch Encounters,* and *The Sasquatch in Minnesota.* Mike did an admirable job of describing several bigfoot encounters throughout north-central Minnesota. Both books quote your standard bigfoot sightings as the creature is crossing a roadway (*The Sasquatch in Minnesota,* Strawberry Lake, 1976, pg. 5). On page seven Quast has a chapter titled, "Indian Legends, Myth or Reality." He wrote about several names that Natives in Minnesota use for the creature including, Windego, Wittiko, Wendego, and Big Man. At one point Quast returned to a Native American family he knew in southern Minnesota and saw them sitting in their yard in front of a fire. He knew "Running Wolf" and met their friend, Ray Owen. Quast showed them a photo of the famous Patterson–Gimlin bigfoot filmed in 1967, and Ray Owen said, "That doesn't look like the ones I see. They're more human. But then, they can look one way to one person and another to another person, however they want to look to get their message across. Whenever they appear, there's a reason, a message." (pg. 13). After a few more statements Owen said, "Then it turned and walked away and vanished into the air. That's the way

376

they can do it. They exist in another dimension from us, but they can appear in this dimension whenever they have a reason to." (pg. 14). His statement that the sasquatch he observed didn't match the Patterson–Gimlin creature is directly in line with the witnesses in Bena, Minnesota (see Bill Bobbilink below), and the vast majority of witnesses in Hoopa and others quoted in this book.

There have been many bizarre coincidences in my life, but none that beats what follows. As I was reading Quast's book, the Native's statements intrigued me, but I was floored by the second paragraph on page 14. "Every tribe no matter where you go knows the big man. Where I felt him the strongest was by Hoopa in California. See, sometimes you can feel him, you can sense he's around. I was driving along when my car died, and when I got out to look at it I could feel him real strong. There wasn't even anything wrong with the car, it started right up again."

I could not believe I had just stumbled onto a paragraph written in 1990 where a Native American from Minnesota revealed his feelings regarding bigfoot and Hoopa. Unbelievable! (Refer to *The Hoopa Project* for details.) Days later I was still in shock over the coincidence of finding this Hoopa notation. This 78-page, bound booklet was part of research materials purchased by NABS in April 2008 from Ray Crowe. The day I wrote the paragraph above was the first day I'd ever heard of the booklet. *The Hoopa Project* had already been printed and distributed by Hancock House Publishing.

One of the last questions Quast asked was, "Could sasquatch be killed if someone got a shot at it?" Owen said, "You could shoot him with a bazooka. You could drop a nuke on him. You can't kill a spirit." (pg. 16, para. 7).

BILL BOBBILINK

After I met with Bob Olson, I spent the night in Deer River and got an early start contacting bigfoot witnesses in the Leech Lake Indian Reservation. The first stretch of roadway I hit was Highway 2, the main road through the reservation. It goes from Deer River west towards Bena and crosses the mighty Mississippi River, but the

Mississippi isn't very mighty up north; it's little more than a creek. This area of northern Minnesota is very wet, almost boggy. Areas have very thick forests that are almost impenetrable. If you like to hunt, the best idea is to set up in an open meadow, or on the fringe of the meadow, and watch for wildlife crossing the area. The meadows that you walk through are treacherous because many of the creeks flow just below ground level and are only covered by light grass. You can easily fall through into the bog. If all of the moss, grass, and creeks don't bother you, the clouds of mosquitoes will, guaranteed.

Bob Bobbilink.

The first stop was the Bobbilink residence along Highway 2 just outside Bena. I had been told that they had a bigfoot visitation on their land. They were hospitable people and were willing to talk even though they were part of the local tribe.

I knocked on the front door and was immediately greeted by one of the Bobbilink girls. I was directed to the living room where Bill was sitting in a large chair. Bill was hard of hearing and he asked me to speak up. I asked him if it was true that he had a bigfoot sighting on his property, and he said it happened in the fall of 2002 at approximately 8:00 p.m. just at dark. He looked out the front of his home for some reason and saw a huge creature standing in the front yard. It was over seven feet tall, completely covered with hair, and appeared very muscular and healthy. As he was looking at it, it started to move around the side of the house towards the backyard. He switched his position and watched the creature from a back window as it slowly made its way into the forest behind his house.

The creature wasn't running, it was moving almost at a walking pace, slightly bent forward, knees bent, and moving with purpose. It appeared not to have a neck and its hair was four to six inches long over its entire body. Bill said he had yard lights on in the front and back and that aided him in viewing the creature. I questioned Bill about the facial structure of the bigfoot, and he answered that it did-

Front of the Bobbilink residence.

The Bobbilink backyard.

n't have hair on its face, but there was hair on almost all other areas of the body. He described the hair as a medium brown in color and consistent over its entire body. He felt that it was a healthy, middle-aged bigfoot in perfect health.

One of the last items that Bill mentioned was that bigfoot had visited his house several times in the past five years and he didn't quite understand why. Bill signed the affidavit, and I took his photo.

Bill's wife had been sitting in the room with us, along with a young girl, Jenny Dick. Jenny was their granddaughter and they were babysitting while her mom was out.

As I was leaving, I turned and thanked the people for allowing me in their home and I shook Jenny's hand. I told her that any time she or anyone in the house had information about bigfoot, or had another sighting of the hominid, to please give me a call. In a very soft but assured voice Jenny asked, "Do you want to hear about my bigfoot sighting?" I was stunned, and her grandmother said, "You'd want to talk to her, she had a very good sighting." I said I absolutely wanted to hear about it, and immediately sat back down in the kitchen with Jenny.

JENNY DICK

I interviewed Jenny in June 2007. At that time she was 11 years old and attended sixth grade in Bena. Her family lived at the Bobbilink residence. The setting of the Bobbilink backyard is important to this story. It sits against an open area that literally goes for 50 miles of nothing but open forests, swamps, bogs, and Lake Winnibigoshish. The lawn area in the backyard is large, close to 60 feet wide and 60 feet deep, with the perimeter surrounded by very lush bushes and trees.

In September 2005 Jenny was still on summer vacation. It was late in the afternoon and Jenny's grandmother gave her a basket of clothes to take out to their backyard and hang on the clothesline. Jenny said she walked out her back door looking at the basket to ensure that she didn't drop any of the clean clothes on the ground. As she walked down the steps towards the clothesline, something made her look up. She saw a huge, hairy creature sitting on the lawn in the shade, directly under one of the lines at the far left corner of the yard and adjacent to the forest line. The creature was sitting with its legs crossed looking up towards the line and twirling a clothespin with one of its extended fingers. The sight of a huge creature sitting on the lawn frightened her, and she dropped the clothes. Jenny yelled to her sister, Barbie, who was just inside the back door that something was with her in the backyard. At this point Bill Bobbilink heard the commotion and ran to the back door and told Jenny to come into the house immediately. She went back into the house, closed the door and they all watched the creature get up and rush straight into the forest, swinging its arms wildly. Jenny thought her grandfather's voice scared the creature, and that's what caused it to leave in haste.

Jenny described the creature as over seven feet tall, very dark brown in color, with hair over its entire body. The face appeared to be half human and half ape, but it didn't have hair in the facial area. There was some hair on the lower chin and dropping down from the head onto the forehead. She described it as a bigfoot that was very healthy and seemed to be quite content playing in the yard twirling the wooden clothespin. She felt bad she had disturbed the creature while it was having fun.

Jenny explained that she, her grandpa, and her sister watched the bigfoot get up and leave the area. I questioned Bill Bobbilink about this incident and he validated what Jenny stated. When Jenny later told her mom what she had seen in their yard, her mom said that bigfoot often came when they wanted something, mainly food. The family put out some fresh fruit that night and it was gone the following morning.

LOCATION

The Bobbilink residence sits just off Highway 2, between Highway 2 and Lake Winnibigoshish. Directly behind the residence is heavy bush with a trail that parallels the highway. Behind the trail are very thick forests and bushes with ponds and swamps.

As I canvassed the backyard area, I noticed there were a lot of berry bushes along the perimeter. The lawn was neatly mowed and there were large trees around the perimeter, which offered ample shade for a creature to sit in the yard and play with the clothespin.

Jenny Dick's story is too unique to have been fabricated. She and Bill Bobbilink signed affidavits to their sightings.

JODY DEVAULT

Part of any research effort is to canvass an area once activity has been confirmed. It was obvious that the region around Bena had bigfoot visitations, and it was worth the effort to knock on doors. Bob Olson had told me of bigfoot tracks being discovered on Six Mile

Jody Devault.

Road just north of Six Mile Lake, so I decided to drive the perimeter of the lake and see what other residents could add to this mystery. One of the homes I walked up to had a very nice woman answer the door who offered to tell her story.

Jody Devault was raised in this area and spent a lifetime walking in the woods and swimming the lakes of the region. She is very knowledgeable about the area and never had any unusual encounters until she was 17 years old.

Her family always lived near Six Mile Lake; it was approximately a two-mile walk from Highway 2 to their residence. Many of her friends didn't like driving the dirt road from the highway to her house, so they would drop her at the intersection and she would walk home. Remember, this is rural Minnesota, where many of the roads are dirt and there is hardly any traffic around lakes such as Six Mile. She had done the walk dozens of times and never had any problems or incidents, except once.

It was a summer night at approximately 10:00 to 11:00 p.m. and

Roadway Jody was walking when she felt she
was being stalked by bigfoot.

Jody was walking alone down her road on her way home. The mosquitoes weren't bad, but they were present. As she started to walk,
she felt that something was different about her surroundings.
Everything in the area was very, very quiet. The normal sounds that
come from the forest weren't present and it was eerily quiet. As she
left the open meadow area near the highway, she entered a narrow
dirt roadway surrounded by towering trees and bushes. She could-
n't see 20 feet into the woods and she could barely see 20 feet down
the roadway in front of her.

About 50 feet down the road Jody started to hear someone walking in the forest on her left, paralleling her. She said it sounded like
a human walking, but every once in a while a very large branch
would crack and break. She stopped and listened for the steps, but
whatever was tracking her also stopped. When she started walking
again, the walking in the forest started too. As she got closer to her
house, the footsteps in the forest seemed to get closer to her; her
anxiety level started to rise. She knew nobody lived in the woods
except her neighbors, and she had heard stories about bigfoot, but
she really didn't believe it would be stalking her. As she got closer
to the house, a horrible smell like rotten meat started to penetrate the

air. Many of the steps in the woods were breaking very large branches, and the cracking sound was very loud. Jody didn't believe a human was following her, but she was positive it was a two-legged creature and not a bear or a mountain lion. As she continued down the road, the smell continued to get worse and the footsteps continued to get louder and closer.

When Jody was within half a mile of her house, she started to jog for 100 yards and then stopped. She heard the footsteps also accelerate and stop just after she stopped, exactly how a human would run. The noxious odor was getting worse and at this point she was very frightened, so she ran all the way home. She made it in her front door and found everyone asleep inside. She took a shower and went to bed.

Reflecting on the incident, Jody now believes a bigfoot stalked her. There have been too many instances indicating that bigfoot calls this region around Six Mile Lake home, at least at some times during the year. Jody said there is absolutely no chance that she will ever make that walk at that time of night again.

DANIELLE DAHL

Danielle Dahl lives with her grandmother in a very beautiful setting adjacent to Six Mile Lake, right next to Jody Devault. She told me there are a lot of big fish in the lake, and it's known to fisherman near and far as a great place to catch a lunker. When I interviewed Danielle in June 2007, she was 19 years old. She had been born in Bemidji, Minnesota and was about to graduate from Cass Lake High School in Bena.

In April 2007, at approximately 1:00 a.m., she was staying up late doing the dishes in the front kitchen area of her grandmother's residence. The kitchen window looks directly at the corner of the rear yard of the Devault residence and there is a yard light directly above the corner where Danielle happened to be looking while washing the dishes. She saw a large creature standing on two feet adjacent to the window. It was dark in color, nearly seven-and-a-half feet tall and large in size. She couldn't say its exact color, but could tell it had hair over its entire body and was definitely not human. She watched in disbelief as the creature took two strides and disappeared around the corner of the house. She checked the area the next day and couldn't find any tracks in the gravel around the outside of the house.

I spoke to both Danielle's grandmother and Jody Devault, and they confirmed that Danielle had talked to them about her sighting, and also that nothing unusual had occurred in the area at that time.

Danielle Dahl.

LOCATION

The spot where the Dahl and Devault residences sit is just down the road from the location where the tracks were found by the grader on Six Mile Road in June 2006. This occurred ten months earlier than

Boggy marshland behind Dahl residence.

the Dahl sighting and less than half a mile from the residence. It would appear that the hominid was in this area for that specific time span. Six Mile Lake is adjacent to the Dahl property and the area behind the residence in a boggy marsh with grass, small bushes, and lots of animals.

Some facts about Six Mile Lake may help with an understanding of the food sources available to a bigfoot. According to the Minnesota Department of Natural Resources, the lake is 1,288 acres and contains black crappie, brown bullhead, largemouth bass, northern pike, pumpkinseed sunfish, rock bass, snapping turtles, tullibee, walleye, white suck, yellow bullhead, and yellow perch. This lake is a veritable smorgasbord of freshwater fish.

I stopped by Cherney's resort on Six Mile Lake and met with the owners, a great group of people. One of the owner's sons was a friend of the grader operator who found the bigfoot tracks, and he took me to the scene of the discovery and explained the precise location of the tracks. The son explained that the area is very wild with a lot of bears, deer, and even a mountain lion, but the resort is so far out in the middle of nowhere that few people want to come out and enjoy what they offer. He also said they are in the unique position of being one of the few businesses allowed to operate in the middle of the Leech Lake Indian Reservation.

SECOND SIGHTING

Danielle had a second sighting in May 2007, just a month prior to my arrival. She was on her school bus at 7:15 a.m. on Highway 2, four miles passed Bena, heading towards Cass Lake. There was almost nobody on the roadway, and nobody on the bus but Danielle and the driver.

Danielle was sitting on the north side of the bus looking out into a passing field. Two hundred yards out into the field she observed a creature walking along on two legs. She was able to see that it was covered with dark brown hair and it matched the description of the creature she had seen next to the Devault residence. She said this was definitely not a human, as it walked just slightly stooped forward and was in an area of marshes and bogs where no human would be walking. She was able to watch the bigfoot for almost three seconds until the bus passed it by. Danielle guessed the creature was over seven feet tall and had a distinct swing to its arms as it was walking along. She confirmed that, based on the photos, film footage, and articles she has read, the two creatures she described and witnessed were both bigfoot.

Danielle signed an affidavit on the two sightings.

REVIEW

The one striking similarity that I found between the Minnesota bigfoot sightings and the sightings in Hoopa and Northern California is the association that witnesses make between the creatures looking half human and half animal. In every instance where there was a sighting, the witnesses made a distinction that what they saw was not human, but specifically stated the creature looked half human. The trait of the creature not having hair on its face is important and aligns with the sighting reports in Northern California.

In the Jenny Dick sighting, a nine-year-old girl walks into her yard to hang clothes and there is a bigfoot sitting on the ground playing with a wooden clothespin. That is priceless. Nobody could fabricate that story, let alone a nine-year-old girl. The creature didn't flee upon seeing the girl, and in fact continued playing. It wasn't until grandpa made his voice heard that it left the area. This confirms the NABS theory that bigfoot is not afraid of children (chapter 3).

Bigfoot has also made an appearance at the Bobbilink property on multiple occasions, something that seems to happen to specific people and residences, but we have no idea why. As reported in *The Hoopa Project,* Inker McCovey and his family have had multiple sightings around his residence in Hoopa. This sighting is also important because three people witnessed it.

The bigfoot sighting in the 1950s that I cited from the BFRO website is important as it fits the profile of the recent sightings in northern Minnesota today, even though it happened over 50 years ago. It supports both our belief that bigfoot is not afraid of children, and also that bigfoot likes berries and can be found around berry bushes (chapter 3).

The extreme weather conditions in northern Minnesota make a bigfoot living in this region a very tough and durable creature. The winter months can have heavy snow and minus-30-degree temperatures, while the summer can have 95 percent humidity and be 100 degrees. Those variations of conditions make the creature a very adaptable mammal, acclimating to conditions that obviously change the hunting and gathering skills needed to survive in this region. This is quite different from the conditions present in Northern California where summer months can have heat as high as 100 degrees, but in winter it rarely gets lower than 20 degrees, 50 degrees warmer than in a Minnesota winter.

This region of Minnesota does not have mountains or ridges. The area is absolutely flat and wet. In every bigfoot sighting or incident I investigated in this area every one was very close to water (chapter 3, water). All of the witnesses were extremely credible and helpful in their descriptions and accounts of their sighting.

The last item of note in this section deals with a Native American/bigfoot connection. The Dahls, Devaults, and Bobbilinks are either 100 percent Native Americans or have it in their blood. All of these incidents occurred on a Native American reservation. The association between bigfoot and Native Americans is very, very strong.

12 OKLAHOMA

Many western U.S. residents are probably pondering why a bigfoot researcher would be driving the plains of Oklahoma looking for bigfoot activity. Good question. North America Bigfoot Search employed the services of Harvey Pratt (Oklahoma Bureau of Investigation Agent) to draw forensic sketches for NABS witnesses. Harvey made two trips to California to meet with witnesses and draw forensic sketches. During Harvey's excursions west, we often spoke about his state and his Cheyenne Arapahoe background. Harvey told me stories he heard about bigfoot in Oklahoma, and the proximity of many of the sightings to his reservation and the community of Concho. It appeared that bigfoot lived in the same vicinity as Native Americans on their reservation.

In late September 2008 I was invited to be a speaker at the Oklahoma Bigfoot Conference in Honobia. Harvey was another of the invited guests and would present with me. As plans started to come together for Honobia, Harvey began to put together a series of witnesses we could interview in the week prior to the conference. The decision was made, the flights were reserved, and I was flying to Oklahoma City.

I landed at Oklahoma City Airport, Harvey picked me up and we took an hour drive to his very rural property. Harvey and Gina live on 75 acres of beautiful rolling hills. Most of Oklahoma is flat, open farmland separated by roads, freeways, and creeks/rivers; no mountains, but many hills. Harvey and Gina were great hosts and we made their house base camp for the week.

Native American reservations in Oklahoma are not the same as they are in California. California reservations have defined borders that act as a property line. The public can drive through the reservation on paved roads and can even stop to visit friends or relatives, but they cannot go off a paved road or walk on rural property belonging to the tribe without a special permit issued by the tribal

Concho pond.

chairman's office. Oklahoma is much different. Many of the tribal members live hundreds of miles from tribal offices, and the tribe owns only a small percentage of the land surrounding offices or casinos.

Concho is the name of the region surrounding the administrative offices for the Cheyenne Arapahoe Tribes. In our travels and visitations with Harvey's friends, family, and acquaintances, there were several stories told about ghosts in Concho and their persistence in the community.

One story involved a local phone installer who was looking for a Concho resident and was told to check the basketball gym. The installer went to the gym and didn't see any vehicles, but went inside anyway. The gym was fairly dark when he entered, but he did find an older, dapper-looking gentleman in a gray suit and a long ponytail standing along the sidelines. The installer asked the man if he had seen this specific individual, and he replied that he hadn't. When asked if he knew where the person might be, he said he had no idea. The installer thanked the man and headed back outside where he found the resident and relayed his story about the man in the gym. The resident told the installer he was nuts—nobody was in the gym; he had just been there. They walked the ten feet back into the darkened gym and nobody was there. It had been a ghost that loiters in different parts of the community and was known to the Concho resident.

Tribal members are often very private people and getting them to talk about bigfoot with outsiders can sometimes be a very tedious issue. It took me almost six months to win the trust of many people in Hoopa *(The Hoopa Project)*, but eventually the barriers were eliminated and the information flowed. In Concho I had a tribal chief knocking on doors and approaching people asking if they would talk—Harvey Pratt. It took us a few days to develop momentum in the community, but once word started to spread that a research team that included Harvey was in the area looking for bigfoot witnesses, information started to come our way.

FARLAN HUFF

Farlan Huff, one of the organizers, invited me to the 2008 Oklahoma Bigfoot Conference in Honobia. Over the course of the two-day conference I spent significant time around Farlan and found him to be a quiet, humble, and very polite man who had a very intriguing bigfoot story. During conference breaks, lunch, and dinner, Farlan told me a story that was very interesting.

Farlan is a 48-year-old semi-retiree and now lives in Wilmington, Illinois, but he was born in Shawnee, Oklahoma (interestingly, in the same building as actor Brad Pitt). He graduated from high school in Maud, Oklahoma, after which he attended Oklahoma State University in Stillwater for two years, majoring in geology. He has been a welder's helper, gang hand, operated a backhoe, and most recently was a refinery technician. He has 23-year-old twins (boy and girl), two stepsons, 17 and 18 years old, and a 13-year-old daughter.

Farlan has always enjoyed the outdoors and going to new spots for different elements of adventure. His family had friends who owned a hunting cabin near Davis, Oklahoma in Murray County. In the spring of 2003 there were no hunters at the cabin, so the friends extended an invitation for Farlan to stay and have a mini retreat. He invited a friend of his to go with him; his friend was six feet, six inches tall, a huge man.

When they arrived at the cabin, Farlan's friend wanted to stay around the cabin while Farlan took off fishing. Returning to the cabin near dusk, Farlan said he had an odd feeling, like someone was watching him while he was walking up the driveway. As he reached the doorway, he found his friend standing in front of the cabin singing loudly, and his friend told him someone had just thrown a rock at him. Farlan laughed at this, because the cabin was in the middle of nowhere and there wasn't anyone for miles. The cabin was part of a 700-acre ranch on small rolling hills with large cedar trees. People don't visit the area often, and the most traffic in the area occurs around hunting season.

Farlan Huff.

In July 2004 Farlan went back to the ranch and took his dog, a small schnauzer. The previous year he had seen a few documentaries about bigfoot in Oklahoma and heard about incidents that people experienced, and this opened his eyes to the type of behavior he should be looking for. He arrived at the cabin with 30 minutes of daylight remaining. He left everything electrical turned off and was putting things away in the cabin when he thought he heard a loud snort at the rear of the cabin. He walked outside and heard sounds like a primate would make. He started to get scared, so he went back into the cabin.

He stood at the front of the cabin and made a "hoot" sound like an owl would make; he got a hoot back. A few seconds after the initial hoot, three other locations near the cabin all started to hoot, almost like a choir. Farlan had planned for this event and had tested his tape recorder at home prior to his departure. He now got the recorder out of his bag and activated the tape, but it didn't work.

As it got dark, more things that were unexplained started to happen around the cabin. He heard a very loud tree knock not far from his location, and he continued to hear odd sounds. He began to think that he might be in over his head, and was definitely scared. He had a shotgun and an SKS (a semi-automatic weapon) leaning against his chair. His dog was cowering under the chair.

A few hours later Farlan heard bipedal footsteps coming up the driveway, just out of his sight. He also heard another biped walk up, and it sounded like it was dragging a stick. At this point he heard something that really bothered him; he heard mumbling, language, a definite language that he couldn't identify. It was definitely two or three different creatures talking amongst themselves. Remember, the cabin was completely shut down, no lights, no electrical, no sounds.

At this point, Farlan said he wanted to wait a few hours to tell me the remainder of the story. I wasn't sure why at the time, but I always try to accommodate witnesses. After the conference was over for the day and other people were not around I heard the rest of his adventure.

He said that what happened next was very unusual, had never happened to him before, and scared the heck out of him. It was after midnight and he had fallen asleep in his chair with his gun lying across his lap. He said he felt like he was waking up, but had a very difficult time opening his eyes as he attempted to move his arms and couldn't. He knew he was awake, and he started to feel his chair and the nearby glass picture window start to vibrate. At this point he was able to open his eyes and he felt like he was in a vacuum of silence, no noise at all. He fell back to sleep and didn't awaken until 10:00 a.m. He walked outside the cabin looking for tracks with his SKS draped over his shoulder, but he didn't find any unusual marks in the dirt. He heard a very loud branch break not far into the forest from his location. He saw something big and dark running down near the tree line about 100 to 150 yards in front of him. He walked closer to the woods and could just see a dark shadow and then heard a loud primate snort again, the same snort he had heard the previous night. Farlan made a point of stating that he can't claim he saw a bigfoot, but whatever was charging through the woods was big, fast, and large, and there weren't bears that large in those forests.

Approximately three weeks after Farlan's incident at the cabin, the owner and his wife made a trip to the location. They did not see a bigfoot, but did hear a vocalization they described it as a "howling roar." They immediately went to their new truck and tried to start it and leave the area, but the truck wouldn't start. The pair then went into the cabin and locked the doors. After an hour of silence, the owners again went back to the truck and this time it started, so they left the area immediately. In prior incidents, the owner had always blamed these odd occurrences on large cats in the area, but not anymore.

CONCLUSION

During my time around Farlan I was impressed with his diligence in telling a story where the facts were understood. He wanted to make it clear that he felt multiple bigfoot were in the area of the cabin when he heard the owl sounds. He had no doubt it was bigfoot mimicking the owls. He also had no doubt that it was bigfoot walking up the driveway while he was in the cabin. Farlan said his dog acted completely out of character during these events, as it was always loud and aggressive even in the face of large animals, but during these encounters the dog was cowering and shaking under his chair. Regarding the paralysis and shaking he felt in his chair, Farlan has no explanation.

One of the last items Farlan and I discussed was the cabin. He explained that 99 percent of the time there is nobody at the cabin; it is vacant. It is only at certain times of the year, usually during hunting season, that people visit and use it as a base camp. The times Farlan chose to visit were times when nobody ever went. Maybe bigfoot didn't expect or appreciate visitors.

PATRICK HOLLIMAN

Oklahoma is a huge state with hundreds of thousands of acres of rolling crops. It is amazing that among the fields and crops there are people having bigfoot sightings. With one farmhouse every couple of miles, it's hard to imagine these people have the time to talk and spread stories about one of the most elusive creatures in the world, bigfoot.

Harvey Pratt and I heard a story about Patrick Holliman, and through a twisted web of contacts we were able to finally meet Patrick at a small diner in an area on the outskirts of Weatherford. We sat in the corner, far away from other guests, as we wanted Patrick to feel completely free to open up and tell his story without concern for outsiders. As he got talking, it was obvious that he and Harvey had mutual friends and relatives, and this brought the interview to an entirely different level.

Patrick is a 48-year-old Cheyenne Arapahoe Native American, but you'd never guess it from his green eyes. He was born in Clinton, Oklahoma and went to high school in El Reno. He didn't play a lot of sports in school, but he enjoyed all the trades: wood, metal, and auto shop. From our initial conversation with Patrick, Harvey and I felt we were sitting with a man who had a lot to say. He was a little shy, very intelligent, and introspective. During his life he had been a hay hauler and ran a wheat harvesting business. At

Patrick Holliman.

the present time, Patrick was the primary care giver to his ailing 87-year-old mother.

Patrick had spent his life roaming the fields, creeks, and rivers of the region surrounding Weatherford, El Reno, and Concho. He knew these areas very well and was quick to answer all questions regarding weather, topography, mammals, etc. He told us he had one unusual bigfoot sighting in his life that he hadn't told many people. He said that the events leading up to the sighting and the region were a little different, and it might be surprising to people that a bigfoot had been in the area, but the sighting is fact.

In July of 1976 Patrick was 16 years old. He and his older brother, Mike, now 53, and his friend, Steve, were going to the south fork of Bear Creek approximately half a mile from the South Canadian

River. The boys were going to explore a very wild area that they hadn't spent much time visiting. They knew the area was remote and that there was an old farmhouse and barn nearby.

Patrick's brother parked their truck near an old bridge that spanned Bear Creek. The boys slowly walked their way up the creek to a point where the forks of north and south Bear Creek meet. This was a hot hike and after a very short time Mike had gotten very thirsty. He decided to head back to the truck and Steve and Patrick would stay behind and check out the abandoned barn.

As Mike went to the truck, he saw a family of small skunks directly adjacent to the front wheels. He made his way cautiously to the front seat of the truck and grabbed his 22-caliber pistol. While Mike was grabbing his pistol, Patrick and Steve were etching their names into a large plank on the second floor of the barn. It was at this point that Mike cranked off two rounds from his pistol in an attempt to scare away the skunks. Steve and Patrick had no idea what was happening with Mike, so Patrick ran over to the opening of the barn that was closest to the truck.

Approximately 15 seconds after the shots were fired, Patrick observed a huge creature walking on two legs down a cattle trail near the area where the truck was parked. He could see that the creature was covered in hair the color of an Irish setter, but because of branches obstructing his view, he couldn't see the face. The hair color was uniform throughout the body. He saw that the creature was casually swinging its arms in the same manner a human would sway its arms as it walked. He only observed the creature for a short period of time as it then disappeared into nearby foliage. The exact location of the sighting was approximately 200 yards upstream from the bridge spanning Bear Creek and that is located half a mile from the Bear Creek confluence with the South Canadian River.

Thirty seconds after seeing the creature walk down the path, Patrick saw his brother casually walking down the same path in their direction. Patrick couldn't believe the proximity of his brother to the creature. He later asked his brother if he had seen or heard anything unusual by the truck, and he said that he had only seen skunks. We asked Patrick what might have caused the bigfoot to suddenly start moving on a hot day in July. He thought the gunshots might have startled or scared it.

Since the Bear Creek event, Patrick has seen many bigfoot specials on television and seen creatures that were described as bigfoot in magazines. He said the creature he observed was close to eight feet tall, massive in size, and it matched the descriptions of bigfoot he has seen on television.

Two years after the 1976 incident at Bear Creek, Patrick was with Mike and another friend as they were driving over the same bridge at Bear Creek. It was approximately midnight on a July night and the boys were going to shoot rabbits. Just as the truck was midspan on the bridge, something hit the side of the truck so hard it made the truck shake. The boys were so scared that Mike accelerated and didn't stop until the boys were clear of the area. They checked the side of the truck and could not find any marks or damage. The boys initially thought that there might have been a loose board on the bridge that lifted up and hit the truck. Days after the event Patrick went back to the bridge and could not find any loose boards. It's still a mystery to everyone involved what might have hit the truck.

In June 1979 Patrick was operating a combine at a farmer's wheat field just northeast of the Bear Creek bridge by maybe half a mile. It was near 10:00 p.m. and he was in the far corner of the field closest to Bear Creek. As he was turning the combine in the corner, the lights of the rig illuminated two bright glowing eyes inside a clump of tall grass just outside the fence line. The eyes were eight to 10 inches apart and five feet off the ground. He knew this wasn't a deer or any other mammal common to the area, and with his past experiences at Bear Creek he was frightened. He pulled the combine up and drove as fast as he could back to the farmhouse. Once back with people, Patrick explained his sighting to his brother and friend who were also working at the site. The other boys wanted to ride the combine back to the corner and see if they could determine the ownership of the red eyes. They went back to the area and made the same turn in the same corner when suddenly both headlights on the combine went out. According to Patrick, Mike got very scared and floored the combine all the way back to the residence. The boys attempted to fix the headlights that night, but could not get them operational. Patrick said it was a new combine and the lights shouldn't have gone out and never had before. The next day the headlights were checked and they were functioning correctly. The lights never went out again.

Harvey worked with Patrick on a sketch of what he observed at Bear Creek. The sketch appears to be of a massive male bigfoot, long hair on the edge of the forearms, no breasts, with the Irish-setter coloring. When the sketch was completed, Patrick said it was exactly as he'd seen, massive and muscular, an obvious bigfoot.

Patrick signed an affidavit.

LOCATION

Patrick escorted Harvey and I to the location of his bigfoot sighting on Bear Creek. The drive reminded me of something out of an old movie that highlights endless fields in Oklahoma. As we were driving down a very dusty, narrow road, we suddenly came upon a small bluff. It was nothing large, just a small bluff that overlooked a valley below. Patrick pointed to the field he had been plowing, the Canadian River, and the drainage area of Bear Creek. As we looked out over the area you could see massive umbrella-tree coverage over the river and creek, with the edge of each equal in altitude to the adjacent fields. We drove further downhill until we came to a small bridge at Bear Creek. Patrick explained the area and the creek, and its isolation. The bank of the creek was muddy under the bridge and we could see many different types of animal tracks, but no bigfoot tracks. We went further up the road and stopped on the other side of the creek in an area with very high weeds and small trees.

As Harvey was talking with Patrick, I walked up the road until I came to a grove of trees on the shady side of the roadway. This area was lush and very green even though it was August. The ground cover almost looked like a manicured lawn. I was looking at the ground when I noticed an armadillo waddling down through the grass. That was my first of many armadillo sightings on this trip. The tracks we observed in the creek, the great canopy over the creek, the thick ground cover and green flooring near the roadway, all point towards significant food source for any bigfoot in the area.

As we were standing in the middle of the road, a man drove up in his truck and engaged us in conversation. I explained our presence and asked if he had any information regarding odd events from the area. He said he was from Weatherford and had permission from local landowners to hunt their farms, and had been hunting this area for over 20 years. He told us that two years ago a local farmer put a foal alone in a small pen surrounded by a tall wire fence. The pen was away from the roadway, unable to be seen by passing vehicles and high enough (seven feet) that predators couldn't enter, but on one of the first nights something got into the pen and carried it away. The farmer knew it had been carried away because there were no marks of a vehicle driving up and there were no hoof marks of it walking away. The farmer was completely baffled by what could have taken it and not consumed it onsite. The foal was never found.

Note: There are many stories in bigfoot research that involve a relationship between horses and bigfoot that we don't understand. At certain times it appears that the horse is almost a pet to the creature, and at others it may be a meal.

The area around Bear Creek and the South Canadian River is desolate and very much alive with animals. There isn't significant hunting pressure having

Patrick Holliman's bigfoot sketch.

a huge impact on the area because there isn't a significant population center for two hours of driving. If bigfoot was going to live in Oklahoma, this was ideal habitat.

FORENSIC SKETCH

Harvey completed a sketch of the creature Patrick saw. Details were slightly lacking because Patrick did not see the creature's face, but the details he did see matched other sketches Harvey has done of bigfoot. Patrick confirmed that Harvey's sketch was an accurate rendition of what he observed.

KEITH LUMPMOUTH

In almost every community that I've visited over the years I've been fortunate to meet an individual who seems to have his finger on the pulse of all bigfoot activity. In El Reno, Oklahoma that person is Keith Lumpmouth.

Harvey Pratt is a chief in the Cheyenne Arapahoe Tribe and over the years has heard numerous stories regarding bigfoot activity. When Harvey inquired about whom he should seek out for more information on bigfoot, everyone told him about Keith. After numerous phone calls and stops at many homes, we were finally able to connect with Keith.

El Reno is a small community approximately three miles from Concho. With many of its store fronts closed, it's obvious the town has seen brighter days, but it has a calm and quaint atmosphere and its large population of Native Americans seems quite content living here.

Keith was very pleasant and seemed happy to meet with Harvey, Gina, and me. We went to the local Christian Elementary School where we were allowed to use a room. The school was a perfect place for the interview; it was quiet, we were able to hook up the computer and printer, and the lighting was good for Harvey to sketch.

Keith stated that he was 44 years old and had been born in Lawton, Oklahoma, the second oldest of five children. While he was still in school his family moved to El Reno where he attended high school and played football and baseball. A short time after graduating, Keith married and had two children that are now 23 and 18. Keith wanted to live the lifestyle he enjoyed the most, being a cowboy and ranch hand, so he spent 25 years working for the Cheyenne Arapahoe Tribal Ranch. He left that job in 2007.

The Cheyenne Arapahoe Tribe has a herd of bison that they raise and later slaughter. During Keith's early years on the ranch they had a constant problem with stray dogs roaming the ranch near

Concho and bothering the animals. People from the outlying communities would drive into Concho and let their dogs run loose and then drive away. The dogs would become pests and it was Keith's job to deal with them. He didn't like killing the stray dogs, but he would routinely shoot 10 to 15 strays a year that were bothering the bison and cattle. The dog problem continued until 1999, when it suddenly stopped. The absence of dogs coincided with witnesses claiming bigfoot sightings in the community.

In early 2000 a resident of Concho, Emmit Redbird, told Keith that several bigfoot were in the area of the Pow Wow Grounds during certain parts of the year. The Pow Wow Grounds are located on the right side of the roadway as you drive from the casino to Concho, and they can be identified by the teepee structures that dot the landscape. Supposedly, when the tribe holds ceremonies at the grounds and tribal music and dance are going on, the bigfoot gather at the far perimeter of the ceremony and watch.

Keith Lumpmouth.

Note: This bigfoot activity of watching a tribal ceremony is something that NABS researchers have heard before. The Yurok, Karuk, and Hoopa in Northern California have all claimed that when they have their Jump Dances near the Klamath River, the bigfoot will stand or sit on nearby mountains and watch. This behavior has supposedly been occurring since the 1800s.

The Cheyenne Arapahoe Lucky Star Casino is located between Concho and El Reno. It's a large, modern facility that sits in the middle of huge fields that the tribe uses for grazing. There are small rolling hills and some creeks and rivers in the immediate area. In approximately 2001, the casino's exterior security surveillance cameras allegedly caught a bigfoot in a grease pit at the rear of the casino. Keith said he viewed the tape and saw a creature almost 10 feet tall with hair over most of its body, walking on two feet like a human. Keith said the bigfoot's height was estimated when it walked under a parking lot light, and the light was later measured. Other bigfoot groups also claimed to have seen this tape and they validated that it was a bigfoot. The tribe has temporarily put a clamp on making any statements about the incident or allowing anyone else to view the tape. They are supposedly afraid of people from the city coming to Concho and searching the area in an attempt to kill a bigfoot, something the tribe definitely does not want to happen. The consensus of opinion is that the tape does exist and it did capture a bigfoot, but the tribe doesn't want this publicized. Keith said that the sighting at the casino does not surprise him because he is aware of others who claimed to have seen a bigfoot rummaging through garbage and dumpsters. It should be noted that the grease pit at the rear of the casino and the associated open garbage bins have been removed and are no longer present at the site.

In October 2004, the tribe sponsored a haunted trail in Concho for Halloween. Keith said they made the trail 500 yards from the road and 200 yards from the railroad tracks, in an area that was very quiet and a little spooky. This immediate area supposedly had bigfoot sightings and sounds, but nothing had been confirmed. The event went smoothly and Keith, his boss, and a small group of others gathered the following day to clean up. Keith's boss didn't get into the field much and people made fun of him for not doing manual labor, so he handed Keith a digital camera and asked him to take a photo of him picking up trash. As Keith turned and quickly took a picture of his boss, he simultaneously realized there was a bigfoot standing 50 yards behind the scene. He immediately raised the camera again to take a second photo, but believes the bigfoot knew what was occurring and purposely turned to the side to only offer a side angle view. Keith described the creature as very large, close to nine feet tall, very well built with defined muscles in the abdomen, and hair on top of its head that stood almost straight up. It had a barrel chest, hair that covered most of its body and was dark in color. Keith guessed that this was the alpha male of the group living in the area. He felt that the creature made no facial expression when he locked eyes with it; he could also see the whites of its eyes. He guessed the creature weighed close to 800 pounds. It quickly left the area after the photograph.

After the bigfoot sighting at the trail, Keith contacted a noted bigfoot organization to report his findings. The group sent out a researcher, and Keith gave him the photo he had of the creature.

This researcher was at a recent conference where I was speaking. Keith had signed an affidavit that I presented to the researcher requesting that he allow me to view and copy the photo for research purposes. The researcher said he wouldn't do it. (Keith claims he no longer has a copy of the photo.) This type of attitude among bigfoot researchers isn't uncommon. Many of these individuals are unwilling to work with other groups, share information, or even comply with a request from one of their own witnesses.

In early September 2005, four bigfoot researchers went to Keith and requested his assistance in locating a group of the creatures. Keith agreed to help them. He took his daughter, Nakayla, and they met the researchers near the outskirts of Concho near the river. They had every conceivable night vision and infrared device known to man. Keith said he pointed them in the right direction and then stayed in the background with his daughter. Keith said he knew the bigfoot would circle back behind the group, and if he stayed far enough back, he and his daughter would have the sighting. After the group walked across a small bridge, Keith heard the creatures behind the researchers. He used his spotlight to illuminate two creatures and observed their bright red eyes. One set of eyes blinked and the other set of eyes moved in behind the first set. Keith and his daughter got within 30 yards of the bigfoot when they both smelled a very pungent odor, something similar to a dirty, wet dog. At this point that Keith felt it was prudent to leave the area. The researchers missed the whole thing.

Keith's third bigfoot incident occurred in September 2006. A fellow ranch worker had found a dead deer hanging in a tree near the gun range in Concho. The deer had been gutted and the guts were lying on the ground nearby. It was obvious to everyone that a bigfoot did not kill and hang a deer with rope, so they all decided to pack up and leave the area. As they started to drive out, Keith saw tracks in the sand of something large that had freshly disturbed the ground. The tracks appeared to be very similar to human footprints, but were 14½ to 15 inches long with a 60-inch stride. The guys followed the tracks to a small nearby hill that was covered with heavy foliage. Keith walked up the hill a few feet and observed a bigfoot partially concealed in a small cedar tree. Keith was only able to see the bigfoot's right hand and right thigh. The hand did not have hair on its backside, but the fingers did. The thumb looked identical to a human thumb, but the hand was huge in comparison to a human's. Keith commented that he thought the bigfoot he observed in this incident was the same one he photographed with the digital camera. When Keith got within 15 to 20 feet of the creature, he felt he was getting too close and told everyone to leave, which they did. He returned several hours later and made plaster casts of the tracks.

Near the end of the interview, Keith commented that he had carried a rifle in his truck for the majority of the time he worked the tribal ranch. He said he initially felt that if he had ever seen a bigfoot he was going to kill it and end the mystery for good. But after he finally saw the creature and observed the many human qualities it has, he now knows he could never shoot it. He is positive that bigfoot can mimic sounds that are identical to coyotes, birds, owls, and other creatures. He also said he has been in the woods around his town and Concho where the sounds evaporate away; there is a total vacuum. All creatures leave the area and it is then you know that bigfoot is in the area.

Keith signed an affidavit.

LOCATION

The tribal ranch land encompasses several distinctly different areas. One zone is very flat, reminiscent of traditional Oklahoma farmers' flat land. The North Canadian River bisects the property and offers substantial cover and refuge for anything inside the boundaries of the river. There are many deer and small game that live in the habitat, and it would be easy to understand how a bigfoot could call this home. Another part of Concho has a body of water that is a cross between a swamp and an overgrown lake. It's difficult to maneuver around the lake, as it is completely overgrown on the fringe. An unusual aspect to this environment is the railroad tracks that go through the middle of the entire area. It would seem that train conductors might get a gook look at bigfoot on rare occasions when it is caught unawares. Yes, I did look to see if there was a train stop or rest zone for the conductors in El Reno, as that would be a great place to interview possible witnesses, but unfortunately there wasn't.

FORENSIC SKETCH

After I finished my interview with Keith, Harvey Pratt started in on his magic. I must admit that every time I see Harvey performing his artistry, I am amazed at how fast and accurately he works. Harvey has a systematic method for prepping the witness and following a uniform list of questions.

As Harvey started in on the sketch, Keith offered a few details about the face that he wanted to change. The bigfoot did not have the abrupt facial qualities of an ape, but the more subtle qualities of an alpha male adult. It looked large, strong, and distinctly male. It did not have as much hair as the Patterson–Gimlin creature, but it did have a beard.

It took Harvey approximately 75 minutes to complete the sketch and then a few minutes to highlight certain colors and spots where Keith had made comments. At the conclusion of our session, Keith

Keith Lumpmouth's sketch.

405

commented that Harvey had done a great job and he was shocked at how accurate the drawing was.

EMMETT REDBIRD

Keith Lumpmouth gave us the name of Emmett Redbird, a resident of Concho who has had multiple bigfoot sightings in the last few years. Keith also felt that other family members might have also had sightings.

We drove to Concho and we were immediately offered assistance from a variety of people about how to locate the Redbird residence. Emmett came to the door and volunteered to talk with us in his front yard. Emmett also brought out his wife, Stella, and his son, who couldn't speak and was obviously suffering from a disability.

Emmett said he had always heard that bigfoot was in the area of Concho, but described serious activity as starting in 2001. He said security guards from the casino would routinely walk around the old school in town looking for bigfoot. In April 2003 at 6:00 p.m., Stella and Emmett were near the old school in Concho when they both saw a large bigfoot standing under a nearby tree. They were 20 feet from the creature and got an excellent view of its face. They described it as walking on two feet, entirely gray in color, and completely covered in hair except on its face. The arms hung lower than human arms, but it had the stature of a tall man. The creature looked much more human than animal. The entire sighting lasted approximately two minutes.

Emmett's second bigfoot sighting occurred six months after the first, and was a little more interesting. Emmett was walking quietly near the exit of the old school, close to the pasture, and he saw a female bigfoot with a small baby on her chest. The female was lying on her back. She was described as having dark-colored brown hair and large breasts. The baby was lying on the mother's chest with the mother's arms wrapped around it. The baby's head was located near the mom's head and breasts. Emmett saw the baby turn its head and look in his direction; its face looked much like a human baby's. It had lighter colored hair that wasn't as thick as the mother's. Emmett felt he shouldn't be in the area and immediately left. He didn't think the female bigfoot saw him.

Emmett told us that his son (approximate age 15) had a disability and couldn't speak clearly, but Emmett could understand what he said. He explained that his son sometimes mowed the lawn on the property, and also routinely walked in the area near the lake and the creek. His son had many instances where the bigfoot came near him while he worked and just stared at him. The boy was afraid of the creatures, but realized they'd never bother him. All of the incidents occurred near the creek and foliage that surrounds the creek at the far side of their house. Many of the descriptions Emmett relayed from the boy matched the description of the creature Emmett had observed under the tree.

The Redbirds.

Stella and Emmett each told me about their cousin, Ronnie Bearbow. Ronnie had lived just down the street from them in the tribal maintenance building. At the end of a long day, a bigfoot would come over to Ronnie's window near where he slept and the creature would knock and scratch on his window. On some nights the bigfoot became so aggressive that it would actually shake the small residence. Ronnie would get so upset at his inability to sleep that he would shoot one round from his rifle into the air to scare the bigfoot away. After the round was fired the creature would leave. Ronnie died several years ago.

Emmett said there were many nights at his house when he could hear the bigfoot yelling at each other, *whoop, whoop, whoop.* The sounds came from various directions, and usually very late at night or early in the morning.

As we neared the end of our interview, I asked Emmett and Stella if there was anything else I needed to know about bigfoot in this area, or if they had any idea why bigfoot stayed in this area over the last several years. In a very matter-of-fact fashion, Emmett stated, "We are probably related to them, this is probably their home too." Here was a Native American Native claiming that his people are related to bigfoot. Not a big surprise to us, but maybe for many other Americans.

Emmett signed an affidavit.

LOCATION

Concho, Oklahoma is the hub of the Cheyenne Arapahoe Tribe. This is where there tribal offices, the Lucky Star Casino, and their herd of bison are located. This is also where the North Canadian River bisects much of their land. The river is large, flows all year, and is home to a variety of birds, mammals, and other wildlife. Concho also has a small lake and very old cemetery.

During my talk with Emmett, he advised that Concho is a haunted city. One story that affected the community involved a Concho resident driving back to the city and seeing a very old man dressed in his finest clothes walking down the highway at noon on a summer day.

Concho water tower.

The Native American had a long ponytail and looked very dapper in his suit. The driver pulled to the road and offered the man a ride, concerned that it was too hot to be out walking. The man got into the car and the driver asked where he was going. The older man stated that he was trying to get back to the cemetery. The driver thought that was a little odd because he hadn't heard about a funeral on that hot day, but he drove the man two miles back to the cemetery, and the man exited and walked towards the rear of the car. The driver turned to ensure the man was off the roadway. He could not see the elderly man and, concerned that he might have fallen, he got out of his car to check on his status. He searched the area completely around the car, but the man had disappeared. This description is identical to the man who was seen in the basketball gym by the phone installer.

FORENSIC SKETCH

Harvey Pratt interviewed Emmett regarding his sighting of the lone bigfoot under the tree. Emmett had stated that he had a good look at the creature's face and felt that he could offer a good description of what he observed. He described an older male bigfoot that was probably on the downhill side of life. This may be compared to a 60-year-old male human, large, gray, and slightly bent over at the shoulders. It took Harvey approximately an hour to complete the

Redbird bigfoot sketch.

sketch. Emmett confirmed it was exactly what he observed standing under the tree and staring at him.

NABS would like readers to take a close look at the Redbird sketch. Our researchers believe there are many human qualities to this sketch, and Emmett's response as to why these bipeds are in this area may be directly on key; they may be related to the Cheyenne Arapahoe tribe.

MARY LONEBEAR

Sometimes you have to go the extra distance to get the big story, and this might be that case. Harvey and I had a report that bigfoot had been harassing a woman in Lindsay, Oklahoma on a fairly regular basis, and it was supposedly still occurring. Our informant was an employee of the Cheyenne Arapahoe Tribal Office in Concho. As luck would have it, the woman who reported the activity happened to be in Concho as we were visiting, and this is how we met Mary Lonebear.

Mary explained that she lived quite a distance from Concho and wasn't going directly home, but she invited us to go to her residence on our own and to search the area for evidence. As Mary and her family would not be home for a few hours, Harvey and I headed over to her place. Harvey has spent his entire life in the great state of Oklahoma and he advised me that it would be an easy trip. Ha! After more than 90 minutes and several phone calls to Mary, we still could not find her house. We finally stopped on a road and, thankfully, Mary drove out and found us.

As we drove up to Mary's residence it struck Harvey and I how isolated she was from any other house. We saw a barn behind the main residence and a creek bed approximately half a mile to the rear of the barn, but it was very isolated.

It was a very warm August day in rural Oklahoma, so we sat in the front yard for the interview and subsequent sketch. Mary was very easy to talk with and was eager for someone to listen to her story.

She started the interview by explaining that she was a 48-year-

old Cheyenne Arapahoe Native American. She was born in Clinton, Oklahoma and graduated from high school in Thomas. Mary was different from many Native Americans; she wanted to travel and see the world, so she joined the United States Air Force (USAF) after high school. Her USAF specialty was Air Crew Life Support. She was assigned to Cannon Air Force Base and that is where she spent a majority of her time. A portion of her tour of duty took her to Australia, Korea, and England, and for two weeks out of every year she went to Las Vegas. Mary has four sons, 21, 22, 24, and 27 years old. In 2002 when she was looking for a location to move her family, she checked the Indian Registry and found the house where she presently lives.

Mary Lonebear.

Here is a chronological list of the bigfoot incidents in the area of Mary Lonebear's residence.

September 2004, approximately two years after the family moved in, the incidents started. Gerald Yelloweagle is Mary's long-term, common-law husband. He was in the backyard with another friend when they observed a nine-foot-tall bigfoot standing in the yard near the barn. The creature was very dark in color and was observed walking from the backyard in a westerly direction and disappearing into an adjoining field.

February 2005, Gerald and two cousins were in the backyard when they saw a bigfoot peer at them from around the corner of the house. The creature was on the side of the house looking in a window where the kids were playing. There are few details available for this sighting because it was dark outside and they only had a short view of the creature. Mary commented that it was very cold outside.

April 2005, at approximately 10:00 p.m. Gerald and Mary were driving home from the city. They both observed a bigfoot standing at the corner of the house as they drove up the driveway. When they got into the house they checked on the kids and found them all in the room on the side of the house where the bigfoot was standing.

July 2006, Mary had a glass front door on her house. During the

(Left) The back of Mary's residence and barn.
(Right) Intersection where accident occurred in
December 2007.

summer months the door would sometimes be left open because of the heat. At approximately 10:00 p.m. one night Mary had just said good night to her son, Randall, and was passing by the front door. Something made her turn towards the doorway where she saw a huge bigfoot standing on the front porch. It was near nine feet tall, had yellow reflecting eyes, and had dark hair covering its entire body. Mary slammed and locked the door and yelled to Gerald what she just observed. Gerald ran outside, but the creature had already disappeared.

January 2007, Randall's cousin was boarding a horse in the barn to the rear of the residence. One night the 30-year-old woman went to the barn to check, water, and feed the horses. She left her young kids in her warm vehicle while she tended to the horses. As she walked towards the rear barn area she saw the horses huddled in one corner of the barn and a bigfoot in the other corner. The woman screamed and ran the entire distance into Mary's residence, forgetting that her children were in her car. Mary and Gerald ran into the barn, but the creature was gone.

February 2007, Mary was in her house with her family when they heard a very loud sound, as though something the size of a car had hit the side of their barn. The family went out to check on the condition of the structure and found their 1200-pound cattle feeder knocked over. It would normally take three full-size men to turn it on its side. They searched the area and didn't find anything in the barn.

December 2007, on a Friday night at approximately 10:00 p.m.

(Left) Creek and embankment with trees.
(Right) Author after hiking creeks and region around
Lonebear property.

Mary went to the store. On the way back from shopping Mary saw an accident that had recently occurred between an SUV and something that was down and not moving on the roadway. The driver of the SUV was a clean-shaven, white male in his fifties, and sitting in the back of an all-black, American-made sedan. Mary also observed a box truck that was similar in design to a U-Haul moving truck, but in very good condition and all shiny black. Mary pulled her vehicle up towards the accident scene to get a closer look at what was occurring. At this point a uniformed officer with no badge but a khaki-colored uniform approached her car. Mary was looking at a tarp that was covering a body on the roadway when a gust of wind blew the tarp off the body. Mary said that she observed a dark-colored bigfoot lying on the road, apparently dead from impact. The officer came to Mary's window and stated, "You didn't see or hear anything, we know where you live, now leave." Mary went straight home and got relatives so they could go back to the scene. Mary returned to the accident location with Chris, Shane, and Shannon and attempted to chase the truck, as it was now heading north on a small country road at over 70 mph. They gave up the chase when the pursuit got too dangerous. Mary said the vehicles were headed in the direction of a nearby military field.

July 2008, Mary's nephew, Bert, was visiting and while sitting in the backyard, he saw two bigfoot walk by the rear barn area. No further description was available.

At the end of July 2008, Gerald's sister and Bert were sitting in the backyard when they observed a bigfoot pick up an old washer that had been placed in the far corner of the yard and roll it down a hillside. They then saw the creature walk to a nearby tree and shake it violently.

Two weeks before we met Mary, she and her family returned home to find their front porch light receptacle torn from the wall. The glass front door was also broken out, but no entry was made to the residence.

After taking copious notes of the bigfoot sightings and incidents that have plagued Mary and her family, I asked her son, Randall, to escort me around the property and down into the creek and river area where they believe the bigfoot live. One of our first stops was the side of the residence where the creature has been seen many times. One of the first items I noticed was a very large and neatly stacked pile of garbage bags 100 feet from the residence. I asked Randall if the pile was ever disturbed or bags had been gone through, and he said it had happened and the family thought it was bigfoot going through their garbage. We continued our walk down through the barn and into the rear area, headed towards the creek while Harvey conducted his interview and sketch with Mary.

Randall and I walked north across two large fields and dropped down into Little Deer Creek. This was running with water in late August. We found numerous deer, pig, and cow tracks. It was obvious that the region had significant wildlife and water for a large mammal to survive. There were many nooks and crannies in the basin to build a large shelter to survive the horrid Oklahoma winters. Randall and I spent almost two hours hiking the region and walking through two different creeks and valleys. It was beautiful country and we never found another human track in the area.

When Randall and I returned from our hike, we were sweating profusely and exhausted. We found that Harvey had almost completed the sketch of the two bigfoot that Mary witnessed, so Randall got me bottled water and I retreated into the residence and wrote the affidavit. While there I met with Mary's boyfriend, Gerald. I asked him about the bigfoot sightings and he confirmed that they had occurred and he validated many of the statements that Mary had made. At this time Harvey and I were more than two hours from base camp and were running late for a dinner meeting. I was able to complete Mary's affidavit, which she signed, but I was never able to complete a form for Gerald.

One of the last statements Mary made to me was in reference to something that continues to ring true with many bigfoot encounters. I could tell by Mary's casual approach to the topic that she wasn't sure if it was relevant to the conversation. She said the family had lost four dogs in the last five years. They completely disappeared, no trace, no signs of a fight, nothing. I advised her that bigfoot has been known to kill dogs that continue to harass them. They sometimes will eat the dogs, but will almost always leave the scene and consume them away from the site where they were captured.

Mary signed an affidavit.

LOCATION

Mary's property is located in the middle of wide open farm fields with no mountains and only a few gently rolling hills. There are two different creeks that run to the rear of the residence in small canyons that have been etched into the land. The canyons are approximately 30 to 40 feet deep with significant foliage that acts like an umbrella over the creek floor. It was easily 20 degrees cooler down in the creek area than up by the fields.

The sighting of the large pile of trash bags adjacent to Mary's residence is reminiscent of a story I related about garbage bags and a bigfoot visitation in *The Hoopa Project*. In the Hoopa incident, a lone female was at home and a bigfoot was seen rummaging through the garbage pile. It would appear that the garbage bags adjacent to Mary's residence are an attraction for the bigfoot and a location where it may find an easy meal.

Mary was concerned that incidents at the house were escalating, and she asked Harvey and me for input on how to stop the visitations. I told her that every time video or game cameras are placed in a location where bigfoot is known to visit, the visitations stopped. I recommended going to the tribe and borrowing three game cameras. I also advised her that bigfoot does not like bright light and that's

415

probably the reason the front porch light was destroyed. We recommended putting up several motion-activated yard lights installed high off the ground with the bulbs protected by wire mesh. We also recommended removing all garbage bags from her yard.

We asked Mary if the tribe or any of her contacts knew who was occupying the residence before they moved in. She said it was abandoned for several years and that the yard and much of the barn was in disarray.

After leaving the residence, Harvey and I drove to the intersection where Mary said she had seen an accident and a dead bigfoot. I went to the farmhouse near the intersection and contacted the owner. He lived in the rear section of the residence, far off the roadway. I informed him who I was and inquired if he had heard about an accident at the intersection, giving him the date in question. He said he hadn't heard about an accident in this area for several years, and he specifically told me that if he had, he would keep to himself and not bother anyone. He then closed the door with the distinct expression he wanted nothing to do with the topic or me.

Mary Lonebear sketches (also shown in color section).
(Left) Bigfoot facial sketch. (Right) The front image is the bigfoot Mary saw at her front door, and is brown. The image behind is golden colored and is the bigfoot Mary and Gerald saw near their home as they drove up the driveway.

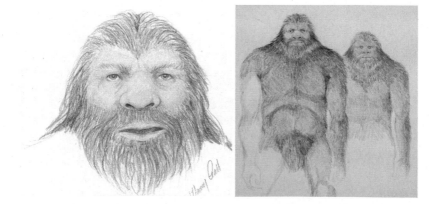

Harvey completed two sketches on one tablet for Mary. The first bigfoot was dark in color (brown) and the other was a golden color, much lighter than the first. Both creatures appear to be mature males with their physical descriptions matching sketches we've made in parts of California and Oklahoma. One sketch was of the bigfoot Mary observed at her front door, and the other was of the creature that she and Gerald had seen standing near their residence as they drove up the driveway.

STUART WHITEHEAD

One of the people I met on the first day of the Honobia Bigfoot Conference in August 2008 was Stuart Whitehead. Stuart is a member of the Choctaw tribe. He initially approached me and asked a series of questions about *The Hoopa Project,* and then spent considerable time talking with Harvey Pratt about his artistic skills. After several minutes lingering near our table, Stuart came closer to me and quietly stated that he had seen a bigfoot at fairly close range. I asked him if he would mind giving me an interview about his encounter and he agreed.

Near the end of our first day I had the opportunity to interview Stuart privately. He said he lived in Lindsay in central Oklahoma. He explained that he was two college credits shy of earning his biology degree and he hoped to finish soon.

Stuart Whitehead with Harvey, working on sketch.

Early in the evening on March 28, 2008, Stuart was driving from his house in Lindsay to his professor's cabin in Pushmataha County, Oklahoma in a 1998 silver Ford Thunderbird. He was driving downhill on a gravel section of the Portland Road when he passed a bigfoot standing on the roadside. As he passed, he made eye contact with the creature and it had a blank stare on its face. Stuart quickly slowed to a stop, grabbed his camera, and returned to the spot where he saw the creature. The bigfoot had already started to walk away, but at one point it turned and faced him, and he got a very good look at its face.

Stuart described the creature as near seven feet tall, weighing 500 to 600 pounds, hair covering almost its entire body, and it possessed a massive and muscular upper body with defined pectoral muscles. Its head appeared to be almost attached to its shoulders, and it didn't look to have a very big neck. The bigfoot had approximately two inches of hair on its body, not really thick, as he could see skin on its chest beneath the hair. The bigfoot stood erect and walked on two feet.

When Stuart initially observed the creature it had a willow branch with attached leaves in its hand and it was dragging the branch and leaves through its mouth and removing the leaves.

Note: I have heard this exact description of eating activity involving an eyewitness account in *The Hoopa Project*. A female firefighter had seen a bigfoot sitting down also dragging a branch and leaves through its mouth.

Stuart said he had seen many documentaries and television specials on bigfoot, and he believes that what he observed on March 28 was a bigfoot. Stuart signed an affidavit.

FORENSIC SKETCH

Dr. Jeff Meldrum from Idaho State University was another of the guest speakers on the conference agenda. Dr. Meldrum had talked to me during the day about Harvey Pratt's sketches and had expressed interest in the process and the accuracy of the drawings. He had never observed a police forensic artist conduct an interview and draw a sketch, so I invited him to witness Stuart's sketch process. At the conclusion of the first day of the conference, Dr. Meldrum,

Stuart, Harvey, his wife Gina, and I stayed in the building and Harvey started the interview and subsequent sketch.

The interview portion involves asking the witness a series of questions about the witness's opinion on a variety of topics related to the sighting, size, weight, color, etc. Harvey obtains an overview description, he asks the witness what type of sketch they feel they can draw (head, body, shoulders, etc.) and then the process of drawing the creature begins.

After approximately 60 minutes, Harvey had completed a facial sketch of the creature Stuart had seen. The creature appeared much like many of the sketches that he has drawn in the last two years. It appeared to be a male, very strong in appearance, and it did have characteristics to its face reminiscent of a human.

Stuart Whitehead's bigfoot sketch.

At the conclusion of the sketch and in the presence of Harvey and Dr. Meldrum, I asked Stuart the following question, "If a photo of the face of the creature you observed was sitting in front of you, and that photo represented a 10 out of 10 in accuracy of what you observed, what rating would you give the sketch Harvey just completed?" Without hesitating Stuart stated, "Ten out of 10; its exactly what I saw."

INA FAULKENBERRY

The 2008 Oklahoma Bigfoot Conference in Honobia had almost 200 attendees. It was a good turnout considering it's a long drive from almost every populated center of the state. We had many people approach us with their bigfoot stories, but none was as unusual or as interesting as Ina Faulkenberry's. She arrived at the conference with her son and another man who was introduced to me as a big-

419

foot researcher in charge of the electronics at the sighting location. I decided to take a break from the day's festivities and went to an outside location to interview Ina.

Ina does not fit the mold of any Native American I have ever met. She is a 61-year-old woman born in McAlester, Oklahoma. She attended high school in Houston and later learned to be a truck mechanic. She spent 12 years fixing trucks. She has also been a maid, and conducted IRS audits. She has been married twice and has one son, Felix. She owns two antique shops in Krebs, Oklahoma and the businesses occupy much of her time.

Ina had a cousin, Raymond Sparks, who owned a ranch in Hartshorne and he regularly invited her up to visit. In 1987 Ina was visiting the ranch and saw a reddish-brown bigfoot on the property. The creature was across a small field and walking towards the forest line from Ina. She wasn't sure what she had seen, so she went to Raymond with her story and asked what it was. Raymond said he had giant bears on the property, but they were harmless. He advised her to leave them alone and they would leave her alone. Ina did some research and learned that there hadn't been bears in those woods for many years. When there were bears in the area, they were very small, not seven to eight feet tall. Since that initial incident, Ina now knows that she was looking at a bigfoot.

Ina Faulkenberry.

Raymond died and Ina inherited the ranch in 1997. It is 600 acres against the Blue Mountains, of which 360 acres are hay pastures, and the remainder are rolling hills with huge pine trees up to 300 feet tall. The property had only been logged once in the early 1970s and only a few trees were taken at that time. The property has an old German cottage as the main residence.

In May 2006, between 4:00 and 5:00 p.m., Ina was exiting her 1994 Silverado pickup truck on the driveway and observed movement in the field near a round bale of hay. Her attention focused on a large buck (deer) standing almost motionless and mesmerized next to the hay; a few seconds later a large male bigfoot stepped from behind the hay and stood directly next to the buck. Ina was standing approximately 30 feet away. The bigfoot had a very human face that was bronze in color. There was no facial hair, but it did have hair completely covering its body, and the hair was messy. The bigfoot was very muscular with a small waist. At this point it looked directly at Ina and then patted the buck on its head. Ina watched as the bigfoot and the buck walked together, in synch, towards the forest line. Ina said the two walked almost as a dog would walk next to its master. She was able to watch this spectacle for five minutes.

Note: There was one other incident where I reported a bigfoot walking with deer in close proximity. That occurred in the Sierras in California and is documented in *The Hoopa Project.*

It was March 2008 when Ina had her last bigfoot incident. It was just starting to get dark and Ina and her son were going out. Just as she stepped outside, she caught a glimpse of a hairy arm making a throwing motion towards the barn. She turned to see a female bigfoot throw something in the direction of the barn. The bigfoot was 20 feet away and Ina could see that she had very large breasts. The female had the face of a black person, no hair on the face but she looked like a man. Ina took a few steps backwards and immediately started to sing a song. The bigfoot squatted and put her hands on her hips as she knelt down. There was an odor of rotten dead flesh in the air that was worse than a burn ward that Ina had once visited. As Ina was backing up, she fell backwards over a log, got back up, and then backed into a tree. Ina stated that, in hindsight, the bigfoot must have thought that she was a very clumsy human. Ina said the bigfoot had very unusual eyes, which she described as elongated cat eyes with a black center and an orange perimeter. Hearing the commotion of Ina falling, Felix came out of the house and he and Ina both got into the truck and left the area, leaving the female behind.

Ina signed an affidavit.

LOCATION

The property is located in eastern Oklahoma and in a general region known for bigfoot sightings and incidents. It's evident from the previous owner's statements that he had encounters with the bigfoot and he didn't want Ina to be afraid. The descriptions given by Ina are highly unusual, but not unheard of. There are several reports I've read that describe a bigfoot as having orange eyes. The description she gave of the bigfoot with the bronze skin is slightly different than many reports we have taken, but the very odd quality of that sighting was that there was no facial hair at all. We have heard this in California sightings but not in Oklahoma.

The most unusual aspect to Ina's entire story is the portion involving the bigfoot and the buck. There are old Native American stories that bigfoot can mesmerize its prey to the point that they are hypnotized. It seemed that the buck was under some type of spell for it to have walked in the manner it did next to what most bigfoot researchers would agree is a predator. The story in *The Hoopa Project* describes a hunter who saw a bigfoot come from behind foliage walking with a group of deer. It is obvious in both stories that the deer were not afraid of the bigfoot, and/or couldn't leave its

Ina Faulkenberry sketches. (Left) The bronze bigfoot that mesmerized the buck. (Right) The black face of the female bigfoot.

422

side. There are also stories of campers in their tents at night who believe a bigfoot is outside and they are so frightened that they cannot move. Earlier in this chapter, Farlan Huff described being in a cabin surrounded by bigfoot and having some type of paralysis so he couldn't move. Could it be possible that bigfoot has the ability to mesmerize or paralyze prey with a type of gas or scent?

Ina had introduced us to an individual who was a bigfoot researcher with another group. He had been to her property many times and had installed a complex system of cameras in an attempt to film the visitors. She said that since the cameras had gone up, all visitations had ceased. Nothing about her lifestyle had changed; the only thing different was the cameras.

Harvey sketched the bronze colored male bigfoot and the black bigfoot with orange eyes. They are very different in appearance and very different from almost anything we've ever seen. The sketches took Harvey much longer than normal because of their unique nature and because he was drawing two different sketches. When Harvey completed the facial sketches, Ina stated they were exactly what she had observed. At this point Felix walked up and also looked at the sketches and confirmed that this was what the bigfoot looked like. It was a bit of a shock to see Ina's son; I didn't even know that he was at the conference until we were completely done with the sketching. I didn't get a chance to obtain his affidavit, but I did get his verbal acknowledgement that the sketches were accurate.

13 THE BIGFOOT/HAIRY MAN/HUMAN CONNECTION

"All men are created equal."
—Declaration of Independence, July 4, 1776

Most avid bigfoot believers look to the Patterson–Gimlin film footage at Bluff Creek in 1967 as the Holy Grail. It shows a large, hairy creature walking on two legs through a creek bed in Northern California. The footage has held its ground in many scientific circles because of advanced motion and kinetic studies that are now available through digital manipulation of the frames; this was not available when the film was shot in 1967.

The creature caught on tape at Bluff Creek had dark hair over its entire body, including its face. It looked much like a giant primate walking briskly across a gravel area, except it walked much like a human. The creature had large breasts and is presumed to be female. This was the only film footage of bigfoot for many years. It is only in the last 15 years that people with video cameras started to catch bigfoot on their personal recorders, and more images of the creature started to be publicly posted. The Patterson–Gimlin footage was taken approximately 25 miles north of Hoopa, California.

In 2007 I brought Harvey and Gina Pratt to Hoopa for Harvey to sketch the creatures that witnesses had observed. Every sketch showed a creature that was similar in size and shape to the Patterson–Gimlin footage, which was not surprising. We also did several facial sketches of the creature where the witnesses were able to get a close view in decent lighting conditions.

It should be stated that residents in and around Hoopa receive cable television, and a main channel is the Discovery Channel, which has aired several bigfoot specials. One would suspect that if witnesses were fabricating their sightings they would describe a creature that matched the Patterson–Gimlin footage. This footage, along with other bigfoot specials, has aired many times on their cable system, and all show the same similar shaped and configured hominid. All of these specials also showed a creature with full facial hair.

However, the sketches made by Harvey with the aid of NABS witnesses had a staggering result. Only one sketch was of a creature with hair covering its entire face, all other drawings were much different. The sketches showed an extremely large creature (seven to eight feet in height) with a large barrel chest, little or no neck, and with facial features that could be called close to human. Many of the sketches had almost no hair on the face, yet every other part of the body contained hair. One sketch (witness Ed Masten) did have very similar features to the Patterson–Gimlin subject.

JOSEPHINE PETERS

Josephine has witnessed bigfoot multiple times and her stories were chronicled in my first book. She also told Harvey, Gina, and me how Hoopa people had a very tough life in the 1800s. She explained that in certain circumstances when members had a child that was extremely disabled, they would leave the child in the hills for the "Big People" to care for. She said this didn't happen often, but she knew that it occurred as early as the late 1800s.

AL HODGSON

I met Al late in 2007 and we quickly became good friends. Al has a wealth of knowledge about bigfoot starting when he became friends with Patterson and Gimlin when they filmed the Bluff Creek creature. Al had also been a longtime friend with Betty Allen, a freelance newspaper reporter working out of Willow Creek for the Eureka *Times-Standard* newspaper. Betty chronicled the events involving bigfoot for many years and believed the hominid roamed the hills surrounding Hoopa and Willow Creek. Betty had interviewed many bigfoot witnesses and made close friends with the Yurok, Karuk, and Hoopa women. The women had confided in

Betty some very deep and dark secrets that Betty had told Al she wouldn't talk about.

Al recalled one incident where his wife, Francis, Betty, and he had driven to Bluff Creek and found bigfoot tracks, whereupon Betty locked herself in the car. When he asked her why she had done that, Betty implied that bigfoot liked women and might abduct a woman if they had the opportunity. Al thought Betty had been told this by the Native American women, and that one of their secrets was that they had been kidnapped and raped by bigfoot.

JACKIE MARTINS

One of the most credible bigfoot witnesses I ever had the pleasure of meeting is Jackie Martins. Jackie is the Hoopa language teacher at Hoopa Elementary School and is a past councilperson for the tribe. In my first book she related a story about a bigfoot sighting that she and her friend, Julie, had while they were driving from the coast to Hoopa for a weekend dance. Both women saw the hominid cross the roadway in front of their vehicle, and both acted completely differently.

Jackie comes from the belief that bigfoot is in the area to take care of the woods and be a friend to the Natives, and not to harm people. If they leave the creature alone, then the creature will leave them alone.

Julie believes that bigfoot is an "Indian Devil" and was screaming this in Jackie's vehicle when the creature crossed the road. Jackie wanted to slow the car and watch the creature as long as possible. Julie wanted Jackie to accelerate out of the area and get as far away from the biped as possible—polar opposite reactions.

LUCY

Lucy was a Yurok Native American living in the community of Pecwan adjacent to the Klamath River. In the late 1800s and early 1900s this was a very rural area and had many incidents that would remind you of the wild, wild west.

The real story about Pecwan and bigfoot involves women. There exists no better story then one that Lucy wrote in her book, *To the American Indian, Reminiscences of a Yurok Woman,* which was originally published in the early 1900s. Chapter nine, "Indian Devils," talks about a group of wild people that lived in the mountains away from Pecwan and were feared by the local women. During the day the men would go out hunting and gathering and the women would stay in small groups near the river, cleaning, fishing, etc. When the men were gone and the women were alone, the devils would kidnap the women, rape them, and sometimes force them to have their children, and rarely would they escape and make it back to the village. Lucy documented one case where a woman was kidnapped and escaped, and made it back to Pecwan. She brought her child home with her, but it was wild and unable to be acclimated to the village, and it escaped and ran back into the woods to live with its father.

The devils were supposedly living in the High Country near the sacred grounds (above Bluff Creek) and they were known to take valuable items when they raided the villages late at night. They were known as very stealthy, almost invisible, and could almost disappear behind large trees. They were also known to escape towards the headwaters of Redwood Creek on the western side of the Bald Hills Mountains, now part of the Redwood National Park. Lucy documented an incident where the Native men of Pecwan once got so mad at the devils that they amassed a large group and went to the headwaters of Redwood Creek and attacked them. They supposedly found many valuable articles in caves, which had been stolen from them.

Additional corroboration on bigfoot abducting Native American women can be found in statements Chief Stokes made to Ray Wallace (see page 457).

To think that Native American women from the area of Pecwan were the only females abducted by bigfoot would be shortsighted. In *The Hoopa Project* I chronicled a story where a young boy was sitting on a toilet and a bigfoot reached through the bathroom window and attempted to grab him (Inker McCovey). Inker had young females living in the house with him and, in hindsight, the bigfoot probably made a mistake and thought Inker was a female.

In "The Track Record" #84 (pg. 12), the following story highlights what can happen when an aggressive bigfoot gets interested in a female. This report was called into Ray Crowe February 3, 1999. It is a believable story because of what I've heard from credible sources that match closely with what is reported here. There have probably been hundreds of attempted abductions of women by bigfoot across the North America. We know that a U.S. Department of Justice survey states that only 50 percent of all rapes are reported by women. We are left to imagine how many reports of bigfoot abductions or attempted abductions go unreported. The story follows.

> What the call was really about though..."do you have many reports of Bigfoot touching people?" This was 30 years ago, and Rod and his wife were visiting cousin Buck, 30 miles south of Chapmanville, West Virginia. At 7 AM in the summer they were in bed, an open window at the bottom of the bed. He was 21 and she 18 at the time. Al [sic] of a sudden something jerked the bed violently, then it was pulled towards the window. A great hairy arm grabbed his wife's leg and tried to pull her out the window. They screamed and beat on the creatures arm until it finally let go and went away. Later, cousin buck noted that boards had been torn away from a nearby steeple, and they thought maybe the creature had been staying in there.
>
> As an aside, Rod mentioned that back during prohibition days, some fellows at nearby Little Hearts Creek, W. Va., that's about 30 miles north of Chapmanville, got chased away from their still by a Bigfoot creature. They jumped a barbed-wire fence and ran off, but the thing chasing them tried to run through the fence. It got tangled in the wire and screamed horribly as the wire kept tangling around it and the barbs cutting deeper.

The abductions of human females by a bigfoot are obviously not isolated to Humboldt County.

The Columbia River has always been a hotspot for bigfoot activity in the twentieth and twenty-first centuries. It's very rare to find reports from the 1800s on any type of bigfoot activity, but remember that "bigfoot" was a term coined in 1958. When you read older articles you need to read for content and description. The article below, from "The Track Record" #85 (pg. 14), describes how the Nisqually people in 1856 described a creature with the exact characteristics of bigfoot. There is also an account of an Indian woman who was abducted by bigfoot along the Columbia River, again in 1856.

"Where the Waters Begin – The Traditional Nisqually Indian History of Mount Rainier," 1994, Cecelia Svinth Carpenter, paperback, 108 pages. Sells for $11.00, in WA .87 tax. The book has many legends of the Nisqually people plus a chapter on The Sasquatch. Some quotes: "The Nisqually people called him Seatco, the demon of the dark forests. They were terrified of him." And, "George Gibbs, writing in 1856, described Seatco this way: 'The Tsiatko are described as of gigantic size, their feet eighteen inches long and shaped like a bear's. They wear no clothes, but the body is covered with hair like that of a dog, only not so thick...they are said to live in the mountains, in holes underground, and to smell badly. They come down chiefly in the fishing season, at which time the Indians are excessively afraid of them...They are visible only at night, at which time they approach the houses, steal salmon, carry off young girls and smother children. Their voices are like that of an owl, and they possess the power of charming, so that those hearing them become demented, or fall down in a swoon.'"

Gibbs, "One Indian woman who lived at Fort Vancouver on the Columbia River told of having been captured by a group of Tsiatkos and taken into the woods. She lived to tell the tale of her adventure. Gibbs also reported that Leschi and Swatiltoh, both Nisqually, were said to have wounded a Tsiatko and tracked him by a trail of blood. Another Indian reported having shot at and wounding a Tsiatko while the beast was carrying off a young girl. Ke-kai-simi-loot, daughter of To-wus-tan, a former chief, and a Nisqually woman, claimed she was descended from four generations of what she called 'skoo-kums.'" 1856

The Western Bigfoot Society had regular meetings and routinely invited special guests to join and make presentations on a variety of topics. In March 2000, Gayle Highpine, a Native American from

Kootenay, British Columbia joined the group to talk about her tribal beliefs and stories, and how they relate to bigfoot/sasquatch. The following story was highlighted in "The Track Record" #95 (pg. 1).

Another Reservation Story

A woman from the Lillooet Tribe in Mount Currie, BC, told Gayle that her grandmother had seen an Indian Maiden that had been kidnapped by a Sasquatch, only to come home heavily pregnant. The child was born alive but died the night it was born. It was said to have been very hairy.

Ms. Highpine's story shows that bigfoot/sasquatch abductions and rapes have occurred in Canada as well as the United States. You would assume that stories like this regarding females in a small tribe would be something that most would be very tight lipped about. The woman probably wouldn't want the entire tribe knowing she was having a sasquatch baby. The number of stories similar to this that exist is much too large for all of them to be fabricated.

BESSY MUNSON

In mid July 2008, over 1,000 fires were burning throughout California in one of the worst fire seasons on record. A freak lightning storm came up from Mexico and brought with it thunder and fires that roared throughout the state. The fires were not confined to one area; they were everywhere. Cal Fire put priorities on fires near communities, and the Big Sur fire had the highest priority. One of the lowest priority fires was burning near Dillon Creek near the border of Siskiyou and Humboldt Counties.

I had just stayed the night in Hoopa and met with their tribal chairman, Lyle Marshall, and presented him with my first book. I expressed my deep appreciation for the access he had given me to his tribe and their lands. Without him the book would never have materialized.

I left Hoopa early in the morning in very smoky conditions from a fire near Burnt Ranch just outside Willow Creek. For the first hour I was driving away from the smoke and into clear weather. I traveled through Weitchpec, Orleans, and eastbound on Highway 96 towards a meeting in Happy Camp. I approached a campground in Dillon Creek and realized there was another fire burning above the Dillon Creek campground and in an area where other researchers had spent two weeks in 1997 studying bigfoot. I stopped and talked to local fire officials and found that strong winds had blown the fire up the previous night and it gained another 4,000 acres. More crews were requested, but few were available. The fire fighters looked tired and I questioned whether they knew they were working in prime bigfoot habitat.

I continued driving eastbound on Highway 96 and passed a residence on the Klamath River side of the roadway. Stacked rocks surrounded the house, some six to seven feet tall. They were very well done. It was a place that I'd always wanted to stop and talk about possible bigfoot habitat, but never had the time. Now I did. I pulled into the driveway and was immediately greeted by a Native American-looking woman who identified herself as Bessy Munson. I explained who I was, what I was investigating, and asked if they had any information about bigfoot. Bessy said her family was Karuk and had always lived on

Joseph Aubrey gold mining Aubrey Creek.

the river, and on their property for almost 200 years, but nobody who lived with her now had ever seen a bigfoot. At this time Bessy's entire family started to surround us and listen intently. Bessy said she had a story about a bigfoot abducting a family member and asked if I was interested in hearing it. I said yes!

Bessy's grandmother's maiden name was Bessy Alberts. She was born in 1884 and is buried on the property where we were talking. In 1899, when she was 14 or 15 years old, she married Joseph Aubrey, a miner much older than her. At this point, Loren Offield, Bessy's brother, brought out an old photo of Joseph mining on Aubrey Creek.

At about the time that Bessy and Joseph Aubrey got married, a cousin, Josephine, came to visit during the warm summer months. During this season the Klamath River recedes enough where it can be crossed in some areas where slack water exists. Behind the Aubrey property, the water of the Klamath runs deep and slow, and Grandma and Josephine were taken across the river to see their neighbors, the Thompsons. Sometime after they hit land, the girls became separated and Josephine was lost. A massive search was initiated and she wasn't found.

The area on the south side of the Klamath River includes the Marble Mountain Wilderness Area, the same area described in "The Hermit of Siskiyou." There are 40 miles of open area, no roads, and no other homes.

Two tense weeks elapsed until Josephine, dirty and tired, wandered into the Thompson residence, and the Thompsons took

Bessy Munson and family.

Josephine back to the Aubreys. Josephine said she got across the river she was immediately abducted by a bigfoot (she didn't use the word bigfoot in 1899). Josephine was returned to her parents and never went back to the Aubreys.

Bessy explained that her grandmother was a Christian woman who never ever lied. She told the story about the bigfoot abduction with the same consistency until her death in 1962. At this point Bessy's brother Loren, sister Betty Mckinnon, and daughter Talonna Nelson were all standing around us listening. I asked Loren and Betty if they had heard the same story from their grandmother, and they said they had. Loren reiterated that his grandma never lied and she believed that a bigfoot had taken Josephine. The grandmother had told the kids that she had seen a bigfoot on the other side of the river on multiple occasions and knew that it was there. It didn't surprise her that a bigfoot would take a young girl.

LOCATION

The Aubrey/Munson property is located on the north side of the Klamath River approximately 10 miles west of Happy Camp. It sits of the fringe of the Marble Mountain Wilderness Area. It's approximately five miles east of the Dillon Creek campground, a location where I've documented a sighting *(The Hoopa Project)* and where there was a major bigfoot expedition led by Dr. Jeff Meldrum. Their residence sits in the middle of a region that has a long history of bigfoot sightings.

If a bigfoot abducted Bessy Aubrey's cousin on the south side of the Klamath River, an entire battalion of Army Rangers could have searched the area and never would have found the girl. There is too much open space, heavy forest, and too many places of concealment, for anyone to be found. The residence sits between two of the largest and most remote wilderness areas in the western United States, Marble Mountain and Siskiyou.

COMMENTS

I found Bessy and her family to be very credible. They could have easily stated that they had seen a bigfoot, heard sounds, screams, etc. They never claimed anything other than what they were told by their grandmother.

This story also supports the hypothesis that bigfoot is attracted to young girls/children and supports the association of bigfoot with Native Americans and reservations. This occurred on the fringe of the Karuk reservation.

EDWARD VON SCHILLINGER

Ray Wallace owned a road construction business that worked in the Bluff Creek region of Northern California and later the Skamania

County area of Washington. Ray was noted for telling great stories, having a special relationship with the local Native Americans, and sometime fabricating tracks around Bluff Creek. In a letter dated December 22, 1997, addressed to Ray Crowe, Ray Wallace talked about the Bluff Creek region and specifically about Hoopa and the region north of it. He also included an affidavit signed by Edward Von Schillinger, dated August 11, 1982, who stated that he was with Ray when Ray had taken 16mm movies of bigfoot catching frogs in Onion Lake near Bluff Creek. He also stated that he had taken bigfoot movies for Ray, and that he had worked for him for five years as an engineer.

Ray said that researchers do many things wrong when going after bigfoot. "I will tell you why no one else hasn't ever gotten a movie of a Bigfoot…is they never rode a horse. As the bigfeet [sic] are not afraid of horse's [sic] so that's how Roger got his movie." The horse riding is an angle that does make sense. Getting the human scent off the ground and the sounds of hooves may not frighten the creature.

Ray claims that he was good friends with Chief Stokes (a local chief of the Hoopa and Yuroks). He also claims that some of the local elders used to call bigfoot *Teeseeatcoes*. One of the most memorable statements that Ray made was, "These Teeseeatcoes used to live with the Indians until one of them carried away an Indian girl at a brush dance at Happy Camp about 200 years ago, or so I've been told. None of the white men could remember the name the Indians gave these giant-sized people Teeseeatcoes," so Jerry Crew gave them the name "bigfoot" because of their large feet.

This is probably the earliest story I've heard about a bigfoot abducting an Indian female. The location of Happy Camp makes sense; that is where a hunter spotted a bigfoot while hunting in the late 1800s. (It's also near the location where a couple claimed they had a niece completely covered with hair that they kept inside. See "Tara Hauki" below for details). This is just one more piece to an ever-evolving puzzle involving bigfoot kidnapping women. The story has validity for our researchers because many elders inside Hoopa have told us that bigfoot sit on the ridges above their jump dances and watch the ceremonies.

The statement that Ray Wallace made regarding the Teeseeatcoes having lived with the Hoopa 200 years ago is compelling. These creatures obviously feel comfortable around the reservation; they continue to be seen there. There is also a physical similarity at some levels; refer to sketches made by Harvey Pratt of bigfoot without facial hair (below and in color section). There is also documentation that bigfoot/sasquatch creatures that lived in British Columbia had been Native Americans that were sent to live away on an island by themselves. They slowly moved throughout the country till they eventually occupied much of the Pacific Northwest.

This is from "The Track Record" #51 (pg. 7, para. 3)

> Michael J. April 16, 1996. (Hair sample included with letter...sample to Dr. Fahrenbach and to Dr. Kojo)...I have made an effort to not embellish or exaggerate. My father told me about gorilla hunts that he went on near Fredericktown, MO. in the mid 1940's, while my father was dating my mother. My grandfather owned a large wooded property about five miles southwest of town. A family of tall hairy ape-people were said to live on his land.

> My aunt was a young girl then. She disappeared for a couple of days. They found her wandering in the woods a couple of days later. She had gray streaks in her hair and said she had been carried off by an ape-man. She is now a nun in (deleted).

ED MASTEN, INKER MCCOVEY & ELTON BALDY

January 7, 2008, 5:15 p.m. I arrived in Hoopa late on a Monday, checked into the motel and went directly to the Hoopa recreation department to meet with Inker McCovey and Ed Masten. Both of these gentlemen had become good friends and had supplied me with significant information about sightings and bigfoot in the area. It had been several weeks since I had seen Ed and Inker, and I was looking forward to catching up on lost time.

I walked into the recreation office and found Ed and Inker talking with another older gentleman. They greeted me and then introduced their friend and tribal councilman, Elton Baldy. Inker indicated that Elton was a good friend and I could talk to him about anything.

I first asked the group if any of them had relatives in Pecwan. Inker said he had almost 30 aunts, uncles, and cousins living there. They all said they had gone to that area many, many times over the years. It was a small town that everyone knew very well.

I advised them that I had heard stories emanating from the late 1800s to late 1950s that tribal women in Pecwan had been forcibly raped by bigfoot. The women were embarrassed by what happened and they didn't want to talk publicly about it. I had been told by more than one source that Pecwan was the center of this activity, but that it had happened throughout the general area of Weitchpec, Orleans, Hoopa, and Pecwan. I asked if they had heard of these stories.

Inker, Ed and Elton nodded simultaneously and stated that it was true. Elton was the first to talk and said that there are very old stories within their tribe that women would be taken and raped by the "big people of the mountains." He stated that "big people" is another older name for bigfoot. He said the stories about the women being taken are probably true. Inker and Ed both agreed that they had heard the stories and that their parents had warned them about the creatures, but stated that they were there to take care of the forests and they would generally leave the boys alone. By coincidence, Inker was almost grabbed by a bigfoot while he sat on the toilet as a five-year-old boy (described in my first book). It's difficult to determine if the bigfoot in Inker's case knew he was a boy because of the physical characteristics of the bathroom and the fact that there were women living in the residence when this incident occurred.

TARA HAUKI

Tara's bigfoot sighting and encounter is chronicled in this book (chapter 8), but she told me of another odd incident that occurred to her while she was attending a bigfoot event. She explained that during a break in the day's activities, an older man and woman walked up and introduced themselves to her. They knew she had bigfoot

436

encounters and felt she was a compassionate person, and they needed to talk with someone about their issue. Tara could tell they were distraught and she offered a sympathetic ear.

The couple explained they were raising a niece at their home at the far northern section of Siskiyou County near the Oregon border. The niece was completely covered in hair and they believed she was somehow related to a bigfoot. They never took her outside, she always stayed in the house, and they felt bad for her existence. Tara did a lot of nodding and sympathetic listening, but in the end the people thanked her for her time and then went on their way in the crowd. Tara believed the people were telling the truth.

BIGFOOT VS. WILD MAN

The term "bigfoot" was generally accepted after 1958 as the name for the hairy giant that roamed the woods. Prior to that there were a variety of words and descriptions used to describe a tall, hairy, two-legged hominid that roamed various parts of North America. The most common names used were sasquatch (in Canada), hairy man, or wild man. Most of these descriptions said that the creatures were huge, hairy, and had a human type of face. The historical accounts go as far back as the mid 1700s.

Many researchers believe that there are two distinctly different creatures that are hairy, tall, and running on two feet through the hills of North America, but they rarely talk about it. They feel it's difficult enough for the public to accept the reality that there might be one creature (bigfoot) living and thriving in North America; the thought of two would be very difficult to accept. Looking back in bigfoot research in the 1940s to the 1960s, there is mention of two distinctly different creatures, one has a hairy face and is very large, one doesn't have a hairy face and is smaller. One of the creatures is forceful and sometimes violent, while the other is passive.

As a research group we do not have expertise outside of North America. Ray Crowe did have contacts throughout the world and regularly ran columns about the bigfoot that roamed throughout other countries. Oleg Ivanov was a Russian bigfoot researcher who wrote a book, *Avdoshki,* a Russian word for bigfoot. He interviewed many people about their sightings and this one in "The Track Record" #72 (pg. 11) was of significance. The sighting occurred in a triangle area surrounded by the villages of Selischi, Gornetskoye, and Poddubbye. It happened in July when a 15-year-old boy went out fishing and was confronted with an avdoshki family. It startled the boy and ran around in front of him in an effort to intimidate, but that's not the important part of this puzzle. The following is a quote from the book that Ray had translated (bolding added for emphasis by this author).

> I will describe the appearance of the Avdoshki. The male was of great height, about seven and a half feet tall. His neck ran together into his shoulders, which were a little more than 3' wide. He had very deep-set gray eyes. His mouth was wide and his lips were plump and brownish. There was nothing striking about his ears. The chin was wide, gradually sloping into his thick neck. The face was smooth with sparse hairs. His nose was not the nose of an animal or monkey. It was the nose of a man, wide, turned up and with large nostrils. **His face was like a man's and the color of the body hair was brown or chestnut.** Under his arms and on his stomach it was lighter. Along the backbone his hair was gray. The female was a little smaller and she had big breasts like pumpkins with brown nipples. The oldest of the children was a male about a foot and a half shorter than his parents and **lighter in color**. The younger one was a female.

The description above is in alignment with our sketches of bigfoot in California. The color of the creatures, and the younger bigfoot being a lighter color, is also in line with our belief that younger creatures have lighter tones. Of special importance is the facial description; it had a human face with little hair. This description and the great majority of sketches coming from our forensic artist fall into a pattern that matches the Russian bigfoot.

    I'll state here that North America Bigfoot Search does not have expertise beyond the Western United States and Western Canada, specifically northwest California, Minnesota, and Oklahoma. We haven't seen any viable photos depicting the hominid in other regions. We do believe that it lives in 49 states, but to say that bigfoot looks the same and has the same physical features in other states would be inappropriate. It does appear that the Oklahoma and Minnesota bigfoot do have many physical features similar to a Western U.S. bigfoot. We believe that the idea there are two distinctly different creatures living in the western United States is interesting, and we will address that through forensics in further studies. We will look at bigfoot/hairy man utilizing accepted forensic methods that have never been applied to bigfoot. You WILL have a clear idea of what this hominid represents and the human qualities it possesses.

REVIEW

Al Hodgson has told us that Betty Allen was afraid to go into the hills if there was evidence bigfoot was nearby. Al has stated that he feels that Betty was scared of being kidnapped by bigfoot.

Ed Masten, Inker McCovey, and Elton Baldy told me they heard the stories about Native American women being kidnapped and raped by bigfoot, and they feel there is validity to the stories.

Josephine Peters talked about incidents involving young disabled children left in the hills around Hoopa to be raised by bigfoot. The "big people" were supposed to care for the children.

Jackie Martins and Julie saw a bigfoot, and Julie used the term "Indian Devil" to describe bigfoot and expressed great fear of the creature. Lucy had a full chapter in her book about Indian devils and how they kidnapped Native American women, raped them, and forced them to have their children.

Tara Hauki met a husband and wife who claimed they were raising a niece with hair over her entire body, and the couple believed the girl was related to bigfoot.

Bessy Aubrey told her family that a bigfoot abducted her niece.

I've outlined six different and distinct stories that exemplify the possibility that bigfoot could have bred with humans during the last 200 years. If we believe this could have occurred, then we should expect changes in the physical features of some DNA lines of bigfoot in the northwest California area. The bigfoot genetic lines that did breed with humans may start to take on human qualities. If we review some of the sketches where there is limited facial hair, you can see human facial features.

Near the end of my stay in Hoopa and while I was accumulating sightings from other regions, I came to the realization there was another skill Harvey possessed that could transcend bigfoot research. Part of Harvey's forensic training is not only drawing sketches from witness descriptions, but also taking those sketches and photos and altering them in a manner a suspect may change his or her appearance in order to avoid apprehension. If a suspect in a bank robbery photo was wearing a beard, Harvey has the ability to take the beard off the suspect's face. Why couldn't we do this with the bigfoot in the Patterson–Gimlin footage and other bigfoot sketches where the witnesses described a full face of hair? Harvey was very eager to take on this "hairy" assignment.

FORENSIC SKETCH ART

My first memory of seeing law enforcement sketch art was in my early days at the Fremont (California) Police Department. I was a sophomore at U.C.Berkeley (1976) when I took a civil service test for the tri cities of Fremont, Newark, and Union City, all located on the eastern side of the San Francisco Bay Area, 30 miles south of Oakland. Those were the days when many young adults wanted to become police officers and this was my first exposure to a civil service test. I was from the San Jose area and wanted to get hired there, but they were under a federal consent decree that they could only hire bilingual (Spanish/English) officers in the late 1970s. When I arrived for the written test for the tri-city police exam I was stunned; thousands showed up, it was a zoo of people. To say that I was distraught when I finished completing the written test would have been an understatement. I went back to school and studying. Four months after completing the exam I received notification that I had passed the written and was now invited to take the physical test. To make a very long story short, I ended up being offered an oral interview for all three departments, I only opted for Fremont. From 2,700 applicants who took the written exam, I ended up number 10 on the hiring list for Fremont. The top four applicants took other jobs and I was hired by the Fremont Police Department in April of 1977.

I spent three years working in Fremont and was fortunate to be surrounded by intelligent and supportive officers. I enjoyed the camaraderie greatly, and I learned a lot about the department and myself. It was during those early years that I saw my first law enforcement sketch of a suspect. I don't remember what department entered the drawing into our bulletin, but it was humorous. I can remember several of us in our briefing room discussing if the suspect was supposed to be white, Hispanic, or black, we couldn't tell. We felt reasonably certain that the sketch wasn't of an Asian, based on the eyes, and that's about all we could surmise.

As crime started to spread north in the Bay Area, San Jose Police started to submit their "Watch Bulletins" to Fremont to distribute to their officers. It was in 1977 when I saw the first San Jose Police Watch Bulletin that contained a "Macrisketch" of a suspect in a robbery. The sketch looked like a piece of art; it was remarkable. Everyone initially doubted that the suspect would actually look like the drawing, but we all agreed that the person who did the drawing was a true artist.

In 1980 I took the test for San Jose Police, was accepted, and resigned from Fremont, joining SJPD in October of 1980. I came as a lateral transfer officer, which meant I didn't have to attend the police academy again, a blessing. I did attend an in-house, six-week academy that taught us the paperwork and policies of SJPD. It was during a tour of the different units in the police building that I met who was behind "Macrisketch."

The far corner room on the second floor of the police administration building held the office of the police artist, Tom Macris. As a group of seasoned officers arriving from a variety of jurisdictions, we had all seen Tom's work, but few had seen the results. The wall of his office had 50 sketches he had completed of various suspects wanted for crimes ranging from petty theft to murder. Next to the sketches were the booking photos of the suspects subsequently apprehended for the crime. Amazing. It truly appeared as though Tom had the suspect sitting in the room while he was sketching the witness description; they were all near-perfect matches. It was on this day in my police career that I forever became a believer in Tom Macris and forensic sketch art.

441

Tom stayed at SJPD for another 10 years and then retired. He was given the task of finding a predecessor to fill his shoes, not an easy assignment. I really didn't follow the process, but I did see the results. I can remember sitting in briefing rooms and seeing Tom's "Macrisketch" replaced by "Zamorasketch." Gil Zamora took Tom's job and immediately started drawing suspects with the same artistic quality as Tom. The booking photos of suspects drawn by Gil were amazingly accurate. The idea that two of the best law enforcement forensic sketch artists in the U.S. had emanated from SJPD was amazing. It can easily be said that Gil and Tom are probably the two officers in San Jose responsible for more criminal apprehensions than anyone who ever worked patrolling the streets.

SKETCH ARTIST

When I originally started to research bigfoot, I knew that somewhere I would utilize a sketch artist to put a current figure and face on the hominid. I knew that Tom Macris was retired, playing a lot of golf, and remodeling his house; drawing sketches of bigfoot wasn't high on his totem pole. Gil Zamora was raising a family, still working for SJPD, and didn't have the inclination to take time away from his lifestyle for a seven-hour drive to Hoopa to draw hominids in the woods. I wanted to find someone with the same artistic qualities as Gil and Tom, yet who had the Native American background to be accepted by the Hoopa people, and in walked Harvey Pratt.

I found Harvey on an Internet search and discovered he had 35 years working for the Oklahoma Bureau of Investigations (OBI), was an assistant director, and was the State of Oklahoma's sketch artist. He had drawn sketches for the Green River Killer, BTK Killer, and a throng of other major cases. His artwork was beyond reproach, stunning, and his law-enforcement-related sketch art was equally as credible as Tom's and Gil's. We had hit a grand slam finding Harvey Pratt!

NABS did a very thorough review of Harvey's sketch and interview abilities and we were thoroughly impressed. It's important for readers to comprehend the skill that's necessary to interview a victim/witness, listen intently to their statements, and then translate their statements and feelings onto paper with the result being a sketch. The sketch that Harvey draws is something that is a very powerful tool for law enforcement. Based only on a sketch of a suspect, a police officer can stop somebody and question his or her identity and the crime. If the sketch isn't accurate, police officers will be stopping innocent people and wasting countless hours chasing a worthless clue. There have been hundreds of suspects arrested based on a police officer stopping a suspect and utilizing the sketch as their only probable cause for the initial stop.

Harvey has drawn thousands of suspects in his 35 years of law enforcement. Included below are a couple of suspect drawings, followed by the actual suspect photo. Each case contains a short description of the incident, the victim, and the suspect. It's important to remember that the skills Harvey utilizes to draw a suspect of a crime, are the same ones he uses to interview a bigfoot witness and extrapolate a description of the hominid to put on paper.

This first investigation involved the disappearance of two young girls from the Oklahoma State Fair and was subsequently investigated by the Oklahoma City Police Department (OPD). OPD developed two female witnesses and sent them to the Oklahoma Bureau of Investigation and Harvey Pratt. Harvey interviewed both girls and, with their assistance, sketched the suspect. In an effort to help with the clarity, and instigate recall for additional witnesses, Harvey did the sketch in color. The drawing led investigators to Roy Russell Long who admitted he was with the young girls and stated he had hired them to unload items from his truck. Since the abducted girls were never found, an Oklahoma District Judge ruled there was insufficient evidence to hold Long, and dismissed the charges. Long later died in a Wyoming prison where he was serving a life sentence for the deaths of two hitchhikers. The drawing was distributed nationwide and was one of the major clues that led investigators to him.

The similarity between the sketch and the suspect are obvious. The beard, glasses, hat style, and the hair actually coming out from the side of the head under the hat is actually replicated in the photo. That is accuracy.

This second case involved a 21-year-old ballerina student at the University of Oklahoma who was abducted from her apartment complex and driven to Lake Stanley Draper. The suspect raped the woman, and then forced her to walk into the lake and he killed her. The body was found washed onto one of the beaches. A witness was driving a vehicle in front of the suspect and found his behavior very odd. The witness continued to look at the suspect through his rear-view mirror for several miles; this was the only view he had of the killer. The witness eventually made it to Harvey Pratt's office where Harvey interviewed him and sketched the drawing based only on this rear-view mirror observation. The suspect was in custody in Oklahoma on a burglary conviction, and DNA taken off the victim tied the suspect to her. When the suspect was brought to trial he had shaved his head in an apparent attempt to change his appearance from when the witness observed him. A jury convicted the suspect of murder, rape, and sodomy.

Roy Long. Photo on left, sketch on right. (Also in color section.

The sketch is a remarkable accomplishment considering the witness was looking only sporadically through his rear-view mirror. The mouth, lips, facial structure and extension of his ears are an excellent representation of the convicted subject, Anthony Castillo Sanchez. This case was highlighted on the television show, *America's Most Wanted.*

After Harvey had finished the first round of sketches for *The Hoopa Project,* I remember sitting in the hotel room with him and his wife, Gina, talking about how different the sketches were from what I had originally thought they would be. Harvey and Gina agreed. We thought they would be much more ape-like, more like animals, yet the consistent theme for all witnesses was that the big-foot they observed was much more human.

Harvey still works for OBI, and Gina Pratt is a law enforcement agent for the Oklahoma Bureau of Alcohol Beverage Control; between the three of us we have over 75 years of law enforcement experience. In talking about the sketches, we all agreed that the witnesses were credible and honest, and most important, believable.

Anthony Sanchez.

In the history of bigfoot research there have been many who have called themselves "researchers" and have made many bold claims and statements. NABS made an assertion early in our formation that one of our internal mottos would be to "promise little and deliver big." After Harvey's first round of sketches we were still hesitant to make any big claims about bigfoot because of the relatively small sampling of sketches from a confined area such as Hoopa. Indicators were pointing specific directions, but it would be best if we held back any public statements that we felt would rock the bigfoot world.

In May 2008, Harvey and Gina came to Northern California for the second round of sketches with an entirely new group of witnesses predominantly from a different region of Northern California. NABS wanted to ensure these sketches were completed prior to the release of *The Hoopa Project* so that no claims could be made that witnesses had seen sketches from our first book and were merely mimicking those. Our publisher (Hancock House) indicated they could have copies distributed in the early portion of summer 2008, so there was a necessity for Harvey to get to California in May.

One of NABS' predominant theories is that the creature has human genetics in its DNA, based on the multitude of reports regarding its mannerisms and appearance. A pessimist might state that this is only confined to a small area of Northern California. We don't think so.

When I finished *The Hoopa Project* there were a few researchers who challenged our results, saying they were only from a very small and confined area, Hoopa. Some challenged NABS to obtain results from a larger area and see if the results aligned with Hoopa. Little did people know that NABS was well on its way to a bigger project.

Immediately after completing Hoopa, we moved into the four counties that surround the reservation (Trinity, Del Norte, Siskiyou, and Humboldt). We had been accumulating bigfoot witnesses from these areas and plans were made to bring Harvey Pratt out for a second round of sketches. Once the second round was completed and the results proved to be very similar to Hoopa, then Oklahoma and Minnesota presented opportunities. Although sketches did not emanate from Minnesota, detailed witness descriptions paralleled bigfoot sketches from California. Oklahoma was the last location that NABS visited, and there obtained a series of forensic bigfoot sketches. At this point in our project I think it would be safe to state that there is a commonality between most of our sketches. Most have a human quality to the face, don't appear to look like any other animal, and appear more human than any other known creature in the world. We believe it would be a mistake to classify this biped as a walking ape; it appears much too human and forensic testing at this time points towards a human element.

In the article below in August 1998 from "The Track Record" #80 (pg. 7–8), Recent Sightings, a husband and wife recounted a sighting near the Canadian border. If not for the location, this could have been in the heart of northwest California.

Sunday, 8/16 Matt and his wife were being watched by a Bigfoot a mile from their home in Northeastern Ohio when they were in the woods engaging in coitus.

The incident caught his interest and he had returned to study the creature. There's a family of four, the old man, gray and 8 foot, he came face to face with Sunday. It was "human" he said. ...no way ape...there was no hair on its face. They were staring at each other, and so intent was the Bigfoot that a deer blundered into him, and he angrily swatted it away. The 8-point buck was knocked off balance. They had been whistling, but when two deer came by and one bumped the creature, it shrieked loudly. The gray creature left 17-inch tracks. Matt lives on 19 acres of wilderness towards Canada and the area has peaches, plums, pears and paw-paws. A neighbor lady complained that a "Big Man" was in her garbage.

The sighting by Matt and his wife is important as it related to bigfoot across the United States. NABS can only document human–bigfoot interaction in California and can surmise from that the possibility of a human/bigfoot hybrid (we believe that once the world knows that bigfoot has committed abductions in California, other victims will come forward from other areas of North America). It would appear from the sighting in Ohio that bigfoot looks much more human in various parts of the U.S. than many believe. This sighting further justifies the position that a wild man/bigfoot creature roams the United States and is much closer to human than animal.

I would also refer to a witness statement on a Minnesota reservation. A young girl saw a bigfoot sitting on her back lawn. She described the creature as having a human-type face with no hair, yet hair covered its entire body. This again confirms a biped that looks very human, not like an animal.

To think that the human qualities in bigfoot are confined to our continent may be a huge mistake. The following article in February 2003, appearing in "The Track Record" #124 (pg. 15, para. 5) may be good evidence that bigfoot/sasquatch/yeti may be much closer to human than many are reporting.

Was it the visitation of the abominable snowman, also called yeti by some? Villagers in this Jammu and Kashmir village are sure it was. At least 20-year-old Raja Wasim has no doubt that it was the Snowman that attacked him. Fondly called "Raju" by his parents, the young man came out of his uncle's home to feed cattle at a cowshed. He heard a strange noise among the greens in the lawn

447

of the house. When he turned around, there it stood: a four foot tall monster covered with dense dark black hair all over, "looking menacingly" at the youth. Claims Raju: "there is no mistake about what I saw. The monster had the face of a man with monkey like features. It was four feet tall but extremely sturdy. It was the snowman. It pounced on me and I jumped back on the veranda shouting for help. My uncle and his family rushed to my rescue and the monster lazily walked away. It was hardly frightened by the encounter.

The important part of this description is the facial features. Rarely have I ever seen a snowman described as having human-type qualities, but rarely has anyone had a close viewpoint. The description and actions of the witness appear credible.

THE SKETCHES

As a team, NABS had discussed at length the first-round sketches found in *The Hoopa Project.* We had looked at the Patterson–Gimlin still frame (creature is turned slightly looking at the camera) of the footage at Bluff Creek and forwarded this frame to Harvey Pratt, requesting that he sketch the face of the creature. His sketch is shown on the next page and in the color section.

Looking at the Patterson–Gimlin (P–G) creature and at the Ed Masten sketch in *The Hoopa Project,* which was the only facial sketch from the Hoopa research that showed hair covering the entire face of the creature, NABS developed a theory.

NABS sent Harvey the Masten sketch he had originally completed and asked him to remove the facial hair from the sketch. This is something that Harvey is routinely asked to do when he has suspects with facial hair or there is a suspicion that the suspect is attempting to alter his or her appearance.

Harvey called me several days later and said, "Dave, you're really onto something here. These things look very human." Harvey sent the sketches to me for comparative analysis.

The sketch of the Masten creature (below) without hair is riveting. Harvey decided to place a human-style haircut on the face without the hair and make the age approximately 60 human years. As anyone can see, the creature without hair looks exactly like a human male, possibly of Native American descent. This one sketch caused NABS to completely re-think our outlook on bigfoot, but one sketch doesn't make or break a theory. Remember, this isn't an American artist taking liberty with a sketch and casually changing the face to accommodate the needs of an organization. Harvey is a professionally trained law enforcement officer who does this with humans on a regular basis. He is the consummate professional when it comes to forensic diagrams. NABS never told Harvey what we were looking for in the sketch; heck, we didn't know and Harvey never asked.

Harvey agreed that the sketch without facial hair did look very similar to a Native American male, and he questioned me about that assumption. I informed him of our findings regarding bigfoot abducting Native American females and he immediately understood why this could have occurred.

A notation on the Masten sketch should be kept in the forefront of the reader's mind: Ed Masten made his sighting of the biped in the Tish Tang valley in 1978, 11 years after the Patterson–Gimlin footage. The idea that genetics had been altered significantly in 11 years was doubtful. I now placed another call to Harvey.

If Harvey could remove the facial hair from a creature described

(Left) Ed Masten "hairy face" sketch. (Right) Same image without hair.

to him by a witness, he surely could perform the same sketch surgery to the creature in the film footage.

Two weeks later, Harvey sent me his results. The first image is the P–G creature as it looked in the film footage, with hair, turning slightly to look at the camera. It shows hair under the eyes, around cheeks and full on the upper brow.

The second image shows the same creature without facial or forehead hair. This sketch is made in one tone, similar to a pencil. It obviously looks like a human, possibly a Native American, possibly a Hispanic or Caucasian with rugged, chiseled features.

This astonishing representation should cause everyone in the bigfoot world to pause and completely re-think the idea that bigfoot is an ape. Remember the old saying, "if it looks like a duck, and it walks like a duck, then it's a _____." We are in no way claiming that bigfoot is human, because all the data isn't in and many more answers need to be addressed. Lets keep reviewing sketches.

Harvey's final sketch of the P–G creature was completed with color and shading (see color section). He again gave it the look of a 60-year-old male with a receding hairline and possessing the same rugged features. Harvey did not alter the creature in any way; he merely sketched the creature the way it appeared in the footage without facial hair.

(Left) P–G sketch with hair. (Right) Clean shaven P–G sketch.

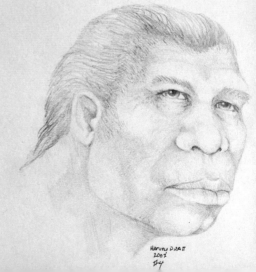

An interesting note on each of Harvey's sketches: it is universally accepted in bigfoot circles that if you are a believer that the P–G footage is of a real bigfoot, then it is also accepted that the creature is a female because of the large breasts (which are obvious on almost every frame of the film). There can be no mistaking that the creature in the film had breasts much larger than any male in the human race could ever possess and still be considered male. If the creature in the footage is a female, researchers must understand that the beauty of a female bigfoot doesn't correlate to the physical beauty of a human female.

Clean shaven PG.

In our continuing effort to review all sketches for the human qualities we reviewed the sketch made by Jennifer Crockett. Jennifer told us several times that she felt this creature was young, acted timid, and never made her feel frightened. She said the face had obvious human qualities that are easily recognizable. This hominid did not have a heavy brow, squared off cheeks, or heavy ridgelines. It had a fairly rounded face with some very soft features. This sighting was made adjacent to the Smith River in far Northern California in Del Norte County very near the Oregon border. Jennifer thought this sketch was an almost exact representation of what she saw in the roadway.

Jennifer Crockett's golden colored bigfoot.

Rose French described the creature she saw while she was watching her son ride his motorcycle on vacation property between Weitchpec and Pecwan, less than 10 miles from Bluff Creek. This is one of the few sketches where the biped didn't have a large, flat nose. This creature also has a round face that was able to express emotion, and soft features. Rose told us that when she saw the creature playfully chasing her son on the motorcycle, it stopped in the middle of the roadway, looked directly at her and gave an expression of great sorrow, as though the game had to end early. She felt the bigfoot was playfully chasing and never had any intention of actually catching the cycle.

After a day of working on a variety of different sketches, Harvey, Gina, and myself had a brainstorming session where we discussed what we had learned from both rounds of sketches. One item that immediately rose to the top of the list was a discussion about the golden or tan-colored bigfoot. Both Jennifer Crockett and Marlette Jackson had described their bigfoot as slight in build and timid, and neither witness felt afraid. The behavior and size of the creature (not fully developed) indicated to us that they were in their adolescent years or were young adults not quite fully mature.

Rob Alley wrote a small segment about a woman that he had known for 10 years. The incident occurred in late December 1999 on Ward Lake Road 7 miles northwest of Ketchikan, Alaska. This is from "The Track Record" #95 (pg. 4).

Rose's young bigfoot.

The woman was driving out early in the morning as she does several times a week to just sit at the end of the road in her truck and sip coffee, her little getaway from a demanding schedule. There was no snow at the time, it had melted a bit, and it was drizzling lightly, still dark out at 5 am. She said, "I had just cleared the windows with a wipe of the wipers and was coming around the right curve past the ward creek bridge and I flicked to high beam, noticing something in the salmonberry bushes on the left. It was tall at first, I thought it was a deer, you know standing on its hind legs, all light brown in color. But I changed my mind pretty quick when it turned its head and shoulders at me. As I drove past it, it turned to its left, and showed a fur covered form, I could only see from the waist up, but it was like a man sort of, about 6 feet tall and probably not over 200 pounds. It was walking when I saw it and after I went past I thought, "Wow I've actually seen one of those creatures they keep talking about." And instead of stopping for my coffee in my truck at the Signal Creek Campground up ahead, I turned back around and pulled over just before getting to the spot I had seen it. I waited for about a half hour but it didn't show itself. I didn't hear or smell anything and I felt pretty safe sitting in my ¾ ton pickup. I didn't go out looking for tracks. That was all I saw.

The description of the creature in Ketchikan matches the description of Jennifer and Marlette very closely. The bigfoot is obviously not fully developed, six feet, 200 pounds, still short, and too slim for a full-size bigfoot. The color also is a match and confirms our suspicion regarding a probable color change as the creature ages. It's interesting that the sighting was made next to a berry bush (see chapter 3, Associations).

From "The Track Record" #99 (pg. 4).

June 25th, 2000, Near Orick, CA off Highway 101.

At 6:10pm while driving on Johnson's Road off Bald Hills Road, witness saw creature striding very far away from the road. It jumped off the embankment. Described as man sized, brown and gold in the sunlight.

Orick is north of Eureka at the fringe of Redwood National Park. The area of the sighting is a very active region for bigfoot sightings. The color of the creature described in the article along with a description as "man sized" is indicative that it is not full-grown. This area is just over the hill from the Rose French and Marlette Jackson sightings.

From "The Track Record #117 (pg 12, para. 5).

> Near St. Louis, MO where there have been two sightings in the past week, of a tan colored Bigfoot, seen by the reporters daughter and grandchildren. Also a possible nest constructed of woven branches and brush had a clump of tan colored hair.

From "The Track Record" #122 (pg. 8, para. 3).

> Vaughn, Montana, October 1976, Gail Kapptie and son found a sheep pen gate pushed in and 18 inch, four toe footprints. A lot of grain was gone from a barrel. That night when the dog was barking, she went out and saw the dog barking at an 8-foot tall, tan, ape-like creature with very long arms standing upright by the chicken coop. It ran off at high speed.

It would appear from "The Track Record" quotes above that tan-colored bigfoot are seen across the United States and are not isolated to the area of the Hoopa Valley.

AGE PROGRESSION

I asked Harvey if he had completed enough sketches of the creature to get a comfort level of what the size and dimensions are for a mature male, versus a young adult male, versus an adolescent male versus...Harvey paused for a few seconds, immediately comprehending where my idea was going. He said he thought he had enough information at that point to comfortably say yes.

I then asked him if he could take one of the younger creatures we've drawn and age progress it in the same manner he would age

454

progress a human sketch, and he said that wouldn't be a problem. We had a brief conversation about which sketch he felt would be most representative and accurate, and be able to age correctly, and Harvey said it would be the Rose French sketch.

At the conclusion of the week's sketches, Harvey and Gina returned to Oklahoma and within 10 days I had my first sketch of an age-progressed Rose French bigfoot.

We believe the first Rose French sketch represented a bigfoot in human age range of 14 to 17 years. This second sketch represents a bigfoot in its mid thirties. The major differences in the second sketch are fuller features, thicker beard, possibly darker hair, and a more full and developed body. Harvey bases his sketches on his age progression of humans, and how that has been validated through the discovery of people who had disappeared and comparing the age-progressed sketches he completed to how they looked when found. He is very accurate in his work.

Two weeks after sending the second sketch, Harvey sent a third sketch. This looked very similar in proportions and features to some of the other sketches he had completed for other Northern California witnesses. He continued with the same hair colors as in the earlier sketches, but that is an uncertainty with bigfoot. With a working theory that they change colors going through adolescence, there is a possibility that Rose's bigfoot may change colors as it ages, similar to a human getting gray hair in later life. This sketch would represent a creature in its mid to late forties in human years. This is a bigfoot that is healthy, full featured, weathered from the elements, and has long hair. In every sketch completed, the witnesses have stated that the creature does not have hair flowing off the forehead into its eyes. Many have stated that it does have hair flowing off the back of the head, neck, and shoulders off the upper back, almost acting as insulation to the head and shoulders. This is, of course, the first area that inclement weather would hit the creature, so having some level of insulation would be important.

Rose bigfoot age progression.

Many months after Harvey and Gina had gone home, I was reading one of Ray Crowe's Track Records newsletters from 1997 (issue #71). In the back of the newsletter was a section titled, "Letters to the Editor," which included a letter written by Ray Wallace, the owner of a road building company that was contracted by the USFS during the late 1950s and early 1960s in the Bluff Creek area. Ray made friends with a lot of people in that area and among his closest friends were the Native Americans living in the Pecwan region. One of their leaders, Chief Stokes, became great friends with Ray. What follows is a statement from the chief about bigfoot and what it has done to his people.

> Chief Stokes says they (Bigfoot) carry away many girls at night-time that we never see again. I asked Tuffy Dowd who is a 35 year old Indian who has graduated from college about this fellow they call Teeseeatcoe (another name for Bigfoot). He said that they are covered with brown fur all over their entire body and the young ones are a brownish light color or tan. He said that they are mean, as they carry away many young girls about 20 years of age or younger, he said. They steal our calves for food, he said that's why no one raises cattle anymore, he said there is a whole family of them living at the headwaters of Blue Creek in an old gold mine.

I can't imagine a better statement validating our belief that bigfoot is a lighter color when young and gradually transitions to dark. Specifically, they are a tan or light brown when they are young, per tribal member Tuffy Dowd. The paragraph is also another confirmation that Native American women in the Pecwan area were being abducted.

NATIVE AMERICANS

During the course of researching this book we came across vast numbers of documents, sighting reports, historical newspaper articles, and witness statements that explain bigfoot encounters and sightings on or near Native American reservations. Our forensic artist has drawn sketches of the biped with and without facial hair. Some of these drawings look identical to Native Americans. We have provided articles that indicate bigfoot hair and blood samples have human DNA.

There have been many researchers, onlookers, and observers from the past who may have heard innuendoes that the bigfoot/sasquatch may be part of a Native American tribe that was exiled. Ray Wallace has stated that bigfoot used to live with tribes along the Klamath River until they had a falling out. An interesting parallel to Ray Wallace's account comes from a Native American writer, Jorg Totsgi. Mr. Totsgi was the editor of "The Real American" and he was a member of the Clallam Tribe of Washington.

On July 16, 1924 Totsgi authored an article for the Oregonian that addressed an attack on a group of prospectors in the Mt. St. Helens area of Ape Canyon. The prospectors had claimed they were attacked at their cabin during the night by a group of creatures that walked on two feet, moved like humans, but had hair over their entire bodies, like human-looking apes. The headlines in Totsgi's article stated, "Big Hairy Indians Back of Ape Tale / Mountain Devils Mystery Grows Deeper / Shaggy Creatures Kill Game by Hypnotism / Ventriloquism Is Used." In the article, Totsgi stated northwestern Indians recognized the big apes reported to have bombarded a shack of prospectors at Mt. St. Helens as none other than the Seeahtik tribe, and that they have kept the existence of the tribe a secret because it is a "skeleton in the northwestern Indians' closet." Totsgi quotes other Indian leaders from neighboring tribes to support his position, such as, George Hyasman (Quinault Tribe), Jimmy James (Lummi Tribe), Henry Napoleon (Clallam Tribe). Totsgi further stated, "Every Indian in the Puget Sound area is familiar with the history of these strange giant Indians." The article covers a variety of aspects of the tribe but leads to the conclusion that the Seeahtiks are a group of huge people with hairy bodies and possess the strange ability to kill game with the use of hypnotism and to call game utilizing excellent ventriloquism. They have a keen sense of smell and excellent night vision. They agreed that the tribe is harmless unless provoked.

Totsgi reports that Indians from a variety of tribes have been humiliated by the Seeahtiks. Many times they steal the tribal women and keep them, only sometimes letting them go. Totsgi writes, "More often she does not, and it is even said by some northwestern Indians that they have a strain of Seeahtik blood in them."

There are many names associated and evolving from Seeahtik including Saskehavas, Seeahtek, Seeahtlk, Seeahtkch, Salatiks, and it eventually evolves to Sasquatch. There are names used in modern Washington State that are very similar to Seeahtik, including Seatac Airport in Seattle.

In the Totsgi article tribal members from different Pacific Northwest tribes stated that "the Seeahtiks," another tribe, are the giant and hairy creatures that live in the middle of the forests; they are sasquatch or bigfoot. A fellow tribal member even implied they are human. The Seeahtiks are different from other tribes because of their appearance and abilities. As they have their own customs, beliefs, and rituals they live separately from one another. It appears from the article that certain tribal members visit the Seeahtik and sustain some type of relationship. These visitations and relationships may account for some of the attempts by bigfoot to walk up to people and attempt communications. While almost all of these attempts have ended with humans running away, it is probable that many of the tribal members of the Clallam and Quinault probably engaged the Seeahtik in some type of communication, a key element for scientists to classify bigfoot as human. Bigfoot is probably confused that people with whiter skin tend to run from them while natives with darker skin are willing to be friends.

Oregonian.

)AY, JULY 16, 1924   PRICE FIVE CENTS

BIG HAIRY INDIANS BACK OF APE TALE

Mountain Devils' Mystery Grows Deeper.

GIANTS SAID TO ROAM HILLS

Shaggy Creatures Kill Game by Hypnotism, It Is Said.

VENTRILOQUISM IS USED

Redmen's Editor at Hoquiam Gives Theory of Reported Attack at Spirit Lake.

"Big Bear" Speaks.

Ciallams Are Killed.

Strange Medicine Used.

Tribe Held Harmless.

APEMAN HUNT BROADENING

Kelso Police Chief and Others Go to Spirit Lake.

APES DECLARED INDIANS
Continued From First Page.

Totsgi article in the *Oregonian*, July 16, 1924.

460

Clue to "Gorilla Men" Found

* * *   * * *   * * *

May Be Lost Race of Giants

Clallam Indians Tell of Eight-Foot Seeahtiks Who Killed Game by Hypnotism—Existence Kept Secret by Other Tribes.

By A STAFF CORRESPONDENT.

HOQUIAM, Wednesday, July 16.— "Mountain Devils" discovered at Mount St. Helens, near Kelso, are none other than the Seeahtik Tribe, said Jorg Totsgi, Clallam Tribe editor of The Real American, an Indian national weekly publication, in an interview here today. "Seeahtik" is a Clallam pronunciation. All other tribes pronounce it "Seeahtkoh."

The Indians of the Northwest have kept the existence of the Seahtiks a secret. Partly because they know no white man would believe them, and the Indian, known for his honesty and truthfulness, does not like to be called a liar, and partly because the Northwestern Indian is ashamed of the Seeahtik Tribe, said Totsgi.

"The 'mountain devils,' or 'gorillas,' who bombarded the prospectors' shack on Mount St. Helens, according to the description of the miners, are none other than the Seeahtik Tribe with whom every Indian in the Northwest is familiar," said Totsgi.

Were Thought to Be Extinct.

"The Seeahtiks were last heard of by the Clallam Indians about fifteen years ago, and it was believed by the present Indians that they had become extinct. The Seeahtik Tribe make their home in the heart of the wilderness on Vancouver Island and also on the Olympic Range.

"As described by the Clallam Indians, the Seeahtiks are seven to eight feet tall. They have hairy bodies like the bear. They are great hypnotists, and kill their game by hypnotism. They also have a gift of ventriloquism, throwing their voices at great distances, and can imitate any bird in the Northwest. They have a very keen sense of smell, are great travelers, fleet of foot, and have a peculiar sense of humor." Totsgi added.

"In the past generations they stole many Indian women and Indian babies. They lived entirely in the mountains, coming down to the shores only when they wanted a change of diet. The Quinaults claim they generally came once a year to the Quinault River, about fall. The Clallams say they favored the river near Brinnon on Hood Canal. After having their fill of fresh salmon they stole dried salmon from the Indian women.

"The Seeahtik Tribe, are harmless if left alone. The Clallam Tribe, however, at one time several generations ago killed a young man of the Seeahtik Tribe and to their everlasting sorrow, for they killed off a whole branch of the Clallam Tribe but one, and he was merely left to tell the tale to other Clallams up-Sound. The Clallam Indians believed that the Seeahtik Tribe had become extinct.

"It is fifteen years since their tracks were last seen and recognized at the Brinnon River. Prior to that time many Clallam Indians have met and talked with men of this strange tribe, for the Seeahtiks talk the strange tongue of the Clallams, which is said to have originated from the bear tongue.

"The Quinault Indians, however, claim that Fred Pope of the Quinault Tribe, and George Hyasman of the Satsop Tribe were fishing about fifteen miles up the Quinault River in the month of September four years ago, when they were visited by the Seeahtiks. The two Indians had caught a lot of steelhead trout which they left in their canoe, and these were stolen by the Seeahtiks.

Possess Hypnotic Power.

"Henry Napolean of the Clallam Tribe is the only Indian who was ever invited to the home of the Seeahtik Tribe. It was while Napolean was visiting relatives on the British Columbia coast about thirty years ago, that he met a Seeahtik while hunting. The giant Indian then invited him to their home which is in the very heart of the wilderness on Vancouver Island. Napolean claims they live in a large cave. He was treated with every courtesy and told some of their secrets. He claims that the giant Indians made themselves invisible by strange medicine that they rub over their bodies, and that they had great hypnotic powers and the gift of ventriloquism.

"Some Indians claim that during the process of evolution when the Indian was changing from animal to man the Seeahtik did not fully absorb the tamanaweis, or soul-power, and thus he became an anomaly in the process of evolution.

"The Indians of the Northwest are of the belief that the 'mountain devils' found at Mount St. Helens are the Seeahtik Indians, as it is generally their custom to frighten persons who have displeased them by throwing rocks at them."

Totsgi interview.

461

Our team has wondered many times why researchers haven't taken the Totsgi claim seriously and attempted to understand his correlations and assertions. NABS has researched this angle of the bigfoot phenomenon at a peripheral level and find it very intriguing. Much of what Totsgi claims is found in many modern day sightings of the creature.

We have investigated several reports of people hearing sounds they recognize, but are out of context where they hear them—children crying, goats bellowing, people talking, owls, and other noises. Bigfoot having the ability to be a ventriloquist and make noises to draw in game or people is consistent with current reports. There have been countless witness accounts of bigfoot having the ability to sound like animals in the forest. In the Mike Cuthbertson encounter (chapter 9) he stated he heard goats, but there were no goats in the area. There are other accounts of bigfoot sounding like owls and other animals and livestock. There have always been rumors that the creature has the ability to simply disappear, tracks that simply stop, etc.

Another of Totsgi's assertions that is very interesting is the bigfoot ability to utilize hypnotism to mesmerize game. NABS reported in *The Hoopa Project* a sighting where deer were reported staying close to a bigfoot in the Sierra Nevada Mountains west of Lake Tahoe. This sighting seems unbelievable due to the numerous reports of people seeing bigfoot killing, feeding on, and stealing deer for food. Why would a deer stay near a bigfoot? Bigfoot is the largest predator in California! NABS has also read reports where witnesses saw a bigfoot in the bushes in Oregon, the bigfoot stared at the witness and a deer then ran directly into the hominid, which it then swatted away as though it was a nuisance. I know this makes no sense when you think in terms of normal wildlife and game, but we know that bigfoot isn't "normal." NABS is keeping an open mind on all aspects of bigfoot until our research clears its own hurdles. The rationale of how a bigfoot "hypnotizes" creatures may simply be a product of a certain smell (we do know that bigfoot has a horrific smell at certain times), pheromone release, or some other paralyzing characteristic that we don't understand.

HUMAN BLOOD/DNA

We can easily point to drawings and make associations about facial similarities between bigfoot and humans. In "The Track Record" #98 (July 2000), Ray Crowe received a letter from a regular contributor to his newsletter. A researcher named "Rita" lived in a remote cabin and utilized a generator for her power. One night she returned to find the generator damaged. The generator was essentially pulled apart, an impossible feat for any human. Since Rita was a researcher, she used her skills to understand how the damage could have been done. She found tissue and blood at the location of the damage. She sent the blood to the University of California at Davis primate center for analysis. The sample came back as human blood! It should be noted that Rita had bigfoot sightings around her property in the past, as is outlined in the article.

> Rita of North CA Sierras tells of a remote cabin, and there's a militar generator used for power. She returned one night and it had been damaged. It was the big 4 cylinder type on skids, water cooled. The wires had been torn apart. They were the 1/2 inch cables, twisted and snapped and pulled apart. She learned that it would take 2000# sq. in. of strength to pull them apart like that...about like two 4 X 4's pulling it between them. There was white hair, tissue, and blood mixed with the wires. Rita sent them to UC at Davis for analysis. They said it was human blood and hair that they couldn't identify.

We believe this is a valuable piece to add to an ever-evolving puzzle. While many researchers are discounting blood and tissue returns that show "human" DNA, we believe this may be exactly the building block of certain bigfoot proof.

Another fascinating article appeared in "The Track Record" #138 (pg. 3). Researchers wrote about an investigation occurring at the Carter Farm, 57 miles south of Knoxville, Tennessee. This is a very well-known location for its long history of bigfoot-associated phenomena. Researchers spent months on the property accumulating evidence, and had sightings at various points in the study. Investigators sent hair samples accumulated during the course of their investigation to Dr. Henner Fahrenbach, the world leading expert on hair identification and a retired scientist from the Oregon Primate Research Center. Dr. Fahrenbach analyzed the hair under a microscope and stated that the hair was consistent with bigfoot hair. The report indicates that the doctor retreated slightly from his position, because DNA testing on hair and scat samples sent to him came back as "human." The article is reproduced below.

> There is an ongoing situation at the Carter Farm that has not been fully exposed to the light of day, although serious efforts have been made. There is a significant amount of circumstantial and eyewitness evidence to support Jan's claim that unusual events occur there. There has been no solid scientific explanation for those events and that evidence. Dr. Fahrenbach has declared the hair to be sasquatch hair based on microscopic analysis, but then has retreated a little because we keep finding human DNA on all the evidence samples, including scat.

> The hair sample evidence, which matches his other sample data base, is the largest collection of hair from a single site that anybody's ever heard of. The scat was deposited in large quantities and in a very non-human fashion, and also yielded bat and dog DNA. Experimental studies in which a scientist attempted to contaminate dog and cat hair (controls) with human DNA did not yield a finding of "human" for those samples; they came up as dog and cat hair. This has led that particular scientist to speculate that the human DNA he has gotten from the hair is, in fact, Bigfoot DNA. He's looking for any possible differences from human (Homo sapiens) DNA, which MUST be there to account for the obvious differences between us and the creatures being described. Fahrenbach is perplexed and does not wish to speculate about the DNA problem right now. He, and other scientists, have expected to see something unique and apelike when Bigfoot DNA is finally analyzed, but that's not what we've got so far.

> I've personally spoken to a dozen witnesses who have described

seeing Bigfoot creatures on the property. I've found numerous stretched and twisted barbed wire fences, four or five dead dogs (including one hanging in a tree), twisted and broken trees, a squirrel skin in a tree, a couple of hundred pounds of human-sized scat in the loft of a relatively inaccessible barn, and a lot of other circumstantial facts that suggest that Jan's story has merit.

What we don't have is a dead or captive Bigfoot or clear, undeniable pictures of any, which is what we need to prove this case (and every other case, too). I think you have an important story to tell, and this case has at least as many important points of evidence as any other, and probably more. But we cannot claim more than that until (a) we get further DNA conclusions to support the idea that the human DNA is actually Bigfoot DNA, or (b) we get perfectly clear pictures of them, or (c) we get one, dead or alive.

NABS researchers believe there is a distinct possibility that bigfoot hair, tissue, and scat samples may contain human DNA. The idea that scientists discount a sample because of this return is amazing. If you believe in the science behind the testing, then you must believe the results. It is true that contamination does occur, but reading the article confirms that even the attempted contamination of controlled samples cannot reap returns as were in the Carter samples.

From the dozens of talks, presentations, and speeches I've done during the last year, there is a segment in the bigfoot/sasquatch world that wants the biped to turn out to be an evolutionary ape. There are many bigfoot/sasquatch books that have "ape" in the title, showing a bias towards this position. We believe that this biped is far from an ape, far more intelligent, and possibly aligned genetically with the Native North American.

MONSTER QUEST

The first season of the *Monster Quest* series on the History channel highlighted a series of incidents that occurred on Snellgrove Lake in Ontario, Canada. The lake is located 250 miles north of Ottawa and is only accessible by floatplane. The closest community is 200 miles away and there is only one cabin on the lake.

The owners gave *Monster Quest* producers the names of several visitors to the cabin who had sasquatch-like activity while they were there. They had experienced wood knocking, sounds, and the cabin was actually shook one night. At the end of one season the owners secured the cabin and left for the year. They returned in 2002 to find that something had entered the cabin and destroyed the interior. The insurance footage of the damage was shown to Dr. Lynn Rogers, a wildlife biologist from Minnesota who is an expert on bear behavior. Rogers viewed the damage and gave the opinion that this wasn't done by a bear, for a multitude of reasons. The most solid reason was that bears in this area of North America would have been hibernating when the damage occurred.

The owners of the Snellgrove cabin had experienced problems at the cabin prior to the break in. They had placed a board with nails protruding upward just at the entry to one of the cabin doors. Whatever had broken into the cabin had stepped directly on the board prior to the entry and bled on the nails and the board.

Dr. Jeff Meldrum (Idaho State University) and Dr. Curt Nelson (University of Minnesota) were invited by *Monster Quest* in August 2005 to go to the cabin and conduct research. They both examined the board, found an outline of a print matching a sasquatch print that was 18 inches long on the board. They also recovered what they thought was tissue, blood, and fibers from the board.

During the *Monster Quest* team's stay at Snellgrove, they had a highly unusual incident occur. On the team's last night they decided to build a large fire in an attempt to attract a sasquatch. At approximately midnight, without warning, something threw a large rock onto the metal roof of the cabin. A crewmember threw a rock into the woods and something threw a rock back. At this point most of the crew locked themselves inside the cabin and the activity subsided. The next day the owner of the lodge arrived and flew the team back to the airport for the flights home.

Once back in Minnesota, microbiologist Dr. Curt Nelson started to conduct DNA tests on the dried blood samples taken from the nail board at Snellgrove Lake. The team also forwarded hair samples recovered from the board to Dr. Lynn Rogers for his analysis.

Dr. Rogers stated that the hair is like no other hair he has seen. It does not match mammals' hair in North America and certainly isn't bear. Dr. Rogers stated, "It looks human." He said that human hair has a medulla, a center core section, but the sample hair did not. The hair had a naturally worn tip, "like it came from a wild human," he said. NABS has a hair sample it recovered from northern Del Norte County that has the exact characteristics as the Snellgrove Lake sample. Everything that Dr. Rogers has cited that is unique about the Snellgrove sample exists in the Del Norte sample.

Once Dr. Curt Nelson was back at his laboratory at the University of Minnesota, he initiated the process of extracting DNA from the dried blood samples. He said that at early stages of testing he found there was an inhibitor in the sample that prevented him from finding any DNA. He continued to work to identify the inhibitor and eventually was able to classify it; the galvanized portion of the board nail was acting to obstruct DNA extraction. Once that was clarified, Dr. Nelson "re-purified and amplified" the sample and found DNA. Dr. Nelson said, "It's identical to human DNA except it had one nucleotide poly-morphism." This is the same difference humans share with chimpanzees. The narrator of the show said, "DNA says primate but not quite human, not quite non-human, one of the base pairs is deviated."

Dr. Nelson explained what needs to be done from a scientific perspective to positively identify the hominid: "The thing we have to do now is look at more DNA. We have to sequence more of it and we have to design primers to amplify different regions of the DNA so we can get sequence across the mitochondrial genome and determine whether or not it is just human DNA, which seems unlikely that something human would step on the board like that."

If a sasquatch were approaching the cabin on a moonless night, its obvious focus would be the interior of the cabin, to know if it was occupied. I doubt that the sasquatch has a habit of looking where it steps when it's entering a danger zone like a human cabin. There are portions of the statements from both Nelson and Rogers that are very compelling. Rogers makes a statement that the hair sample is very similar to a "wild human." Nelson's conclusions on the blood sample show it could be human, even though it came from a bloody footprint 18 inches long and identical to a bigfoot/sasquatch print.

The Snellgrove Lake research is very compelling. Academic professionals conclusively have shown that DNA and physical examinations of evidence taken from a suspected sasquatch footprint and hair sample could possibly be human.

Note: In the summer of 2008 I made several presentations at bigfoot conferences throughout the western U.S. At two of these conferences Dr. Jeff Meldrum was also a guest and we had several conversations about Snellgrove Lake. He had completed his second trip to the lake and had the opportunity to be flown to a village of Native Canadians that was relatively close to Snellgrove. They had told Dr. Meldrum that sasquatch was a part of their culture and they had seen it for years, and some had sightings fairly recently—another example of the bigfoot/sasquatch relationship with Native North Americans.

HYPERTRICHOSIS

This is a human condition in which people are covered completely with hair, including all parts of the body that normally would not be affected. Called the werewolf disease in some circles, there are significant populations of these affected individuals in Europe and South America. It can be caused by a recessive gene that is passed to the next generation only if both parents possess the gene. There is no guarantee that the offspring of the parents will be affected with the condition, but there are other extenuating conditions that affect the patient and the amount of hair they may have.

At the point where our team noticed a human quality in our sketches, we started to question whether bigfoot may be a victim of the hypertrichosis condition. We have shown that when hair is removed from the bipeds, their facial features closely match those of humans. While there are physical differences between bigfoot and North American caucasians, we must keep in mind that there are also substantial differences between different races throughout the world: height, weight, frame build, etc.

Is it possible that bigfoot is a victim of the hypertrichosis condition and this is the reason they possess far more hair than a normal North American?

BIGFOOT LANGUAGE

One major stumbling block for many researchers to allow that bigfoot may be human is the assumption that bigfoot does not use language. During an email exchange with many of the top researchers in the world, the suggestion was raised to submit a tape recording of purported bigfoot speech to a linguist for expert opinion.

The witness to the bigfoot sounds was Ron Morehead, who reportedly recorded a conversation between two bigfoot at a very remote location in the Sierra Nevada Mountains in California. Many witnesses have said the tape sounds like two Japanese people speaking in a garbled dialect, thus the tape is called, "Samurai Chatter."

The tape was submitted to R. Scott Nelson who is a retired crypto linguist from Naval Intelligence, has worked in the Naval Security Group, and is a two-time graduate from the Naval Defense Language Institute. He is also a two-time graduate of the U.S. Navy Cryptologic Voice Transcription School at Naval Security Group Detachment. He has logged thousands of hours of voice communications as a cryptologic interpreter.

Mr. Nelson conducted a comprehensive study of the tape and the result was a far-reaching study that shook many bigfoot researchers' basic beliefs. I won't discuss all of the details of the report (it's long) but you can read it in its entirety on our website at www.nabigfootsearch.com. Nelson goes into many technical aspects of the language he reviewed, but the following is understandable.

> All this may be delving a bit too deeply into Psycholinguistics for our purpose, but since the creatures in our study are using language, these speculations may serve to alert us to the homo-centric tendency to classify Bigfoot into one group or another. Is Bigfoot an animal, or is he Human? Is he or is he not sentient? Does he think in linear patterns or more holistically? Here we

469

must not forget the *tertium quid* that Bigfoot may be very different from any creature ever classified. We cannot assume that he has not developed a graphic system for expressing language, simply because we have not discovered it.

Near the end of the report, the following is included:

. The intonation contour, stress pattern and speed (approximately twice the speed of human speech) at which the vocalizations are delivered makes it impossible for humans to understand. In addition to this, the rate of discourse, or the speed of exchange of conversational turns is such that the creatures are virtually "stepping on" each other in their responses. This also makes it impossible, in real time, to distinguish the utterance of one creature from that of another. The conclusion that must be drawn here is that the creatures mentally process information at a much higher rate than humans do, or at least they are able to communicate their ideas much faster. Some might argue that the creatures are able to do this because their thoughts are much simpler, but I think this would be a very homo-centric way of looking at this issue.

The bottom line in Nelson's report is that there appears to be language being used by the bipeds—a monumental turning pint for bigfoot research! As you can imagine, many researchers immediately called for another study, more expertise, etc. Since NABS does not have expertise in the criteria, we will refrain from commenting other than to say we believe that bigfoot has too many human qualities to ignore. We are not geneticists, but we do have a high level of common sense and we'll let the results lead our research.

THE BIGFOOT BULLETIN

In the 1960s and 1970s an Oakland, California native, George Haas, produced "The Bigfoot Bulletin." It was a monthly newsletter that is similar to "The Track Record," but decades before Ray Crowe had produced his. George was a great researcher and someone who had the ability to communicate with other investigators/researchers and keep everyone's egos in moderation, not an easy task in the bigfoot world.

On January 2, 1969, George wrote an extensive article about the opinions of bigfoot researchers in California and what they perceived the biped to be. This is page 2, paragraph 3.

Most of us in the Bay Area (San Francisco Bay Area) feel that we are dealing with a creature that is more than a "mere animal"; that we are dealing with a being very close to man, a sub human perhaps, or even a subspecies of the genus *Homo*. However, whatever the Bigfeet turn out to be, we feel that they already compromise an endangered species. Considering the current rape of our National Forests, the continued shrinking of our wild areas and the increasing invasion by backpackers and others of what used to be truly "Wilderness" areas, we hope that eventually we can see laws passed preferably on the national level, to stringently protect them and areas set aside for their preservation.

NABS believes that Mr. Haas was on the right track on many levels of his research.

The mission statement for NABS is to acquire enough evidence to assist in the development of legislation to protect bigfoot.

I can absolutely guarantee that when we first started this effort we never had any idea we would be heading down the path we are now on. We were always pushed towards the belief that bigfoot was some type of ape, gorilla, maybe further down the chain from human, but never, ever did we think it would have the significant human physical qualities that we have found. This is not to say that we didn't do our research and read the countless witness statements and hunter sightings where they stated that the creature looked too human to shoot; "The Hermit of Siskiyou" is a good example.

With every assumption there needs to be rational thought, and we think we've laid a path that everyone can follow. Bigfoot may well have been a creature much further away from human genetics than anyone could have imagined countless years ago. It appears that something happened. Maybe a rogue bigfoot couldn't find a female bigfoot, so he abducted and raped a human female; or, as in the John Lewis story, a female bigfoot abducted a human male. This behavior may have been communicated to other bigfoot and so began a pattern, as bigfoot found they could easily overcome human resistance. The offspring that resulted from the fornication probably died more times than lived. There must have been occasions that the child survived, as was recounted by Lucy in her chapter on "Indian Devils." The child still had too many genetic ties to the wild bigfoot to live within the confines of a Native American village, as Lucy wrote in her book. There have been other accounts of offspring of a human/bigfoot interaction where the child survived, usually dying before its 30th birthday.

BEHAVIOR

I think it's important to review bigfoot behavior and see what parallels can be made between bigfoot and human behavior.

Language: Refer to the report by R. Scott Nelson.

Tools: There are many accounts where people have seen a bigfoot throw a large rock and kill a deer. I have personally read several accounts of this same behavior, and it seems to be one of their preferred methods for bringing down other game. Rocks are also used as an intimidation tool to scare away humans, as is recounted throughout this book and *The Hoopa Project.* There are also many occasions where bigfoot bang rocks or wood together either for communication or for intimidation.

Thief: There exist other accounts of a bigfoot stalking hunters and waiting for them to shoot a deer. Once the deer is down, the bigfoot runs up and steals the kill. This tells us that bigfoot understands hunter behavior, stands back, watches the stalk, the kill, and then takes his dinner (*The Hoopa Project,* Damon Colegrove). This behavior is not for the timid. Bigfoot obviously understands that a rifle is a potentially life-threatening weapon; yet, they continue to follow and stalk hunters and take advantage of their kills. One of the many examples of a bigfoot stalking a hunter and ending up the hunted is Greg Fork's story of the young hunter who was scared by the bigfoot. There are many, many others.

Visitations: Bigfoot likes to visit campsites, predominantly when younger women are present. They've done nothing overt towards women in the last 50 years that has been reported, but they are inquisitive. They seem to understand when people are sleeping and they seem quite capable of picking the tent with the young female. A search of a variety of online bigfoot files shows many examples of campers in tents hearing and seeing a bigfoot just outside their sleeping area. In a few isolated examples, the bigfoot will actually open the tent door and look inside.

Scare But Don't Hurt: I have personally investigated many incidents where bigfoot throws huge boulders into close proximity of witnesses (*The Hoopa Project,* Hank Masten). They have also been known to shake trees/bushes violently and bellow, all in an apparent effort to frighten (*The Hoopa Project,* Corky Van Pelt). Many accounts also indicate that small stones are thrown at people while in the woods, sometimes while in tents. This behavior is interesting, as the stones are so small they never cause injury, but they do cause concern and send a specific message, "I am here." Bigfoot obviously has great aim with large stones and boulders and could easily kill a human if they wanted, but they haven't. I couldn't find one instance where someone thought a bigfoot threw anything with the intent to injure or kill.

INTELLIGENCE

If anyone believes that bigfoot is not intelligent, I'd like to meet with them. How many creatures are left in the wild where the best footage anyone can find of them is over 40 years old and is less than one minute long? That's the Patterson–Gimlin footage. There is footage that has more frames of a creature in the Marble Mountains, but the subject is so far away that physical features cannot be determined. This footage has been questioned as to its authenticity.

It would appear that bigfoot has the ability to be quite elusive.

It may understand human behavior better than we understand its behavior. Its ability to stay away from humans, to avoid fatal attacks where bigfoot bodies are discovered, and to conceal their deceased, purposely or not, is amazing.

Yes, there are unconfirmed reports of people saying that bigfoot bury their dead, either in rivers or under huge boulders. Perhaps bigfoot watched and adopted the tradition from Native Americans. It is obvious from our studies in Oklahoma, Minnesota, and California that bigfoot has stayed near reservations for hundreds of years. There are multiple reports of bigfoot watching jump dances and Native America ceremonies. They are probably watching us more than we are watching them. Could they have adopted some of our customs by watching us?

As humans continue to seek isolation and remote living, land and free space is being taken away from bigfoot. This does not appear to be causing a lack of bigfoot sightings. They are still being seen, but "where, when, and how" are the million dollar questions. Some researchers believe that bigfoot has been incredibly lucky for decades, and this is a major reason why they haven't been discovered and placed under some type of surveillance. We believe that the main reason a bigfoot colony, family, or even a single bigfoot hasn't been located for long-term surveillance is because there has never been a well-funded group working on this on a full time basis with paid professional researchers. Hopefully, that day is on the horizon.

The accounts we've quoted in this chapter lead us to believe that bigfoot is much smarter than most researchers are willing to admit. It is possible there are specific creatures roaming the woods that are much more human than others. The case can also be made that there are two distinctly different bipeds walking the woods of the Pacific Northwest—bigfoot and wild man—bigfoot being the creature that appears more animal and wild man being more human. There have also been discussions about witnesses finding human-like footprints with four toes and others with five toes. The toe disparity may be associated with genetics and the evolution of the species from bigfoot to wild man; maybe it's a genetic deformity from breeding within a family. We don't know. We do believe that it is time for researchers to open their minds, professors to open their doors, and academic institutions to open their wallets to serious, prolonged research into this phenomenon. NABS will continue our quest to understand the biggest mystery in the woods of the world.

NABS believes the study of bigfoot should be under the appellation "cryptoanthropology" (a term we first heard used by Dmitri Bayanov) and hope that long-time researchers will take a second look at all aspects of bigfoot and adopt this title for future research. While NABS has spent considerable time investigating all aspects of the bigfoot phenomenon, we admit that this is merely a well-researched hypothesis that deserves further inquiry. We hope that this new research designation encourages more anthropologists to delve into the bigfoot arena.

By the time you read this book, NABS will have a policy in place to never call bigfoot a creature again. We believe bigfoot should be called a tribal member, indigenous person, or Native American. Let your conscious be your guide.

We are well into our next phase of research. Keep in touch! www.nabigfootsearch.com

INDEX

479